QUANTUM MEASUREMENT

This book adopts a novel, physics-first approach to quantum measurement, using physical experiments as the basis to describe the underlying mathematical formalism. Topics covered include weak measurements, quantum measurement reversal, quantum trajectories, and the stochastic path-integral formalism. The theory of quantum measurement is also covered in detail, including discussion of how it can be tested and demonstrated in a laboratory: how to build quantum-limited amplifiers, fundamental noise limits imposed on measurement by quantum mechanics, and the design of superconducting circuits. This text is an excellent introduction for students with a basic understanding of quantum mechanics wanting to learn more about measurement theory, and the inclusion of a wide selection of end-of-chapter exercises makes this book ideal for emerging courses on the topic. Key chapters introducing the foundations of quantum computing and the history of measurement theory are equally accessible to a broader, less specialized audience.

ANDREW N. JORDAN received his B.S. from Texas A&M and his Ph.D. in physics from University of California–Santa Barbara. He serves on the faculty at the University of Rochester and Chapman University where is he is co-Director of the Institute for Quantum Studies. Andrew has received the NSF CAREER award and a Simons Fellowship in theoretical physics for his work.

IRFAN A. SIDDIQI received his A.B. from Harvard University and his Ph.D. in applied physics from Yale University. He is a member of the Physics and Electrical Engineering faculty at the University of California–Berkeley and faculty scientist at Lawrence Berkeley National Laboratory. Irfan is a fellow of the American Physical Society and has received numerous awards including the George E. Valley, Jr. Prize, the John F. Keithley Award, and the University of California–Berkeley Distinguished Teaching Award.

QUANTUM MEASUREMENT

Theory and Practice

ANDREW N. JORDAN

Chapman University and University of Rochester

IRFAN A. SIDDIQI

University of California–Berkeley

Shaftesbury Road, Cambridge CB2 8EA, United Kingdom

One Liberty Plaza, 20th Floor, New York, NY 10006, USA

477 Williamstown Road, Port Melbourne, VIC 3207, Australia

314–321, 3rd Floor, Plot 3, Splendor Forum, Jasola District Centre, New Delhi – 110025, India

103 Penang Road, #05–06/07, Visioncrest Commercial, Singapore 238467

Cambridge University Press is part of Cambridge University Press & Assessment,
a department of the University of Cambridge.

We share the University's mission to contribute to society through the pursuit of
education, learning and research at the highest international levels of excellence.

www.cambridge.org
Information on this title: www.cambridge.org/9781009100069

DOI: 10.1017/9781009103909

© Cambridge University Press & Assessment 2024

First published 2024

A catalogue record for this publication is available from the British Library

A Cataloging-in-Publication data record for this book is available from the Library of Congress

ISBN 978-1-009-10006-9 Hardback

To our wives, children, and families. Marian, Thomas, Juliana, Catherine, James, Terence, Sylvia, Peter and John. Lutfur-Rehman, Qudsia, Ambreen, Talha, and Mariam.
For all you have done to support us in our work

Contents

Preface

The subject of quantum measurement is now over a hundred years old, and yet we are still discovering new facets of it. Remarkably, despite the chaotic historical origins of quantum theory, its pillars have withstood the scientific test of time. While the mathematical formulations of quantum measurements have been demonstrated to be correct, they do not tell you "how it works." Over time, we have achieved a deeper and deeper understanding of this physics, with an increasing set of phenomena associated with it. Indeed, in past decades with the introduction of ever-improving quantum technology, there has been a quiet series of groundbreaking experiments that have pushed our best theories of quantum measurement in new directions. These discoveries have allowed physicists to experimentally probe the most fundamental aspects of quantum measurement and, for the first time, to "look under the hood" to understand the dynamics and statistics of the quantum measurement process. These findings help to elucidate many of the puzzling features of quantum physics: What is wavefunction collapse? How can quantum-limited measurements be made? How can quantum systems be monitored continuously? What is the connection between physical reality and information? Can we describe the measurement quantum mechanically? And so on.

However, in our experience, most physicists' understanding of measurement is stuck somewhere between 1920 and 1980. Even the most mathematically advanced treatments of quantum physics have a very crude understanding of how quantum measurements work, either in theory or in practice. The three cardinal properties of textbook measurements, of being projective, irreversible, and instantaneous, are idealizations and are usually only approximations to experimental reality. How to generalize and adapt the basic concepts of measurement to more realistic situations was unexplained until fairly recently. This text will guide the reader through the theory of generalized quantum measurements, the role of measurement strength and time, as well as mathematical formulations of continuous measurement. We will also explain how these new types of quantum measurements are designed and carried out in different fields of physics.

A number of books have been published that discuss quantum measurement and related fields: see Mensky (1993); Carmichael (1993); Braginsky and Khalili (1995); Percival (1998); Breuer and Petruccione (2002); Gardiner and Zoller (2004); Haroche and Raimond (2006); Barchielli and Gregoratti (2009); Wiseman and Milburn (2010); Jacobs (2014). We mention in particular the classic text by Braginsky and Khalili, focusing mainly on interferometeric light techniques, and of great practicality to the Laser Interferometer Gravitational-wave Observatory (LIGO), as well as the book of Wiseman and Milburn, focusing on the mathematical development of the theory of quantum measurement, particularly for atomic and optical systems, complemented by the book of Carmichael, all pioneers in this area. The book of Jacobs, as well as that of Breuer and Petruccione, address a wide variety of topics on this and related subjects in great detail. We also mention the helpful introductions of Jacobs and Steck (2006) and Brun (2002).

Sometimes books tend toward the extremes of the mathematical or the encyclopedic and usually are sparse on the experimental aspects of the subject. In addition, there are many philosophical texts on the subject of quantum measurement, of varying degrees of quality. Popular books (and movies!) on the subject (that we will not cite here) can be cringeworthy products written by novices who know just enough to be dangerous, or by scientists that have an axe to grind – usually related to issues of philosophy. Additionally, there have been several new experiments pushing the boundaries of our understanding and technical ability, as well as new theoretical formalism and insights. Most importantly, many of the subjects discussed here do not appear in any book we know of. This situation has convinced the authors that we are at a time when a new book on the subject, aimed at a wide audience, is needed. Our approach here is to write an accessible scientific book on the subject that is grounded in experimental findings.

It is our intention that much of this book can profitably be read by advanced undergraduates with some knowledge of quantum physics. The main audience is graduate students, postdocs, and other scientists wanting an accessible introduction and limited survey of the field. The literature is too vast for a complete treatment, so we focus on what we feel are the essential aspects of the field and give ample references to the original literature for further learning. Our goal is to concatenate the theory canon into a streamlined tool chest for quantum "mechanics" and to give experimentalists the techniques and methods needed for carrying out a wide variety of quantum measurements.

Information is the most spiritual thing in the physical world.

- Andrew N. Jordan

Quantum mechanics is the science of describing a vector that lives in an imaginary world, can only be reconstructed through our eyes in certain directions, and even those glances have already shaken it from its slumber.

- Irfan A. Siddiqi

Laguna Beach, CA   May 13, 2022

Andrew N. Jordan (left side of both photos) and Irfan A. Siddiqi (right side of both photos), celebrating at the Montage Hotel in Laguna Beach, CA, after finishing this book.

Acknowledgments

A complete list of thanks to all those supporting this book is impossible, so we highlight just a few people who were most closely involved.

Thanks go to Philippe Lewalle, who helped make some of the figures for the book; to Sasha Korokov, for many discussions about quantum mechanics; and to the late Markus Büttiker, for his mentorship. The excellent cover illustration of an artistic rendering of a superconducting quantum circuit was made by graphic artist Peter Jordan. We thank our colleagues for many discussions over the years that have helped refine our understanding of the field and spurred on the discoveries described in this book, including Joe Eberly, Justin Dressel, and Howard Wiseman. It is a pleasure and privilege to acknowledge our students and postdocs over our careers, who taught us as much as we taught them.

We owe a huge thanks to the government agencies and foundations that have supported our research reflected in this book. In particular, we thank the US Army Research Office, especially T. R. Govindan and Mike Metcalfe. Other supporting institutions are the National Science Foundation, the Department of Energy, the US Air Force Office of Science, the John Templeton Foundation, the Simons Foundation, and our affiliations: Chapman University, the University of California–Berkeley, and the University of Rochester.

We thank Justin Dressel, Michel Devoret, Alexandre Blais, Nico Roch and Kevin O'Brien for providing critical comments, and Daniel Esteve, for urging us to write this book. Thanks also go to Sarah Carranza, for her help in the final stages of preparing the book for publication. Finally, we thank the Cambridge University Press team for their support in the production and publication of this text.

List of Symbols

α	Coherent state amplitude
$\mathbf{A}$	Vector notation
χ	Dispersive frequency shift
Δ	Tunneling rate or Detuning
δ	Dirac or Kronecker delta function
ϵ	Energy asymmetry
η	Efficiency
Γ	Rate
$\hbar$	Reduced Planck constant
κ	Generalized eigenvalue or Cavity escape rate
$\langle \ldots \rangle$	Expectation value
λ	Eigenvalue or Wavelength or Coupling constant
μ	Magnetic moment
ω	Frequency
$\overline{\overline{A}}$	Matrix notation
Φ	Magnetic flux
ϕ	Phase
Φ_0	Magnetic flux quantum
Π	Projection operator
ψ	Quantum state
τ_0	Correlation time
τ_m	Characteristic measurement time
Θ	Temperature
ξ	Langevin random variable
$*$	Complex conjugate
$\dagger$	Hermitian conjugate
A	Amplitude

B	Magnetic field
C	Capacitance or Bhattacharyya coefficient or Concurrence
c	Speed of light in a vacuum
D	Displacement operator or Bhattacharyya distance
d	Degree of decoherence
dW	Wiener increment
E	POVM element or Electric field
e	Electron charge or Euler's number
E_C	Charging energy
E_F	Fermi energy
E_J	Josephson energy
F	Force
G	Conductance or Gain
h	Planck constant
I	Electrical current
L	Inductance
n	Number of Cooper pairs
P	Probability
p	Momentum, or Probability density
r	Measurement result
R_q	Resistance quantum
S	Noise spectral density or Scattering matrix
T	Duration of time
V	Voltage or Potential
W	Wiener random variable
x, y, z	Bloch coordinates
Y	Spherical harmonic
Z	Impedance or Partition function
$\bar{A}$	Average of A
$\mathcal{H}$	Stochastic Hamiltonian
$\mathcal{L}$	Lindbladian or Lagrangian density
$\mathcal{M}$	Murity
$\mathcal{N}$	Wigner–Smith time delay matrix
$\mathcal{Q}$	Accumulated charge
$\mathcal{R}$	Signal-to-noise ratio
$\mathcal{S}$	Stochastic action
$\mathcal{T}$	Time-ordering operator or Transmission
$\hat{\Omega}$	Kraus (or measurement) operator
$\hat{\rho}$	Density operator
$\hat{\sigma}$	Unnormalized density operator or Pauli operator

$\hat{\Theta}$ Time-reversal operator

$\hat{a}, \hat{b}, \hat{c}$ Bosonic annihilation operators

$\hat{H}$ Hamiltonian operator

$\hat{L}$ Lindblad operator

$\hat{U}$ Unitary operator

$\hat{X}, \hat{P}$ Quadrature operators

$\hat{O}$ Operator or Observable

1

Introduction to Quantum Physics and Measurement

1.1 Prologue

Quantum measurement sometimes has a bad reputation. No less an authority than J. S. Bell once wrote an article entitled "Against 'measurement'" (Bell, 1990). Philosophers of physics are not usually fond of it either, usually only referring to it indirectly as the "measurement problem." Practitioners of the physics of quantum measurement are sometimes derisively labeled as "instrumentalists." Why then should we devote a whole book to the subject? The simplest answer is that the word *measurement* is essentially a shorthand for how we interact with the world. It is, strictly speaking, the only way we can gain information about the world, and only from the data we collect can we begin to formulate theories about the world around us.

In the context of quantum mechanics, as students and researchers of the field, we spend so much time dwelling on the mathematics of the subject – quantum states, observables, Hilbert spaces, symmetries, equations of motion, and so on – we sometimes forget that the results of experiments and every lived experience we have of the world involves none of those things directly. It only involves measurement. Without measurement, none of those mathematical objects would have any scientific meaning. It is this tendency to sweep measurement under the rug, as well as our role as an observer, that has, in our view, held back the further elucidation of quantum physics. Thus, while some physicists think of wavefunction collapse as the "ugly scar" on an otherwise beautiful theory (Gottfried, 1989), it is only by taking its scientific role seriously that we can make further progress in the understanding of quantum physics.

While both philosophy and mathematics will arise in this book, it is our goal to have a "physics first" approach to this subject. In the founders' era of quantum mechanics, direct experimental evidence of many of the predictions of quantum mechanics was lacking because the necessary technology did not exist. This resulted in most of the "experiments" from that time being *Gedankenexperiments*, that is, thought experiments about what might happen. In the subsequent decades, the situation has changed radically. Scientists and engineers have developed many new instruments and physical systems to be able to explore and test new phenomena and extreme limits of the theory to unprecedented levels. There are many new deep and exciting theoretical predictions and concurrent experiments in this subject that have come up in the past decades. In the chapters that follow, we will guide you though some of them. By putting experimental phenomena first, followed by a mathematical description, we will gradually gain an intuitive understanding of how quantum measurement works as an operational science.

Philosophical matters are nevertheless important. The past half-century or more has led to a sprouting up of numerous "interpretations" of quantum mechanics. A clear understanding about the various mathematical objects and physical predictions in the theory can lead to new understanding. These interpretations are a metanarrative, riding over the theory and attempting to give further meaning to the theoretical content as well as draw out possible philosophical implications. These interpretations are too often an ex post facto forcing of physical phenomena into contorted mental constructs. This brings us to our maxim regarding interpretations of quantum mechanics: the best interpretation is a fruitful interpretation. It is only by leading one to new insights and discoveries that an interpretation becomes more than a stumbling block. In the epilogue that concludes this book, we will give an outlook on the metaphysical understanding of the field and how interpretation can guide us in physical discoveries.

1.2   The Era of the Founders: 1920s–1950s

Quantum theory developed in the 1910s with the introduction of the Bohr model of the atom, and quickly entered its modern form in the 1920s with the complementarity principle of Bohr, the uncertainty principle of Heisenberg, and the wave equation of Schrödinger. Measurement entered quantum mechanics via a postulate which stated that interrogating a system results in a final state that is a single stationary state, or eigenstate, of the measurement apparatus. A system may at any time prior to this be in a superposition of such eigenstates, and measurement collapses the wavefunction into a well-defined outcome whose probability does not evolve in time. One notes that it is therefore not possible to directly access the full wavefunction of a system in a single measurement, and measurement necessarily causes disturbance. What then does the wavefunction actually represent?

While we are more comfortable now with the idea that all accessible information can be packed into a complex wavefunction that can be extracted in certain allowed combinations using mathematical operators, the interpretation put forth by Max Born in around 1926 in a series of papers that linked the modulus squared of the wavefunction with the probability density to find the system it described was a tremendous step forward in linking abstract quantum quantities with measurement outcomes (Born, 1926). Born noted that the coefficient associated with a given eigenfunction, when the wavefunction is expanded in the basis of the measurement apparatus, represents the probability amplitude that such an outcome would be observed. Born received the 1954 Nobel Prize in Physics "for his fundamental research in quantum mechanics, especially for his statistical interpretation of the wavefunction."

The measurement postulate, as originally conceived, is perfectly compatible with all past and future experimental results tested thus far. Nonetheless it is ad hoc and leaves much to the imagination with respect to the details of the measurement process itself. While a system evolves according to the Schrödinger equation, measurement was put forth as an "instantaneous" collapse of the wavefunction – a process that inherently assumes that information extraction is too fast to access, and is at odds with the fact that any physical apparatus has a characteristic measurement time. Moreover, there are many experimental situations that are not naturally suited to this canonical description. For example, the spontaneous emission decay of an atom ultimately results in the occupation of the ground state whether the atom was always in the ground state from the start or arrived there after a photon emission. More subtly, the measurement rate from the point of view of an external observer is not on equal footing for these two outcomes. For a relaxation event, a rapid quantum jump is the signature of the process, whereas if no such jumps are detected, we must wait much longer than the average decay lifetime to safely conclude the atom was originally in the ground state.

As such, since the early foundations of quantum mechanics, there was extensive discussion of the role of measurement, which was quite important in the thinking behind the Heisenberg uncertainty principle (Heisenberg, 1985). One of the most influential and systematic early thinkers in this area was Hungarian-American John von Neumann. In his classic text *Mathematical Foundations of Quantum Mechanics* (Von Neumann, 1955), originally published in 1932, von Neumann advanced an important line of research taking seriously measurement as a physical process. He recognized that an instantaneous measurement of, for example, energy would run afoul of the time-energy uncertainty principle, and that therefore it is not possible to carry out an arbitrarily precise measurement in a very short amount of time. This led him to develop a dynamical model of the measurement process, whereby the information about the system of interest can be extracted by coupling it to an auxiliary degree of freedom, or meter.

The meter, or probe, variable could be treated within a fully quantum mechanical framework, and then itself be detected using a recording device via the projection postulate. This procedure will be referred to as von Neumann's measurement model. It is an important advance because the combined system/meter/recording device is given a dynamical description, with its own Hamiltonian and equations of motion. This model can be seen as the precursor to later mathematical models of quantum dynamics, including decoherence theory, weak measurement, and quantum trajectories that will be addressed in later chapters of this text.

Nevertheless, the concept of instantaneous wavefunction collapse sparked much debate, including the celebrated paradox put forward by Einstein, Podolsky, and Rosen (EPR). A pair of entangled objects cannot, by definition, be described by concatenating independent descriptions of each constituent piece. Correspondingly, measurement outcomes of each element of the pair must be correlated, seemingly implying that if entanglement exists between physically separated objects, then quantum mechanics is capable of instantaneous, nonlocal influences. This led to the EPR paradox (Einstein et al., 1935), which raised the apparent incompatibility of this notion with locality since the information needed to generate the correlations associated with quantum mechanics would need to be exchanged faster than the speed of light if the two parts of an entangled pair are spacelike separated. This led EPR to conclude that the quantum description of reality as given by a wavefunction is not complete. Without further thought experiments that could prove the existence of quantum entanglement while simultaneously establishing the inability of any classical theory to predict these unique measurement outcomes, and without the experimental tools to access the EPR regime, this dilemma of, as Einstein put it, "Spooky Action at a Distance" lay dormant and unexplored until the arrival of John S. Bell (Fig. 1.1).

1.3 The Era of Bell: 1960–1970s

Alternative Formalisms to Cope with Reality

During the Second World War, science was mostly on hold. Notable exceptions are, of course, the Manhattan Project and development of the atomic bomb, harnessing nuclear physics for the war effort. Following that period, research in quantum mechanics was focused mostly on relativistic generalizations and on the growing field of particle physics. The 1960s was the golden age of field theory, giving rise to the systematic categorization of the many different particles that were being discovered, leading to the so-called standard model of particle physics. The existence of quarks proposed by Murray Gell-Mann and George Zweig in 1964 was verified with the discovery of all the quark species, the last being the top quark discovered

Figure 1.1 From left to right: Niels Bohr and Einstein debating the nature of reality. John Bell at the board, decorated with his famous experiment. Reprinted with permission from the Niels Bohr Library & Archives, American Institute of Physics, and CERN.

in 1995 at Fermilab. The final missing particle, the Higgs boson, proposed independently in three papers in 1964 by Brout, Englert, Higgs, Guralnik, Hagen, and Kibble, was finally discovered at the CERN particle collider in 2013.

While there were important contributions to quantum mechanics in this period (we mention, for example, Feynman's path integral formulation in 1948 (Feynman, 1948) and the Aharonov–Bohm effect in 1959 (Aharonov and Bohm, 1959)), there was not much activity on what we may call the "fundamentals" of quantum mechanics. However, some scientists were unhappy with the philosophical implications of the dominant quantum interpretation, the "Copenhagen interpretation," most systematically described by the Danish physicist Niels Bohr. They were following in the footsteps of Albert Einstein, who, although a co-founder of the field, was deeply disturbed by the intrinsic randomness of the theory, and the lack of a clear "realist" view of what quantum theory describes. In classical physics, we encounter the notion of randomness only as a lack of complete knowledge – for example, whether a flipped coin lands "heads" or "tails" is simply our practical inability to quickly integrate Newton's equations from the given initial condition. The mathematical description of classical probability theory is given in Appendix A. However, in quantum theory, to the best of our knowledge, probability theory is inescapable. Identically prepared initial conditions can give rise to different final outcomes. Thus the concept of preexisting properties before measurement is generally rejected in quantum theory. This concept of a system's property, like position, momentum, energy, and so on, existing without any reference to measurement or observers is what we mean by "classical realism."

One of the more influential scientists of this era was David Bohm (1917–1992), who advanced another view of quantum mechanics. Bohm, together with Ahanonov, reformulated the EPR argument (originally made for continuous position and momentum variables) in terms of measurements of components of spin-1/2 particles. They pointed out that currently it was practicable to test it experimentally only in the study of the polarization properties of correlated photons (Bohm, 2012; Bohm and Aharonov, 1957). Bohm had interpreted quantum physics in 1952 as describing an underlying real particle degree of freedom that was "guided" by an unobservable wave, sometimes called a "pilot" wave (Bohm, 1952a; Bohm, 1952b), called by Bohm a "hidden variable," drawing from the earlier "hidden parameter" of John von Neumann (1955). This view of quantum mechanics is mathematically the same as that of Madelung's hydrodynamic formulation of quantum mechanics (Madelung, 1926; Madelung, 1927) and philosophically similar to the early views of de Broglie (1927). It avoided the "no-go" hidden parameter theorem of von Neumann by also giving the detector an active role to play in the particle dynamics. The Bohmian interpretation had the feature of giving a clear mental picture of the particle degree of freedom, but suffered a number of drawbacks. When applied to more than one particle, such as in the Einstein, Podolsky, Rosen thought experiment (Einstein et al., 1935) (questioning the completeness of quantum mechanics), the pilot wave must act instantaneously across space and time in order to "guide" the particle to the correct detector result. That feature, together with an infinite number of variant theories possible, and difficulties in making a relativistically invariant version (which was already well established within standard quantum mechanics), led most physicists to discount it. However, the fact that such a reinterpretation of quantum theory was possible in the first place was an advance in foundations of quantum mechanics and motivation for further thought.

Bell Correlations, Testing Differences with Hidden Variable Theories

The issue lying in the background is whether quantum mechanics, like statistical mechanics, is an emergent theory describing a yet more fundamental physics of unobserved degrees of freedom, the abovementioned hidden variables. Further, if these hidden variables, exist, the thinking goes, perhaps they will restore our view of objective reality – where physics describes properties that exist before we measure them. The existence of the Bohm reinterpretation led John S. Bell (1928–1990) to think more broadly about the problem and to see what properties *any* theory that involved this type of hidden variable would have. In a groundbreaking paper published in 1964 in an obscure journal *Physics*, which only published four volumes between 1964 and 1968 (Bell, 1964), Bell published his inequality. The result of the inequality puts up a scientific conflict between a class of "hidden variable" theories and quantum mechanics. The scientific nature of this inequality is quite important – it is not an interpretation of quantum theory, but an experimental test that these hidden variable theories must satisfy. The basic setup is that two particles, called S_1 and S_2, are spatially separated such that no communication between them is possible that is faster than the speed of light, and then each is measured to collect data on its properties by two different measuring devices. Bell put in the ingredients that a naive scientific realist would want in a theory for this simple case:

- That the properties of each particle, called A and B, to be measured have definite, preexisting values, determined by the hidden variable.
- That statistical outcomes of measurements are simply the result of averaging over the hidden variables (which are produced randomly at the creation of the two-particle system with some unspecified distribution), as is the case in statistical mechanics.
- The "vital" assumption is that the result B for particle 2 does not depend on the detector setting for particle 1, nor result A for particle 1 on the detector setting of particle 2. Bell quotes Einstein: "But on one supposition we should, in my opinion, absolutely hold fast: the real factual situation of the system S_2 is independent of what is done with the system S_1, which is spatially separated from the former" (Schilpp, 1949). The spatial separation should be sufficient that no causal influence can be transmitted faster than the speed of light, in accordance to the principles of relativity theory.

Some modern commentators also argue there is an implicit assumption of freedom of choice; the experimenters operating the measuring devices of particles 1 and 2 are free to change their settings at will (there is no superdeterminism). There are also more exotic hidden assumptions like no retrocausality.

In his six-page paper, Bell goes on to show that this class of "local realistic" hidden variable theories (local because of the third assumption, realistic because of the first) make definite predictions about the statistical result of measurements made with different settings of the two measurement devices: a correlation function of the measured data has certain bounds on its value, no matter what the distribution of hidden variables. Consequently, any experiment that violates the Bell bound on local realistic hidden variable theories will rule such theories out. Furthermore, quantum physics can violate that bound. It should be noted that not all hidden variable theories fit into this category. For example, the hidden variable theory of Bohm contradicts assumption 3, allowing nonlocal influences.

After Bell's paper was published, the next obvious question was to test it: which was right, the above hidden variable theories, or quantum mechanics? An important step was the introduction of a modified inequality in 1969 by John Clauser, Michael Horne, Abner Shimony, and R. A. Holt, which was the test usually implemented in the experiments (Clauser et al., 1969). There was a series of optical experiments testing the preceding inequality, focusing on the polarization degree of freedom of two entangled photons. In particular, we mention the 1972 experiment of Freedman and Clauser (1972), which shows Bell's inequality was indeed violated. Interestingly, the 1973 experiment by Holt and Pipken using atomic mercury to produce two-photon events at Harvard University (Holt, 1973) showed that Bell's inequality was *satisfied*, casting doubt on quantum mechanics! Clauser repeated the experiment (Clauser, 1976) and found, together with the independent experiment of Fry and Thompson (1976), that Bell's inequality was indeed violated in this type of system.

This created a series of Bell-type experiments that continue to the present day. Of great interest is the possibility of ruling out "loopholes" in the experiments, whereby some experimental imperfections or design flaw may not truly rule out local realistic hidden variables. We mention in particular the influential 1981 and 1982 experiments of Alain Aspect (Fig. 1.3) and coworkers Grangier, Roger, and Dalibard (Aspect et al., 1981; Aspect et al., 1982), the latter of which was the first to implement fast random switching of the polarization analyzers, faster than the light travel time between the measured systems. This ruled out the possibility of the setting of one polarizer being able to causally influence the outcome of the other (nonlocal) photon. Physicists also started violating Bell's inequality over longer and longer distances, such as the work of Nicholas Gisin's group showing violation using Swiss Telecom lines between two villages in the Geneva vicinity, separated by 18 km (Salart et al., 2008), testing the speed of the "spooky action at a distance." Recent notable experiments along these lines are a series of the experiments in 2015 designed to rule out two loopholes (also closing the "fair-sampling" loophole) in the same experiment (Hensen et al., 2015; Giustina et al., 2015; Shalm et al., 2015). An event-ready Bell experiment was subsequently made (Rosenfield, 2017). These experiments are the most aggressive tests of local hidden variable theories to date and all showed that Bell's inequality remained violated. Consequently, we are stuck with quantum mechanics unless we want to break one or more of the assumptions of Bell's theorem. For these experiments the 2022 Nobel Prize in Physics was awarded jointly to Alain Aspect, John F. Clauser, and Anton Zeilinger "for experiments with entangled photons, establishing the violation of Bell inequalities and pioneering quantum information science."

Decoherence Became Fashionable with Zurek and Kraus

It is clear that dissatisfaction with ideas of quantum measurement is behind much of the work of Bohm, Bell, and others. This idea crystallized into what is sometimes called the "measurement problem." The basic point is to decide when we should describe the quantum dynamics with a unitary Schrödinger equation, and when we should describe it with the projection postulate of wavefunction collapse. This problem was, of course, well known to the founders of quantum mechanics and is also sometimes called the "Heisenberg cut" – where should we cut off when to describe physical systems as having classical properties versus describing them as being in a coherent quantum superposition?

This question brings us into the 1980s, where a school of thought arose that we could solve this measurement problem with the process of *decoherence*. The basic idea runs as follows: We know that in reality quantum systems are not isolated, and interact with their environment. This environment can also be described quantum mechanically. However, it is impractical – perhaps even impossible – to control every aspect of it, so it is natural to average over the dynamics of the uncontrolled and unobserved environment. Rather than use a pure quantum state, the concept of the mixed quantum state of the subsystem is then of great utility, which is a statistical mixture of pure states. Mixed states include pure quantum states, but also include classical statistical mixtures – for example, how we describe the results of a classical coin flip where the randomness is simply a reflection of our ignorance of the "microstate" of the coin. Technically mixed states are described with density matrices, which generalize the concept of state vectors. Mixed states are reviewed in Appendix B.

The idea of decoherence is that the environmental interaction degrades the quantum coherence of a pure state, converting it into more palatable classical statistical mixtures. In describing the dynamics of a system of interest with an environment, the effect of the simplest type of coupling is to add a phase to the off-diagonal density matrix elements (defined in a basis specified by the environmental coupling). This phase depends on the state of the environment, so that the dynamics of the subsystem should be averaged over the fast environmental dynamics. The result of this is that the off-diagonal density matrix elements are suppressed, resulting in a purely classical statistical mixture. The basic insight advocated by physicists such as Wojciech Zurek (1981, 1982; see Fig. 1.2) was that the environment dictates the manner of decoherence the system experiences, and quantum measurement should be seen as a kind of decoherence. At the very least, this explains why coherence is difficult to maintain in large quantum systems, or systems not isolated from their environments. This approach, fully within the standard canon of quantum mechanics, was very popular at the time.

The shortcoming of the decoherence worldview is that no measurement – in our sense of learning something about the system – ever takes place until the very end when all coherence is removed. The theory describes statistical ensembles, not measured data. The fact that decoherence theory does *not* explain individual measurement results shows it is not a fundamental theory of measurement, and

Figure 1.2 Hiking photos of Karl Kraus (left) and Wojciech Zurek (right), reclining. Reproduced with permission from wikimedia.org under CC-BY SA, and from W. Zurek.

more is required despite common impressions from the recent literature. We will see in later chapters that when the environment is monitored, a (stochastic) pure state description of the dynamics can be restored, bringing us back to a collapse process.

At the same time, Kraus (1981, 1985) and Kraus et al. (1983), continuing in the line of Sudarshan et al. (1961) and Davies (1976), were also working on the theory of open quantum systems and developed the concept of a quantum process, or quantum dynamical system. This is closely related to a dynamical description of the open quantum systems of Lindblad (1976), as well as Gorini et al. (1976). We will return to these concepts in the coming chapters. While correctly capturing the dynamics of open quantum systems, including the decoherence process, these approaches do not fully capture the informational aspects of quantum theory, which brings us into the Quantum Information age. Let us first take a tour through some of the classic experiments that helped to form our understanding of the field.

1.4 Classic Experiments: 1970–1980s

The last quarter of the twentieth century was a tremendously active period in which the seeds for another quantum revolution were sown, particularly as new experimental tools and techniques became available. These developments coincided with new theoretical ideas put forward by Bell and others, and their confluence led to experimental tests of uniquely quantum mechanical effects. While many notable experiments should be highlighted in a thorough review of quantum milestones, we will discuss only a few exemplars that have had a profound impact in the field,

Figure 1.3 From left to right: Alan Aspect excitedly lecturing, Serge Haroche giving instruction, Leonard Mandel looking relaxed. Reproduced with permission from wikimedia.org and wikipedia.org under CC BY-SA, and Marlan Scully.

affirming the basic postulates of quantum theory and laying the foundation for modern quantum technologies in communication, computation, and sensing. In particular, we detail advances in photonics, atomic physics, and superconducting circuits, along with the basic configurations of these classic experiments.

Photons and Atoms

The invention of the LASER (Light Amplification by Stimulated Emission of Radiation) played a tremendous role in expanding the bounds of quantum measurement. Invented in 1960 by Theodore H. Maiman, this optical device was an advance on the MASER built in 1953, the microwave frequency version, invented by the group of Charles Townes. Townes, together with Basov and Prokhorov, won the 1964 Nobel Prize in Physics "for fundamental work in the field of quantum electronics, which has led to the construction of oscillators and amplifiers based on the maser-laser principle." With this source of coherent photons with narrowly defined optical frequency, one could precisely excite desired atomic transitions with great specificity as well as produce sparse sources of emission where the granularity of individual light quanta were readily observed and the statistics of their generation and detection were imprinted with the signatures of quantum theory.

The experiments to test Bell's inequalities, see Fig. 1.4, performed by Alan Aspect and co-workers (Aspect et al., 1981, 1982) were truly paradigm shifting. Aspect sought to realize the theoretical experiment proposed by Bell with as few modifications as possible – a single source producing entangled, polarized photons, captured by two distant detectors whose mutual output correlations could be quantified to prove the persistence of quantum interactions, even at a spacelike physical separation.

The contribution of Aspect's team. – Aspect's work incorporated several technical advances, which allowed the violation of Bell's inequalities to be verified with many standard deviations of precision. While, as mentioned earlier, several beautiful experiments aimed at observing EPR correlations preceded Aspect's work, they were subject to the challenges of technical noise and drift. Furthermore, in the detectors, the measurement outcomes for two different polarization states were not on an equal footing. A particular polarization direction was transmitted and no signal was expected for the orthogonal one, making it difficult to distinguish between a missing photon's "click" in the presence of poor measurement efficiency and the true absence of an absorbed photon. These new experiments used a radiative cascade decay in calcium-40 pumped by a krypton laser to produce a bright entangled photon source that produced 50 pairs per microsecond and enabled an experimental run to be completed in 100 s, avoiding slow drift in the apparatus. The dual channel polarizers employed were cubes comprised of two prisms coated with thin film dielectrics so that parallel polarization was transmitted and perpendicular polarization reflected; both outcomes could be detected using two photomultiplier tubes, essentially creating a Stern–Gerlach analogue for photons.

In its early version, the polarizers were stepped through relative orientations to calculate the correlations between detection events; Bell predicted that entanglement between the photon pairs would lead to stronger correlations than can be obtained with any classical system with a varying degree of departure from classical statistics depending on the angle between the detectors. Indeed, the predictions of Bell were unambiguously observed. However, one could argue that since the directions of the polarizers were fixed before the generation and emission of the photon pairs, classical information exchange could be responsible for such observations. Aspect, Dalibard, and Roger (Aspect et al., 1982) partially addressed this loophole by not selecting the orientation of the polarizers in the detection system until after the entangled photons were generated, and distancing each detector 6 meters from the source. Each dual channel polarizer was now replaced by a pair with two different fixed orientations, and an incoming photon was directed into one or the other depending on the state of an acousto-optic switch that flipped every 10 ns; see Fig. 1.4(a). This time, as well as the internal 5 ns delay in the atomic cascade, was much shorter than the 40 ns photon transit time, thereby spacelike separating the detectors and supporting the notion of the nonlocality of quantum entanglement. While not completely free of all loopholes, this work elevated quantum entanglement from a theoretical notion to very tangible property that could be measured in any experiment conceived to date.

The contributions of Mandel's team. – In the Aspect experiment, a single photon is incident on a dual-channel polarizer and will traverse one path or the other. Another distinct facet of quantum measurement science is observed when two identical photons are incident on the two input ports of a 1:1 beam splitter, as shown in Fig. 1.5(a). Each impinging photon can be either transmitted or reflected, giving

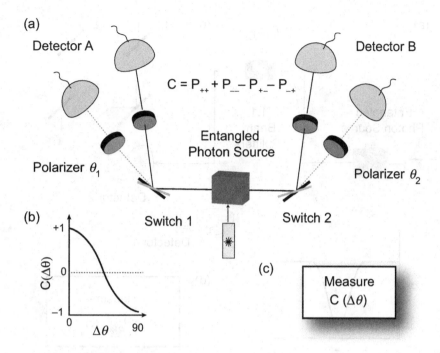

Figure 1.4 (a) A Bell's inequality setup involves shuttling the two constituent photons, produced from an entangled photon source, through two spacelike separated detection arms, each with a variable polarizer and a photomultiplier tube for detection. To address possible local communication loopholes, a fast switch is used to randomly select between two possible polarizer orientations after an entangled photon pair is produced. (b) The correlation coefficient, C, is plotted as a function of the relative polarizer angle, $\Delta\theta$, between the two arms. (c) The quantity to measure is the correlation function of photon coincident counts.

rise to four possible outcomes for two-photon incidence. Surprisingly, rather than being equally distributed among these four cases, the two photons preferentially bunch. An ideal beam splitter would not have any memory of which outcome actually occurred, requiring us to add the probability amplitudes of all four possibilities, shown in Fig. 1.5(b) on an equal footing to predict the output state. Note that the two cases corresponding to both photons being reflected or transmitted, labeled 3 and 4 in the figure, respectively, are physically indistinguishable for fully identical particles. Additionally, an ideal beam splitter is unitary, and thus reflections on one side of the beam splitter acquire a π phase shift. As such, the probability amplitudes for both particles being reflected or transmitted destructively interfere and cancel each other. This means that identical photons that spatially overlap when they enter the two input ports of a beam splitter will always exit together through the same output port, as depicted in cases 1 and 2. This effect was observed by Hong, Ou, and Mandel (HOM) (Hong et al., 1987) at the University of Rochester

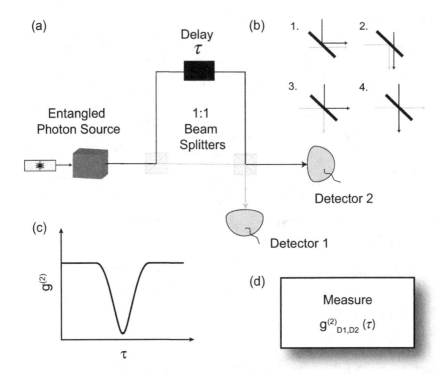

Figure 1.5 (a) The Hong, Ou, and Mandel (HOM) effect is seen as the bunching of output photons when an identical pair is incident on an ideal 1:1 beam splitter. (b) Of the four possible photon trajectories (1–4), shown in the upper right, the wave amplitudes corresponding to both photons being reflected (3) and transmitted (4) have equal magnitude and opposite sign, and thus interfere destructively, resulting in both photons always exiting the output beam splitter through the same port (1,2). (c,d) A time delay τ is inserted in one arm to measure the second-order correlation function $g^{(2)}$ as a function of temporal coincidence. The dip is when the timing of the photons arriving at the beam splitter is perfect.

in 1987, and is manifest as a dip in the coincident count rate observed in two detectors monitoring the two output ports of the beam splitter; see Fig. 1.5(c,d). The HOM dip, named after the trio that discovered it, is complete (i.e. the correlation function goes to 0) for perfectly identical photons and degrades in proportion to the degree of photon distinguishability. Notably, for the case of classical light described by a coherent state, a dip is also observed but only drops to 0.5 at best. Thus, the HOM effect is indeed quantum mechanical and has a distinct measurement signature. It is used extensively to verify the purity of single photon sources, and also underlies the basic operating principle for producing an effective, measurement-mediated gate interaction in linear optical quantum computing, such as in the Knill–LaFlamme–Milburn (KLM) protocol (Knill et al., 2001).

The contributions of Haroche's team. – Around the same time period, another fundamental quantum state was being pursued: the Schrödinger cat state. The classic thought experiment of Erwin Schrödinger (1935) has a feline cohabiting a box with an unstable radioactive atom in a superposition state of having decayed and not decayed. This indeterminacy is transferred to the cat which is then simultaneously dead and alive. Serge Haroche and coworkers, including J. M. Raimond and M. Brune at ENS Paris, developed a series of experiments (Brune et al., 1990, 1996) where single photons could be trapped in an extremely high-finesse cavity through which individual atoms could be transited, as shown in Fig. 1.6. In such a cavity quantum electrodynamics setup, described by the model of Jaynes and Cummings (1963), the itinerant circularly polarized Rydberg atom (a highly excited atom, with its electron(s) having a large principle quantum number) traverses a microwave cavity and is a sensitive probe of the intracavity photonic field. Haroche and the team established a Schrödinger cat state of order 10 photons in the cavity and reconstructed its dynamics by probing it with a series of atoms sent through the cavity one at a time. Physicists now had a powerful laboratory for dissecting decoherence in a small, well-controlled bath. The physics of light–matter interactions beautifully showcased in these experiments directly influenced the field of circuit-based cavity quantum electrodynamics that would follow two decades later, and introduced many techniques currently employed in modern neutral atom quantum computing. For these and related accomplishments, Serge Haroche together with David J. Wineland won the 2012 Nobel prize.

Superconducting Circuits

During the 1980s, another set of outstanding fundamental questions in quantum theory were being pursued in parallel. Quantum mechanics has its roots in the physics of light and atoms, and in particular the phenomena of photon emission from solids and gases. Does the theory apply to electrical circuits? Pioneering research to answer this question was undertaken by Anthony Leggett, John Clarke, and their collaborators. Leggett, recipient of the 2003 Nobel Prize in Physics for his work on the theory of superfluidity, has written extensively on the topic of macroscopic quantum coherence (Leggett and Garg, 1985), and quantitative measures of how macroscopic a system actually is, which continue to be a topic of contemporary research. On the experimental front, a team of researchers at University of California, Berkeley in the laboratory of J. Clarke realized a set of experiments that put electronic circuits on the same footing as their microscopic cousins, establishing a powerful testbed for testing fundamental concepts in quantum theory in an open quantum system (well coupled to a complex bath), and a versatile toolset for future Quantum Information processing hardware.

The Berkeley experiments employed a Josephson tunneling junction to realize a quantized nonlinear oscillator. Superconductivity, in a very basic hydrodynamic

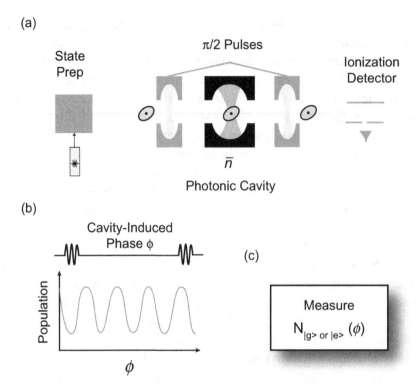

Figure 1.6 (a) In an atomic cavity quantum electrodynamics experiment, itinerant single atoms traverse a central microwave cavity populated on average with n̄ photons, and a strong light–matter interaction results in entanglement. This imprints a phase shift ϕ on the effective two-level atomic wavefunction which can be probed by Ramsey spectroscopy. (b) Applying a pair of $\pi/2$ pulses at the input and output of the central cavity followed by state-selective field ionization detection to measure (c) the number of atoms (N) in a particular state results in an oscillation of the average population of either the ground or excited state with ϕ. Diminishing contrast signals the presence of decoherence. Schrödinger cat states can be readily formed by adjusting the interaction time of the photonic field with a single atom such that, when the latter is produced in an equal superposition of two Rydberg states, it becomes maximally entangled with the former after exiting the cavity.

description, results from an effective pairing interaction between single dressed electrons that transforms a normal metal into a condensate that can be described by a quantum mechanical amplitude and phase. When two superconductors are separated by a barrier, be it a normal metal or an insulator, then a phase difference can be established between the two superconducting electrodes that determines the transport properties of the junction. Brian Josephson, recipient of the 1973 Nobel Prize in Physics for the effect that bears his name, showed that the current flowing across such a junction is proportional to the sine of the gauge-invariant phase difference across the device, and the voltage is proportional to its time derivative.

Figure 1.7 A smiling Anthony Leggett (left) and an earnest John Clarke (right) in front of the blackboard. Photo Credits: University of Illinois/L. Brian Stauffer and University of California/S. Wittmer.

These equations describe the motion of a fictitious particle in a cosine potential whose coordinate is given by the phase difference across the junction (Josephson, 1974).

In an experiment carried out by Devoret et al. (1985), the transition from the superconducting state, characterized by zero DC resistance, to the voltage-state was used as a measurement probe of when the fictitious particle describing the junction dynamics was static and confined to a single well of the cosine potential or if it had escaped, resulting in a time evolution of the phase difference at a finite rate, as illustrated in Fig. 1.8(a,b). By applying a static current to the junction, the potential well was tilted, effectively lowering its barrier and eventually allowing the particle to escape under the influence of thermal and quantum fluctuations. At sufficiently low temperature, the transition rate to the voltage state, Fig. 1.8(c), became independent of the physical temperature of the junction, indicating that the dominant source of fluctuations was quantum zero-point motion which drove a quantum mechanical tunneling process rather than a classical Arrhenius type of activation over the top of the potential barrier. This result was a dramatic demonstration that a collective macroscopic variable can behave in full accordance with the postulates of quantum mechanics.

An important practical difference between photonic/atomic and superconducting circuits, however, is that the latter are inherently open quantum systems that readily couple to both the vacuum and the imperfections of a solid-state environment. In a second experiment, the Berkeley team excited the Josephson junction oscillator with a microwave field to drive the system to an excited state which would then rapidly tunnel to the resistive state of the junction and produce a readily detected voltage signal (Martinis et al., 1985). A critical technological advance in this experiment was the use of highly attenuating filters, specifically made of metallic copper powder, that would attenuate microwave signals over a very wide

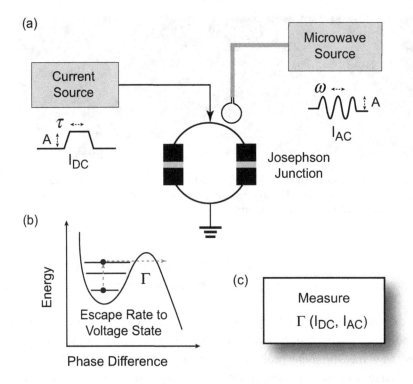

Figure 1.8 (a) A superconducting quantum interference device is shown that contains two Josephson junctions. The Josephson junction is a superconducting nonlinear electrical oscillator whose dynamics can be described by (b) the motion of a fictitious particle through a periodic potential. A static current tilts the potential wells and microwave pulses drive resonant transitions. By using a combination of both of these excitations, the energy level structure of the oscillator can be ascertained by (c) measuring the escape rate Γ of the junction out of the potential well via macroscopic quantum tunneling as the signature that the external microwave drive has excited the system out of a lower energy level.

range of frequencies and effectively shunt the junction with a high impedance. Without such control of the dissipative environment surrounding the junction, the linewidth of the individual quantized levels within a well of the periodic junction potential would be too broad to allow for their spectroscopic isolation. These experiments directly mapped the specific energy levels of the junction, establishing that a Josephson junction is essentially a nonlinear quantum oscillator with distinct particle-in-a-box type energy levels. Several subsequent experiments with D. Esteve and C. Urbina further elucidated the critical role played by the impedance shunting the junctions on its dynamics (Esteve et al., 1986; Urbina et al., 1991). A single junction was later used as a phase quantum bit, or "qubit" (Martinis et al., 2002) – the elementary unit of quantum computation, and circuits with different shunting elements and junction count gave rise to the current flora of charge- and flux-controlled superconducting qubits (Clarke and Wilhelm, 2008).

1.5 The Quantum Information Era: 1990s–Present

With the notion of quantum entanglement on a much firmer experimental footing, ideas related to possible advances in computation and metrology based on quantum coherence began to take shape. The concept of projective measurement and its use in secure communication was appreciated earlier, with the celebrated work of Bennett and Brassard appearing in 1984 (Bennett and Brassard, 1984; Bennett and Brassard, 2014). The BB84 protocol used the fact that measurements of the polarization state of a polarized photon will collapse into a definite outcome, faithfully representing the preparation of the photon, when the detector orientation is aligned with the initial polarization of the photon, or will result in a probabilistic outcome when the measurement and polarization axes are not commensurate. Thus, the choice of initial polarization and measurement detection constitute a key to securely encode and decode sensitive information, with the added bonus that the action of an eavesdropper could be readily discerned by the observation of incorrect measurement statistics.

Additional developments in quantum sensing took place during this period that took advantage of non-classical manifestations of light such as squeezed states (Walls, 1983) where fluctuations are asymmetrically distributed between two quadratures, such as photon number and phase, allowing one to surpass the so-called standard quantum limit of measurement sensitivity. Analogous quantum behavior was later demonstrated in matter, for example in the spin degree of freedom of trapped atoms and ions (Wineland et al., 1992). Extensions to many-body quantum states such as NOON states (Kok et al., 2002) have also been proposed, and represent a route to obtaining quantum enhanced scaling (Heisenberg limit) in measurement sensitivity, which in quantum metrology represents an improved scaling in which the measurement sensitivity increases with the resource of a measurement such as the time of the experiment or the number of photons used, and is a topic of contemporary research.

The concept of a quantum computer that was more powerful than classical computers on an algorithmic level was a huge advance. Quantum algorithms have been developed that can perform certain tasks faster than the best known classical algorithms. The most promising algorithms (such as Shor's factoring algorithm (Shor, 1994)) give an exponential speedup, while others (such as Grover's search algorithm (Grover, 1996)) give a polynomial speedup. We will give a more indepth discussion of this area later in the chapter. While also utilizing quantum coherence for enhancements over classical technology, computing has additional requirements, most prominently the need to both control and measure entanglement over an extant array of quantum objects, with numbers ranging anywhere from a few hundred to a million. Quantum state measurement in such an architecture is challenging, as one cannot simply decouple from the surrounding environment to maintain coherence, as such attenuation would reduce the signal-to-noise-ratio, and clever schemes are needed to effectively filter out noise channels that cannot be probed by the experimentalist. Much work has been carried out on a variety of platforms, including nuclear magnetic resonance systems for ensemble-based computation, and by ion trap–based architectures. Constructing high-efficiency amplifiers and detectors with near-quantum-limited sensitivity has been critical for progress in solid-state and photonic quantum computing platforms. Superconducting nanowire and Josephson junction devices have provided access to single-shot projective measurements, and sufficiently high-quality weak measurements to execute both proof-of-concept quantum algorithms and new classes of fundamental tests of quantum theory.

In addition to a readout resource, measurement can also be integrated directly into the computational step of a quantum processor. In particular, the effective nonlinearity associated with the measurement process can be used to generate entanglement between two qubits traveling through linear guided wave elements, and can also execute one-way or measurement-based quantum computing (Raussendorf et al., 2003) with specific, highly entangled cluster states such that a destructive measurement deterministically yields the results of a desired computation. Such methods are actively being pursued in photonic architectures where the weak interaction between photons and losses in nonlinear elements makes conventional quantum mediated by direct photon–photon interaction a challenge.

Technology, Foundations, and Philosophy

In the preceding section, we discussed the advent of quantum information science as a branch of information theory and computing, discussing the concept of quantum advantage, or "supremacy." It is nowadays possible to find the view of quantum computing as a respectable, trendy, cutting-edge, hard science, while research in quantum foundations is practiced by philosophers of physics, at best, or charlatans, at worst. However, it is our strong belief that the field of quantum computing and the field of quantum foundations are really two sides of the same coin.

Indeed, while Richard Feynman is credited with the basic idea of quantum simulation, pointing out that a quantum system is best suited for simulating other quantum systems (Feynman, 1982), the modern field of quantum computing may be said to begin with a conversation between Charles Bennett and David Deutsch, of whom the latter proposed the general formulation of a quantum computer (Deutsch, 1985), and in a further development of that paper, together with Richard

Jozsa, developed the first quantum algorithm that showed a computational advantage compared to its classical counterpart (Deutsch and Jozsa, 1992). This was two years prior to Peter Shor's celebrated factoring algorithm paper (Shor, 1994), and four years prior to Grover's search algorithm (Grover, 1996).

It is fitting that Deutsch helped to start this field. An eccentric genius, he has never held a conventional faculty job (or indeed, any job), but continues to the present day to hold courtesy appointments and stay home writing books. He is also passionate about topics of fundamental importance. In Deutsch's first paper on the topic, he repeatedly discusses the Everett many-worlds' interpretation in the presentation of his ideas as the only possible interpretation. We bring it up here to stress that quantum computing began as an exercise in quantum foundations research, by a scientist who has never had a job. Progress in quantum mechanics, as in science in general, can only be made by pushing the limits of a theory, both practical and conceptual. By asking questions that have never been asked, proposing new kinds of experiments, and searching for new phenomena in the lab, we can hope to come to a deeper understanding of the physical world. Sometimes that can bring up bizarre ideas, and work in quantum foundations continues to the present day.

In the context of quantum measurement, while its origins must be clearly understood to underpin the foundations of the theory, it can also be used as a resource for quantum information tasks. Measurement can be – and often is – used to prepare quantum states before a quantum information task begins. It can be used to monitor and correct errors in the context of a quantum error correction protocol. It can even be used to generate entangled states between distant quantum systems. The trick is to do it right – measurement is no longer simply "looking"; what you choose to learn changes the system being measured, and by learning selectively and carefully, it becomes another tool in the box of the quantum mechanic.

1.6 Generalized Measurements

Quantum mechanics is a theory of probability. As such, it is important to understand the mathematics of probability and different ways to understand probability as a concept. In classical probability theory, we learn simple principles; that, after defining a set of possible events, by observing the frequencies of the events, one can define an empirical probability distribution in the limit of many events.

Reasoning about Ignorance: Classical Probability Theory

The mathematics of probability is the systematic reasoning about degrees of ignorance, such as the probabilities of disjoint events are to be added, while the probabilities of two independent events are to be multiplied. We can also define the notion of joint and conditional probabilities when reasoning about multiple events.

The conditional and the unconditional probabilities are related by the law of total probability, which we will discuss in Chapter 3. The conditional probabilities can be related to each other by Bayes' rule, which provides a powerful way to reassign probability assignments based on the availability of new data – this concept will also be explored in more detail in Chapter 3. We review the mathematical formulation of probability theory in Appendix A. There are two main schools of thought about probability theory – the first is the "frequentist" school. This approach to probability is that probability is only a meaningful concept in the first situation outlined above – as the limit of large data concerning the frequencies of outcomes. The second is commonly called the "Bayesian" school of thought, which can best be summarized as "probability is not real," and is merely a reflection of our beliefs about a given situation. The reader can then well appreciate that, if there is such a diversity of thought about the meaning of classical probability theory, quantum mechanical schools of thought about the interpretation of the theory are even more varied! We will give a reflection on them in the epilogue at the end of this book.

Quantum Mechanics as a Generalized Theory of Probability

The above examples of classical information theory have quantum generalizations. One of the fundamental principles of quantum mechanics, the uncertainty principle, gives a limit to our ability to reduce the joint uncertainty of complementary variables. This illustrates that notions inherent in classical probability theory must be suitably generalized when applying statistical reasoning to quantum mechanical systems. Quantum systems do not have only probabilities of outcomes; rather we assign wavefunctions or state vectors to the system (or more generally, density matrices) as a generalized notion of probability. These state vectors encode a richer space of possible measurement outcomes. There is more information contained about the system than simply the probabilities of the outcomes in a fixed basis; the quantum states also contain the coherences between the states that are important to keep – in particular when explaining a sequence of measurement results or measurements in a variety of different bases.

When we confront common examples of measurements applied to quantum systems in the lab, we quickly realize that the idealized wavefunction collapse assumption breaks down, and a new concept is necessary. The simplest way of generalizing this concept of wavefunction collapse is to draw on the previously mentioned model of von Neumann, and to introduce another system into the analysis that plays the role of the measuring device. This device need not be described classically, and indeed should be treated fully quantum mechanically. If we allow both entities to interact with each other over some period of time, then the resulting joint state will become entangled. We cannot correctly describe the measurement outcome statistics of one without referencing the other. Viewing one of the systems as the "meter" allows us to apply the projection postulate to that subsystem, giving us the probability of the readout apparatus when the system is also not directly measured. In this way, we have indirect access to the system of interest. By examining the meter outcomes, we can use our understanding of the correlations between the system and the meter encoded in the entangled state to make inferences about the quantum state of the system – even if it is previously unknown to us. The way we can assign probabilities to the meter results, as well as how the quantum state is disturbed by the measurement process, can be formalized, and this will be discussed extensively in Chapters 3 and 4. This is usually described in the literature with a poor choice of name – "positive operator-valued measures," or POVMs. We will usually just refer to this formalism as generalized measurement. Once the formalism is in place, the role of the meter can be abstracted away, and we discuss the system again in isolation under the influence of these new kinds of generalized measurement; however, we should not forget about the other part of the story and that there are other degrees of freedom that are ultimately responsible for these disturbances.

Breaking the Cardinal Properties of Textbook Measurements

Generalized measurements lead to many interesting phenomena that break the cardinal properties of textbook measurements: its projective nature, its irreversible nature, and its instantaneous nature.

On breaking projection. – One of the first consequences of the previously mentioned generalization is the concept of the information versus disturbance trade-off. By varying how long the system and the meter interact with each other, the amount of entanglement can be varied from zero to a maximum. The act of then measuring the meter gives rise to two corresponding effects – the first is the amount of information we can learn about the system of interest from the meter, and the second is the degree of disturbance we give to the system's quantum state. On one side of the limit is textbook wavefunction collapse – we obtain perfect knowledge of the property of the quantum system we are measuring when there is maximal entanglement between system and meter, with an associated projection of the quantum state onto the eigenstate of the observed property. On the other side of the limit is when there is no entanglement between the system and meter, and consequently no knowledge of the properties of the system is obtained. The system state is completely unaffected by what happens to the meter. We can move continuously between these extremes by varying the degree of entanglement between system and meter. The greater the knowledge we obtain about the system, the larger the associated disturbance is.

A very interesting limiting case here is the concept of a "weak measurement." In this case, we consider a limiting process where there is only a tiny amount of entanglement between system and meter. The resulting measurement of the meter is mostly uncorrelated with the state of the system – but a tiny amount of correlation remains. By repeating the process many times with identically prepared systems, and averaging the results of the meter over that ensemble, it turns out that the average system property can be extracted, even though the disturbance to the system state is arbitrarily small in each run of the experiment. This limit is useful to formalize, since we can discuss certain counterintuitive properties of quantum systems in the limit of negligible disturbance – essentially taking a peek into the workings of the quantum world without much of the associated baggage of projective measurements.

On breaking irreversibility. – We have already discussed how the projection property must be generalized into a partial wavefunction collapse. Another property that falls is the irreversible nature of measurement. Let us recall the words of Niels Bohr: "It is imperative to realize that in every account of physical experience one must describe both experimental conditions and observations by the same means of communication as one used in classical physics. In the analysis of single atomic particles, this is made possible by irreversible amplification effects – such as a spot on a photographic plate left by the impact of an electron, or an electric discharge created in a counter device – and the observations concern only where and when the particle is registered on the plate or its energy on arrival at the counter" (Bohr, 2010). Bohr lays the source of classicality in the measurement result at the feet of irreversible amplification. In his famous article "Law Without Law," John Wheeler expresses this idea with poetic flare: "We are dealing with an event that makes itself known by an irreversible act of amplification, by an indelible record, an act of registration" (Wheeler and Zurek, 2014). He credits the use of the word "indelible" to the Dutch physicist Belinfante (2016).

One great discovery in the study of generalized quantum measurement is that just because the act of amplification is irreversible and associated with an indelible measurement record does not mean that wavefunction collapse is. We have seen that wavefunction collapse can be seen as lying upon a continuum, depending on the strength of the entanglement between system and meter. For this reason, we can consider a sequence of two generalized measurements – one that partially collapses the state, and another that "uncollapses" the state to restore it. The necessary condition for the second uncollapsing measurement is that the net information acquired about the system in both measurements – each of them indelible – is zero. That is, while both measurements are the result of amplification, the right combination of measurement results leaves the quantum state of the system untouched – as if it had never been measured in the first place. Remarkably, this is true even if the observer has no idea what the original quantum state is. We will go into greater detail about this startling effect in Chapter 8.

On breaking instantaneity. – Once we have the concept of a sequence of two generalized measurements, it is a natural step to the concept of a continuous measurement. This is the last, and most dramatic, paradigm break from textbook measurement theory, where we learn that the measurement process is instantaneous. While this idea of instantaneous collapse is latent in the writings of Einstein, it is made explicit by Heisenberg: "There is then a definite probability for finding the photon either in one part or in the other part of the divided wave packet. After a sufficient time the two parts will be separated by any distance desired; now if an experiment yields the result that the photon is, say, in the reflected part of the packet, then the probability of finding the photon in the other part of the packet immediately becomes zero. The experiment at the position of the reflected packet thus exerts a kind of action (reduction of the wave packet) at the distant point occupied by the transmitted packet, and one sees that this action is propagated with a velocity greater than that of light. . . " (Heisenberg, 1949). Heisenberg's idea that the wavepacket reduction is immediate (or instantaneous) has become part and parcel of textbook quantum mechanics.

However, it has gradually become clear to physicists that all measurements take some time to occur. This naturally leads to the question "How long?" concerning the associated wavefunction collapse. The answer to this question simply depends on you, the experimenter. It can take as long or short a time as you like, in principle. More formally, we can distinguish two main types of time-continuous measurements, "quantum jumps" and "quantum diffusion." They are both characterized by acquiring information about the quantum system at some rate, leading to the dynamics of the quantum system no longer described only by the Schrödinger equation, but by another type of dynamics we will examine in the following chapters. We can consider this type of continuous measurement as being mathematically described as a repeated sequence of weak measurements where the measurement strength limits to zero as the time window of the description is decreased. Like all stochastic processes, quantum diffusion is only a quasi-continuous process in reality, holding when there is a wide separation of timescales between the system dynamics and the dynamics of the detector. What is remarkable about the resulting formalism is that the underlying quantum state of the system can be "tracked" in time as the data stream comes in from the detector. These so-called "quantum trajectories" can be followed and validated even for a single run of the experiment, despite the fact that they cannot be predicted in advance. The mathematical description of this set of phenomena goes beyond both projection operators and the Schrödinger equation. We must describe the combination of continuous, coherent, and nonunitary dynamics of the quantum state. These topics will be explored in detail in Chapters 5 and 6.

1.7 What You Will Learn in This Book

This text assumes some familiarity with quantum mechanics and builds on it. You will learn about the topics we introduced in this chapter, along with other related topics. While the text is laid out in a pedagogical way from cover to cover, the reader who wants to learn immediately about a certain topic can consult the chart that follows for a list of key topics and chapters in the text, as well as a flowchart for following a sub-thread of the topics, shown in Fig. 1.9. The flowchart is organized in three main topics – Fundamental Concepts, Advanced Topics, and Applications and Implementations. Depending on your interest, you can skip to the chapter of interest and follow the other order of study.

Fundamental Principles

In Chapter 2, we will review basic facts of quantum measurement that are usually discussed in basic texts on quantum mechanics. These include a motivating experiment – the Stern–Gerlach effect – and discussions of measurement results, statistics, the Born rule, and wavefunction collapse. Chapter 3 takes a step beyond textbook measurements and introduces generalized measurements, beginning with the motivating experiment of an optical polarization measurement with a calcite crystal. In this case, wavefunction collapse is imperfect, and we discuss how to describe and predict the statistics of outcomes and how to assign postmeasurement states. This topic is closely related to Bayesian probability theory, and we discuss a "Quantum Bayes Rule." In Chapter 4, we take a limit where the coupling of the measurement apparatus to the quantum system is very small, and in this limit, we discuss weak measurements and weak values. The latter involves a sequence of a weak and a strong measurement. We discuss generalizations of these effects using the concepts in Chapter 3, and introduce generalized eigenvalues of quantum observables that can exceed the eigenvalue range, expanding the concept of observables in generalized measurement theory.

Advanced Topics

This section moves away from discrete sequences of measurements and dives into quantum measurement as a dynamical process. Chapter 5 considers the case of *diffusive* continuous measurements, where the measurement outcomes and quantum state dynamics is analogous to a Brownian noise process. We motivate this type of measurement by considering the example of a double quantum dot quantum system being measured by a quantum point contact. The intrinsic shot noise of the measurement naturally brings about an effective time-continuous measurement. A second example of a superconducting circuit made from a Josephson junctions readout with a microwave-frequency electromagnetic wave is also discussed in detail. The mathematics of a formally continuous quantum trajectory theory is then pedagogically built up, resulting in the stochastic Schrödinger equation,

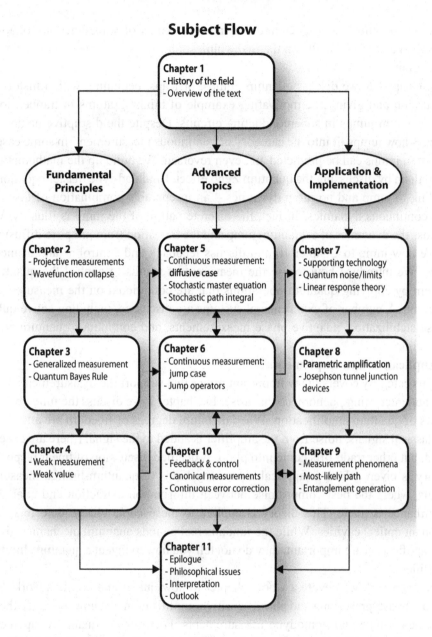

Subject Flow

Chapter 1
- History of the field
- Overview of the text

Fundamental Principles

Advanced Topics

Application & Implementation

Chapter 2
- Projective measurements
- Wavefunction collapse

Chapter 5
- Continuous measurement: diffusive case
- Stochastic master equation
- Stochastic path integral

Chapter 7
- Supporting technology
- Quantum noise/limits
- Linear response theory

Chapter 3
- Generalized measurement
- Quantum Bayes Rule

Chapter 6
- Continuous measurement: jump case
- Jump operators

Chapter 8
- Parametric amplification
- Josephson tunnel junction devices

Chapter 4
- Weak measurement
- Weak value

Chapter 10
- Feedback & control
- Canonical measurements
- Continuous error correction

Chapter 9
- Measurement phenomena
- Most-likely path
- Entanglement generation

Chapter 11
- Epilogue
- Philosophical issues
- Interpretation
- Outlook

Figure 1.9 Possible reading flow of this book. The chapters are listed vertically by themes. Arrows indicate possible reading order of the chapters, depending on reader expertise and interest.

the stochastic master equation, and the stochastic path integral. We also discuss experimental data and its comparison with this theoretical formalism. These experiments allow us to peer into the inner workings of wavefunction collapse, giving an empirical handle on the many philosophical issues that arise in quantum measurement.

In Chapter 6, we discuss quantum jumps, or leaps, beginning with a historical discussion and giving the motivating example of blinking atoms in trapped ions and quantum jumps in superconducting circuits. Despite the disruptive name, we discuss how jumps fit into the category of continuous measurement. In some cases, quantum jumps can be predicted and even reversed. We build up the mathematical formalism by discussing the quantum Zeno effect, Lindblad-type master equations, leading to jump and no-jump dynamics – an inseparable combination of discrete and continuous dynamics. In fact, the discrete nature of the jump is illusory. We discuss the dynamics of a quantum jump and the transition from jumps to diffusion.

We now jump to Chapter 10 and discuss feedback and control as an advanced topic. We introduce how to use the measurement results to control the quantum system by applying quantum operations that are conditional on the measurement outcome. A number of experimental systems are discussed, including active qubit phase stabilization, adaptive phase measurements, and continuous quantum error correction.

Application and Implementation

This category covers many important topics that support the measurement process, and interesting phenomena that arise. In Chapter 7, we discuss the fundamental limits of quantum amplification. When quantum degrees of freedom are amplified to classical signals, noise from the amplifier is added to the signal (there are exceptions, but other trade-offs come into play). A detailed discussion of linear response theory is given, which is applicable to many kinds of quantum-limited measurements, where the best compromise between information extraction and quantum disturbance is made. This theory is applied to mesoscopic charge detectors and resonant optical cavities. While the fundamental bounds quantum mechanics gives to amplification are important, they do not tell you how to invent a quantum-limited amplifier.

In Chapter 8, we devote a whole chapter to how quantum amplifiers work. We discuss phase-preserving and phase-sensitive amplification and how to realize these with heterodyne and homodyne measurements. Focusing on quantum superconducting circuits, we discuss three-wave and four-wave mixing, and the different types of circuits to build in order to realize amplification, and how it can go wrong.

Chapter 9 explores many interesting phenomena that arise in the physics of continuous quantum measurement. We discuss measurement reversal (or "quantum uncollapse"), the most likely path of continuous quantum measurements, the joint simultaneous measurements of noncommuting observables, and entanglement of distant quantum systems by continuous measurement.

The book concludes with Chapter 11, where we give in an epilogue a more philosophical reflection on the state of the field. We discuss what it all means, where the field is going, how quantum computers are the ultimate test of quantum mechanics, and speculate on a future post quantum science.

2

Projective Measurement

2.1 The Stern–Gerlach Experiment

In 1922, Otto Stern and Walther Gerlach published what has come to be seen as a paradigmatic experiment on quantum measurement (Gerlach and Stern, 1922). It revealed the existence of the spin degree of freedom, although it was not recognized at the time. The Stern–Gerlach (SG) experiment sent a beam of hot silver atoms through a region with spatially varying magnetic field; see Fig. 2.1. The silver atoms formed a film on a collecting surface (and were originally detected when the smoke from Otto Stern's cigar turned the film black!). The beam of silver atoms was split apart into two regions that we now interpret as "spin up" and "spin down." The silver atom has one valence electron, and the valence electron's spin is responsible for this effect.

We identify the two important degrees of freedom as the quantum spin variable, and the transverse position of the atom after it has traversed some distance. The interaction between the spin and the diverging magnetic field comes from the magnetic dipole moment of the electron μ. In a constant field, the magnetic moment experiences a torque that orients it with the magnetic field, whereas in a gradient field, the moment experiences a force, the direction of which depends on the orientation of the moment. More formally, the Hamiltonian is of the form

$$H = -\mu \cdot \mathbf{B}, \tag{2.1}$$

with a magnetic field $\mathbf{B}$. Classically, we expect a force in the (say) z-direction,

$$F_z = \mu_z \frac{\partial B_z}{\partial z}, \tag{2.2}$$

when the magnetic field gradient is also in the z-direction. Here the subscripts indicate the component of the vector.

The SG experiment resulted in two distinct spots on the far screen, rather than a continuous blob, as would be classically expected from a random distribution of magnetic moment components. The experiment was very challenging, requiring an

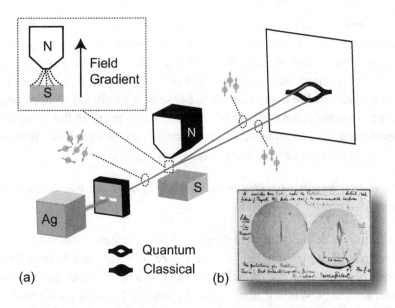

Figure 2.1 (a) In the original Stern–Gerlach experiment, a beam of silver atoms is generated in an oven and is collimated as it traverses through a magnetic field gradient. The quantized spin of the silver atoms' electrons interacts with the field, resulting in a force which causes the atoms to separate into spin-up and spin-down species, as recorded on a piece of photographic paper downstream. Classically, one would expect a random distribution to be peaked at the center. (b) Stern and Gerlach sent their results on a postcard to Niels Bohr, reproduced here with permission from the Niels Bohr Archive.

alignment of the slits and pole pieces to an inaccuracy of no more than 0.01 mm to generate the deviation of 0.2 mm finally observed by Stern and Gerlach (Friedrich and Herschbach, 1998). While the team sought to prove that the orbital angular momentum is quantized, as put forward by Bohr, additional analysis led them to the conclusion that in this case they were dealing with an underlying quantum degree of freedom (spin) where the magnetic moment takes on two values, $\pm\mu_z$. Indeed, if we rotate the SG device so the field gradient is in another direction (say x), then the two spots of silver atoms also appear, but now split in the x-direction. Technically, the value of μ_z can be related to the Bohr magneton, $\mu_B = e\hbar/(2m_e)$, where m_e is the electron mass, e the electron charge, and $\hbar$ the reduced Planck's constant, $\hbar = h/2\pi$. Planck's constant is given by $h = 6.62607015 \times 10^{-34}$ J s and is the characteristic physical constant of quantum physics, having the dimensions of an action – energy multiplied by time. The relation is $\mu_z = -g_s\mu_B/2$, where the electron g-factor $g_s \approx 2.002319$ comes from relativistic corrections (Odom et al., 2006). More generally, we can relate the magnetic moment of the electron to its spin operator $\mathbf{S}$ as $\boldsymbol{\mu}_s = -g_s\mu_B\mathbf{S}/\hbar$.

Quantum mechanically, the spin has a quantum description with a state vector $|\psi\rangle$, and for such a two-state system, it can be represented in its own basis. We call it the $|+\rangle, |-\rangle$ basis when the magnetic field gradient is oriented in the z-direction, so any allowed quantum state has the form $|\psi\rangle = a|+\rangle + b|-\rangle$, where a, b are complex coefficients. We now consider the case of a single silver atom prepared in this state. Sending it through the SG apparatus will result in the atom arriving at the developing plate in either the high $(+)$ or the low $(-)$ position. Which one? This is one of the most puzzling facts in quantum physics:

Repeated experiments with identically prepared initial conditions can give rise to different final results.

One definition of insanity is doing the same thing over and over again and expecting different results. Quantum mechanics produces different results from the same initial conditions all the time, which may explain why so many practitioners of the subject are crazy. While it may seem like science loses all power to describe and predict in such a situation, we can adopt a less exacting standard than perfect prediction and be content with statistical predictions. That is, you do not try to determine where the atom goes every time; rather you are content to simply predict the probability of the result. Quantum mechanics is this kind of theory – it only gives us the odds of events happening. In the context of the SG device, quantum theory says that the probability P of the atom landing in the high $(+)$ or low $(-)$ position is given by

$$P_+ = |\langle+|\psi\rangle|^2 = |a|^2, \qquad P_- = |\langle-|\psi\rangle|^2 = |b|^2. \tag{2.3}$$

Here $|\ldots|^2$ indicates the squared magnitude of the complex numbers. We have adopted the Dirac notation, where bras $\langle\phi|$ can be combined with kets $|\psi\rangle$ to produce bra-kets, $\langle\phi|\psi\rangle$, a complex number. This notation is wide-spread and we encourage the unfamiliar reader to consult any elementary text in quantum mechanics for more details. This complex number may also be viewed geometrically as the *projection* of a vector in Hilbert space $|\psi\rangle$ onto a dual vector $\langle\phi|$, as we will discuss in more detail later.

Similarly, if we were to rotate the SG device in the x-direction, we would define the basis vectors in this direction as $|+x\rangle, |-x\rangle$, which are related to $|+\rangle, |-\rangle$ by a unitary transformation. The same state discussed earlier can be equivalently written as $|\psi\rangle = c|+x\rangle + d|-x\rangle$, where c, d are complex coefficients related to a, b by the unitary transformation, which involves the relative orientation of the rotated SG device to the original SG device. With this new representation of the quantum state, the probability assignments that correspond correctly to the experimental data are also of the form (2.3), but with the new coefficients c, d taking the place of the old coefficients a, b.

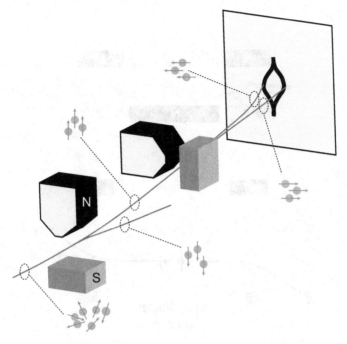

Figure 2.2 In a concatenated sequence of Stern–Gerlach-type experiments, each set of magnets induces a measurement along its respective field gradient. A random beam is first separated into components that align or anti-align vertically. If one of these components is measured by a magnet orthogonal to it, it is then decomposed into a new set of eigenstates that are oriented horizontally.

A crucial part of quantum formalism is to predict what happens to the quantum state post-measurement. In this case, the state of the silver atom is now found in either the high (+) or the low (−) position, and we say the spin state is *collapsed* to either the state $|+\rangle$ or $|-\rangle$, defined by whatever position we place our SG device in. This is nothing more than readjusting our state assignment to reflect what we know to be true. If the electron is found to have spin $|+\rangle$, then we must assign the state to now be $|+\rangle$ (see Fig. 2.2).

Sequences of SG Devices

We note in the previous section of this chapter that the measurement process is crucial to the probability assignments. Instead of detecting the position of the atom, suppose we consider a sequence of two back-to-back SG devices. If the second device has reversed the direction of the magnetic field gradient with a longer extent and we carefully arrange the spacing between the two devices together with the final detection screen, then tracking the dynamics of first the $|+\rangle$ state and then the

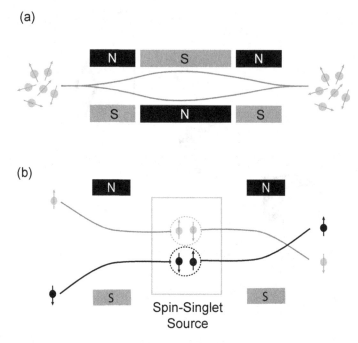

Figure 2.3 (a) Feynman proposed an arrangement of magnets where a beam of spins is separated and then recombined, demonstrating how entanglement can be manipulated before an irreversible projection onto the measurement screen takes place. (b) If a spin-singlet state is produced where the spin species are always anti-correlated, this property is detected when subjecting each component of the beam to a Stern–Gerlach-type measurement. This arrangement is central to many Bell-test experiments.

down $|-\rangle$ state reveals that the up (down) push of the first magnetic field gradient is compensated by the down (up) push of the second magnetic field gradient. The net result of this is nothing! Consequently, detection of the final position of the silver atom does not give any information to us about the spin state of the electron, so no spin measurement occurs. A more clever design described by Feynman (see Fig. 2.3(a)) consists of three SG magnets, two identical ones bookending a central magnet with twice the spatial extent but opposite polarity. The net result of this device is not only to split the beams and to bring them back together, but also to let them continue on their way (Feynman et al., 2011).

More generally, we can consider an arbitrary sequence of SG devices, where we can make a measurement of the spin state at each stage of the experiment by inserting a *beam block* after the SG device. This has the effect of removing the spin state we no longer wish to consider from the experiment, and then we can direct the remaining part of the beam into the next SG device. The next SG device can be oriented at some other angle relative to the first SG device. In this way, we can make a whole series of measurements, or we can also use the first measurement to prepare a known state to be analyzed by the second SG device and detection process.

Let us consider, for example, a general initial state, followed by a z-oriented SG device with the $|-\rangle$ state blocked, followed by an x-oriented SG device, followed by a detection screen. The first device will result in finding the atom passing the beam block with probability $|a|^2$, and in producing the state $|+\rangle$, as discussed in the previous section. The second device will find the atom in the position on the final screen corresponding to either state $|+x\rangle$ or state $|-x\rangle$. Using our probability assignment rule (2.3), we find the results:

$$P_{+x} = |\langle +x|+\rangle|^2, \qquad P_{-x} = |\langle -x|+\rangle|^2. \tag{2.4}$$

The experiment shows these probabilities are both $1/2$, so both outcomes are equiprobable. By rotating the SG devices in different directions, the coefficients determining the probabilities can be mapped out. For example, the x basis states are expressible in terms of the z basis states as

$$|\pm x\rangle = \frac{1}{\sqrt{2}}(|+\rangle \pm |-\rangle), \tag{2.5}$$

which are consistent with the previously noted probability assignments but contain more information, such as the fact that while the SG-x device will always find an $|+x\rangle$ deflected in the $+x$ direction on the screen, a SG-z device will find it deflected either up or down, half the time. From this kind of reasoning, the entire Hilbert space structure of quantum mechanics of spin-1/2 particles can be built up.

2.2 Measurements on Multiple Systems

Quantum mechanics becomes even more interesting when we consider the physics of multiple systems. Here we consider a variation on the SG device, where we have two SG devices, but also two particles. Suppose the two particles are both spin-1/2 for simplicity but emerge from a common source, which creates *entanglement* between them, a quantum mechanical correlation that we will now describe. Suppose the creation of this pair of particles conserved angular momentum. We would then expect that if one particle has a magnetic moment pointing up, the other should be pointing down. However, we have no preference as to which should be which. So, we write down a quantum state for both possibilities, $|+\rangle \otimes |-\rangle$, and $|-\rangle \otimes |+\rangle$. Here the $\otimes$ symbol indicates a direct product of two quantum states belonging in different Hilbert spaces. The first register indicates the first particle, while the second register indicates the second particle. Quantum mechanically, we are allowed to have coherence between these two states, so any combined quantum state of the form

$$|\psi\rangle = a|+\rangle \otimes |-\rangle + b|-\rangle \otimes |+\rangle \tag{2.6}$$

is possible. In the equiprobable case, we have that $|a|^2 = |b|^2 = 1/2$, but a and b may have a relative phase that is physically important. Let us take for simplicity the so-called *singlet* state,

$$|\psi_S\rangle = (1/\sqrt{2})(|+\rangle \otimes |-\rangle - |-\rangle \otimes |+\rangle), \tag{2.7}$$

and suppose this is the state prepared before the experiment begins. The singlet turns out to have total spin 0, so it is a natural product of a decay process that conserves angular momentum from a spinless initial state.

We now consider two SG-z devices on either side of the source, so that by momentum conservation, if one spin enters the first SG device, the second enters the second device (see Fig. 2.3(b)). The question we are considering is where the spins land on the two different screens. One spin-1/2 particle must be registered in each for a successful detection. As in the single-particle case, the magnetic field gradient exerts a force on the particle, so if each particle can be registered high (+) or low (−) (corresponding to spin up or down), then there are in general four possibilities: ++, +−, −+, −−. What we find from experiments is that for the singlet state, results ++ and −− never occur, and we only see +− or −+. That is, whenever we see one SG device register "high," we know *for sure* that the other SG device will register "low," but we cannot say in advance which is which. But we can say that, by projecting on the bras $\langle -| \otimes \langle +|$ or $|\langle +| \otimes \langle -|$, half the time it will be one way, and the other half of the time it will be the other.

One of the amazing things about quantum mechanics is that we can keep this exact same source with the same singlet state, rotate the SG devices to be SG-x devices, and ask what happens. The quantum calculation is straightforward. We start with the singlet state (2.7), using the conversion from x into z basis states (2.5), to find, after some algebra,

$$|\psi_S\rangle = (1/\sqrt{2})(|-x\rangle \otimes |+x\rangle - |+x\rangle \otimes |-x\rangle). \tag{2.8}$$

This is the same singlet state as before! The only difference is an overall sign that does not affect the probability assignments in any basis. It turns out the singlet state has the remarkable property that no matter what basis it is expressed in (among the infinite number of choices), it remains invariant, up to an overall phase. The physical consequence of this is that no matter how the SG devices are turned, the same conclusions we first drew remain true: the spin orientation will always remain perfectly anti-correlated, but with no way of predicting which device will get the "high" register and which will get the "low" one. This type of experiment is behind the reasoning of Bell's inequality, discussed in the first chapter.

2.3 Mathematics of Projective Measurement

Now that we have physically motivated some of the basic phenomena of quantum measurement, we will give a more complete mathematical treatment to present the general structure of the theory more abstractly. We have been describing the simplest kind of measurement – the projective measurement. This corresponds to an idealized limit where perfect knowledge about the quantum system is gained by some means. It can be direct interaction, auxiliary system interaction, or inference about the system in some other way. The name comes from the description in terms of Hermitian (self-adjoint) *projection operators* $\hat{\Pi}_j$, where j indexes the number of possible outcomes that measurement can take. In the example of the SG device in the previous section, the single-particle case corresponds to two possible outcomes, whereas the two-particle case corresponds to four possible outcomes. A projection operator has the property that

$$\hat{\Pi}_i\hat{\Pi}_j = \hat{\Pi}_i\delta_{ij}. \tag{2.9}$$

Here, δ_{ij} is the Kronecker delta function – it takes a value of 1 if $i = j$, and is 0 otherwise. The projector set also obeys a completeness relation,

$$\sum_i \hat{\Pi}_i = 1, \tag{2.10}$$

where **1** indicates the identity operator. Here we consider the simplest case where the number of different detector outcomes equals the dimension of each projector $\hat{\Pi}_i$. The operators are defined by their action on quantum states. We let $|\psi\rangle$ be a Dirac "ket" residing in a Hilbert space, as mentioned in Section 2.1. This quantum state represents the quantum mechanical description of the system of interest. Suppose now a measurement is made, where a number of outcomes are possible. The probability of outcome j, P_j, is generally given by

$$P_j = \langle\psi|\hat{\Pi}_j^\dagger\hat{\Pi}_j|\psi\rangle = \langle\psi|\hat{\Pi}_j|\psi\rangle, \tag{2.11}$$

where we have made use of the fact that the projection operators are self-adjoint. We see immediately that the completeness relation (2.10) of the projectors implies that the probability distribution (2.11) is normalized ($\sum_j P_j = 1$), provided the state is properly normalized. It is convenient to write the projectors in terms of an outer product of some states $\{|j\rangle\}$ of the system Hilbert space with themselves as a rank-one operator,

$$\hat{\Pi}_j = |j\rangle\langle j|, \tag{2.12}$$

where $\langle j|$ is a "bra" dual-representation of the same ket. The projection property
(2.9) implies that the state set $\{|j\rangle\}$ is orthonormal and consequently forms a basis
in the system Hilbert space. If we express any state $|\psi\rangle$ in this basis, it takes the
form

$$|\psi\rangle = \sum_j c_j |j\rangle, \qquad (2.13)$$

where the complex coefficients c_j can also be expressed as $\langle j|\psi\rangle$, by the orthonor-
mality of the state set $\{|j\rangle\}$.

We can understand why the operators $\hat{\Pi}_j$ are called projection operators by
considering their action on the general ket $|\psi\rangle$,

$$\hat{\Pi}_j |\psi\rangle = |j\rangle\langle j|\psi\rangle = \langle j|\psi\rangle \, |j\rangle. \qquad (2.14)$$

We notice that the operator takes whatever state we begin with and collapses it to
the ket $|j\rangle$, the associated eigenstate of the projection operator. Furthermore, the
state is no longer normalized, but has a complex coefficient, $\langle j|\psi\rangle$, which is the
inner product of the state with the 'ket j, which is the projection of the state onto
the measurement eigenstate. Its absolute square corresponds to the probability of
the result j, as in Eq. (2.11). That is,

$$P_j = |c_j|^2 = |\langle j|\psi\rangle|^2. \qquad (2.15)$$

How do we know which projection operators to use? The basis of measurement
is specified by the settings of the measurement apparatus. The dimension of the
system Hilbert space is set by the number of possible outcomes of a measurement
on the system (assuming no degeneracies and a maximally informative type of
measurement, a topic which we will return to later). Physically, this is set by the
type of coupling arranged between the system of interest and the measuring device.
That is, the basis of the coupling sets the basis of the projectors. We will see this
more explicitly in the next section.

In the examples in the previous sections, the projection operators corresponding
to a $+z$ measurement or a $-z$ measurement of the SG-z device correspond to

$$\hat{\Pi}_{+z} = |+\rangle\langle+|, \quad \hat{\Pi}_{-z} = |-\rangle\langle-|, \qquad (2.16)$$

whereas the projection operators for a $+x$ measurement or a $-x$ measurement of
the SG-x device correspond to

$$\hat{\Pi}_{+x} = |+x\rangle\langle+x|, \quad \hat{\Pi}_{-x} = |-x\rangle\langle-x|, \qquad (2.17)$$

and so on. In the multiparty case we have parties A and B, so our sample space
is expanded to four events with two indices, i,j. We can generalize the preceding
definition and define the two-party projectors $\hat{\Pi}_{ij} = |i\rangle_A \, {}_A\langle i| \otimes |j\rangle_B \, {}_B\langle j|$, where
$i,j = +, -$. This analysis can be extended to an arbitrarily large number of systems
of varying dimension.

Projection on Eigenstates of an Observable

Very often, we can associate the projectors of the measurements with eigenstates of a Hermitian operator, corresponding to the quantity that is being measured. This operator is typically called an "observable" $\hat{O}$ in quantum mechanics and can be represented as

$$\hat{O} = \sum_j \lambda_j \hat{\Pi}_j, \qquad (2.18)$$

where λ_j are the eigenvalues of the operator, and the set $|j\rangle$ are the eigenvectors of the operator. The Hermitian nature of the operator implies that the eigenvalues are real, and the eigenvectors form a complete and orthonormal set of states as anticipated on the Hilbert space of the system. The state collapse discussed in the previous section implies that any postmeasurement state is an eigenstate of the measured observable associated with the eigenvalue λ_j. Exercise 2.5.2 proves that the set $\{\lambda_j\}$ are the eigenvalues of $\hat{O}$.

The *expectation value* of the observable is the average value, taken over the set of all allowed outcomes, in the initial state $|\psi\rangle$,

$$\langle \hat{O} \rangle = \sum_j P_j \lambda_j = \langle \psi | \hat{O} | \psi \rangle. \qquad (2.19)$$

Here, the expectation value may be interpreted as a weighted average of the eigenvalues of the measurement with respect to the probabilities P_j given in Eq. (2.11). Indeed, in individual realizations of the measurement process of the observable corresponding to $\mathcal{O}$ on identically prepared states, each outcome j occurs, associated with the system eigenvalue λ_j. Statistically, by averaging over this process, we obtain the expectation value. Similarly, higher-order moments of $\mathcal{O}$ may also be considered in order to capture the fluctuations of the process,

$$\langle \hat{O}^n \rangle = \sum_j P_j \lambda_j^n, \qquad (2.20)$$

which you will prove in Exercise 2.5.3.

A measuring device can resolve the physical values of the eigenvalues, such as a Stern–Gerlach device being sensitive to the value of the magnetic moment of the atom. As such, the measurement device resolves the system by its observable eigenvalues, sorting the system accordingly. As such, the postcollapse state corresponds to the eigenstate of the observable that is realized.

We now give some simple illustrations of these general properties for spin-1/2 systems. The observables corresponding to the spin components are written as $\hat{S}_i = (\hbar/2)\hat{\sigma}_i$, where $i = x, y, z$, and $\hat{\sigma}_i$ are the Pauli operators, named in honor of the Austrian physicist Wolfgang Pauli. The total spin has value of $\hat{\mathbf{S}} \cdot \hat{\mathbf{S}} = \sum_i \hat{S}_i^2 = \hbar^2/4$ and is fixed. If we suppose the spin-1/2 particle is in a uniform magnetic field $\mathbf{B}$, then the Hamiltonian is of the form (2.1). Aligning the z-direction along the magnetic field direction, this takes the simplified form $\hat{H} = -\mu_z B \hat{\sigma}_z$. Consequently, the eigenvectors of the Hamiltonian operator are given by the $|+\rangle, |-\rangle$ kets of the $\hat{\sigma}_z$ eigenkets, and the energy eigenvalues are given by $E_\pm = \pm \mu_z B$.

Other common example of operators in continuous variable systems are the position and momentum of a particle. Their expectation values in a quantum state $|\psi\rangle$ are given by $\langle\psi|\hat{x}|\psi\rangle$, $\langle\psi|\hat{p}|\psi\rangle$ and variances $\Delta x^2 = \langle\psi|\hat{x}^2|\psi\rangle - \langle\psi|\hat{x}|\psi\rangle^2$, and $\Delta p^2 = \langle\psi|\hat{p}^2|\psi\rangle - \langle\psi|\hat{p}|\psi\rangle^2$. These satisfy the well-known Heisenberg uncertainty principle, $\Delta x \Delta p \geq \hbar/2$.

Projection on Subspaces of an Observable

In the preceding section, we focused on the simplest case of projection on eigenstates of a Hermitian operator, and we implicitly assumed that all eigenvalues were distinct, the most common situation. As we mentioned in the previous section, the detector physically measures the system by the response of the system according to its observable eigenvalues. Suppose now there is a degeneracy in those eigenvalues, for example we let $\lambda_1 = \lambda_2 \neq \lambda_3, \ldots, \lambda_N$. Then, the observable can be written as

$$\hat{O} = \lambda_1(\hat{\Pi}_1 + \hat{\Pi}_2) + \sum_{j=3}^{N} \lambda_j \hat{\Pi}_j. \qquad (2.21)$$

Suppose the detector shows a result indicating the system takes on the eigenvalue λ_1, but gives no way to further resolve any information within the $\{|1\rangle, |2\rangle\}$ subspace. In that case the appropriate projection operator for the system is neither $\hat{\Pi}_1$ nor $\hat{\Pi}_2$, but rather $\hat{\Pi} = \hat{\Pi}_1 + \hat{\Pi}_2$. The operator $\hat{\Pi}$ is also a projection operator, as you will prove in Exercise (2.5.7).

It is now of great interest to investigate the postcollapse state of the system, given by

$$\hat{\Pi}|\psi\rangle = (\hat{\Pi}_1 + \hat{\Pi}_2)|\psi\rangle. \qquad (2.22)$$

Taking the decomposition of the quantum state given in Eq. (2.13), we find the result

$$\hat{\Pi}|\psi\rangle = \sum_j c_j \hat{\Pi}|j\rangle = c_1|1\rangle + c_2|2\rangle. \qquad (2.23)$$

Consequently, the physical (renormalized) state is given by

$$|\phi\rangle = \frac{c_1|1\rangle + c_2|2\rangle}{|c_1|^2 + |c_2|^2}. \qquad (2.24)$$

That is, the state does *not* collapse into any particular eigenstate of the measurement operator. Rather it falls into a superposition of the kets in the subspace defined by the degeneracy in the eigenvalue of the measured observable.

Let us consider an example of a rigid rotor of fixed length and moment of inertia, tethered at one end. It is prepared with a quantum wavefunction $\psi(\theta, \phi) = \langle \theta, \phi | \psi \rangle$, where θ, ϕ are the polar and azimuthal angles. We consider a measuring device that is able to measure the total angular momentum $\mathbf{L} \cdot \mathbf{L}$, but not any component of angular momentum. We will now find the event probability and postmeasurement state, given that the detector returns a value of $l = 1$, where l is the orbital angular momentum quantum number. That is, the eigenvalue of the Hermitian operator $\hat{L}^2 = \hat{\mathbf{L}} \cdot \hat{\mathbf{L}}$ is given by $\hbar^2 l(l+1)$, where $l = 1$ is the result.

We recall from quantum treatments of angular momentum that, in such a situation, several different values of the azimuthal quantum number m_l are permitted, where these are the eigenvalues of the z component of the angular momentum, $\hat{L}_z$. In this case, m_l can take the values $\{-1, 0, 1\}$. Denoting the eigenkets of $\hat{L}^2, \hat{L}_z$ as $|l, m\rangle$, we define the three projectors $\hat{\Pi}_{1,1} = |1, 1\rangle\langle 1, 1|$, $\hat{\Pi}_{1,0} = |1, 0\rangle\langle 1, 0|$, $\hat{\Pi}_{1,-1} = |1, -1\rangle\langle 1, -1|$. Drawing on the treatment earlier in this section, the correct projection operator for the $l = 1$ output of a detector that only detects total angular momentum is then given by

$$\hat{\Pi}_{l=1} = \hat{\Pi}_{1,1} + \hat{\Pi}_{1,0} + \hat{\Pi}_{1,-1}. \tag{2.25}$$

To find the probability of this outcome occurring (of all the possible values, $l = 0, 1, 2, \ldots$), we can use our result (2.11), so the probability $P_{l=1}$ is given by

$$P_{l=1} = \langle \psi | \hat{\Pi}_{l=1} | \psi \rangle. \tag{2.26}$$

This result can be further simplified by inserting complete sets of states $\int d\cos(\theta) d\phi \, |\theta, \phi\rangle\langle \theta, \phi| = \mathbf{1}$ and expressing the result in terms of integrals of the wavefunction and the spherical harmonics, $\langle \theta, \phi | l, m \rangle = Y_l^m(\theta, \phi)$:

$$Y_1^{-1}(\theta, \phi) = \frac{1}{2}\sqrt{\frac{3}{2\pi}} e^{-i\phi} \sin\theta, \quad Y_1^1(\theta, \phi) = -\frac{1}{2}\sqrt{\frac{3}{2\pi}} e^{i\phi} \sin\theta,$$

$$Y_1^0(\theta, \phi) = \frac{1}{2}\sqrt{\frac{3}{\pi}} \cos\theta. \tag{2.27}$$

The full expression is left to the reader. We can move on to finding the (unnormalized) postcollapse state as

$$|\phi\rangle = c_{-1}|1, -1\rangle + c_0|1, 0\rangle + c_1|1, 1\rangle, \tag{2.28}$$

where we define the coefficients using the extension of the result (2.23) to a three-state system,

$$c_m = \int d\cos(\theta)d\phi\ Y_1^m(\theta,\phi)^*\psi(\theta,\phi), \tag{2.29}$$

where $m = -1, 0, 1$. As we saw previously, the postcollapse state indeed has a definite value of l but resides in a quantum superposition of the three possible m values consistent with the measured l value.

2.4 Continuous Variables

While the previous discussion is focused on systems with a finite Hilbert space, it is possible to generalize the properties discussed there to continuous variable systems. Suppose we wish to measure a continuous variable like position or momentum. We define the eigenbasis of such an observable as $\{|z\rangle\}$, with associated projectors $\hat{\Pi}_z = |z\rangle\langle z|$. The continuous observable $\hat{O}_c$ can then be written as

$$\hat{O}_c = \int dz |z\rangle\langle z|\ \mathcal{O}(z), \tag{2.30}$$

where $\mathcal{O}(z)$ are the continuous eigenvalues of the operator. We must integrate over the index, rather than sum. We can handle the generalizations of Eqs. (2.9, 2.10) by replacing sums by integrals and Kronecker delta functions by Dirac delta functions. Consequently, we have

$$\hat{\Pi}_z\hat{\Pi}_w = \hat{\Pi}_z\delta(z-w), \tag{2.31}$$

and

$$\int dz\hat{\Pi}_z = 1. \tag{2.32}$$

Here the integrals are taken over the suitable range for the system of interest. If we consider the examples of position or momentum measurement, and write the projectors as outer products of position or momentum eigenstates, then the preceding properties can be understood as being identical to the statement that the inner product of these states with themselves yields a Dirac delta function (mathematically this is a distribution, strictly speaking).

2.5 Discussion of the Cardinal Properties of Projective Measurement

We conclude this chapter with a brief discussion of the cardinal properties of projective measurements that we have learned.

(i) The projective nature of the measurement gives maximum distinguishability of the different eigenvalues of the observable being measured. It is also maximally disturbing. No matter which initial state you start with, the projection operator collapses them all to the eigenstate of the observable, $\hat{\Pi}_j|\psi\rangle \propto |j\rangle$, unless $|\psi\rangle$ has no component along $|j\rangle$, in which that measurement outcome will not appear.

(ii) The irreversible nature of projective measurement can be seen both mathematically as well as physically. The aforementioned projection property is a many-to-one mapping of quantum states to quantum states: Every possible state that is not orthogonal to $|j\rangle$ is reduced to that single state. Such an operation has no inverse in the mathematical sense. Physically, if we try to time-reverse the measurement process operationally, we run into problems. Consider the setup of the SG-z device and run it backwards. Suppose we put in the spin state $a|+\rangle + b|-\rangle$ and end in the state $|+\rangle$, corresponding to the atom deflecting up and being measured there. Now let us try and go backwards. If we time-reverse the momentum of the atom in a spin-up state, we have a left-moving, spin-down atom (since spin is an angular momentum, its time-reverse flips sign, like momentum). Now the atom will track the same course backwards, but the end of its course (corresponding to the beginning of the time-forward process) will not recover the initial situation. Even after again time-reversing at this point to try and recover it, we find the spin state is simply $|+\rangle$ and not the coherent superposition we began in.

(iii) Its instantaneous nature seems to hinge on what happens at the screen (or beam block). If at any point in time before detection of the atom's position, the process does indeed appear to be time-reversal invariant. If we describe both the orbital and spin degrees of freedom, then before detection, we have the state

$$|\Psi\rangle = a|+\rangle \otimes |\text{deflect up}\rangle + b|-\rangle \otimes |\text{deflect down}\rangle. \tag{2.33}$$

If we instead take this state and time-reverse it, we will then reverse the direction of the particle's momentum and spin direction via the time-reversal operator $\hat{\Theta}$, an anti-unitary operator, such that both momentum kets $\hat{\Theta}|p\rangle = |-p\rangle$ and spin direction kets $\hat{\Theta}|S_j\rangle = |-S_j\rangle$ are inverted. In this situation, the original state is recovered after running the process an equal amount of time and again time-reversing! Hence the collapse of the wavepacket appears to happen instantly upon reaching the detecting screen or beam block, resulting in the reduced state. This reasoning supports Heisenberg's argument mentioned in Chapter 1, that the detection of a particle at one place after a scattering event immediately suppresses the amplitude to find the particle anywhere else, no matter how far apart these possible locations are.

In the chapters to come we will see how all of these properties can be broken.

Exercises

Exercise 2.5.1 Consider a Stern–Gerlach apparatus with a fringing field of 100 gauss per cm. What length of the magnet is required to have a spacing of 1 mm on the screen placed 40 cm from the device? The oven temperature is approximately 250 degrees centigrade (use the typical velocity from this oven for a silver atom). First solve the general problem, and then put in specific values.

Exercise 2.5.2 Using Property (2.12) of the projection operators and the orthonormality of the set $|j\rangle$, prove λ_j are the eigenvalues of operator $\hat{O}$.

Exercise 2.5.3 Show that the formula (2.20) is correct, given the properties (2.9) of projection operators.

Exercise 2.5.4 From the two properties of projective measurement operators (2.9, 2.10), show that the projection operators may be written as the outer product of a complete set of orthonormal basis states with themselves.

Exercise 2.5.5 Express the projection operators for $\pm x$ spin-1/2 measurements (2.17) in the z-basis.

Exercise 2.5.6 Consider a spin-1/2 system and a general SG-θ device that can project the spin onto an arbitrary angle. Recall that if we rotate a spin-1/2 system about a direction described by the unit vector $\hat{n}$ characterized by the polar and azimuthal angles, β and α respectively, we can describe the "up" spinor in the z-basis as

$$\chi = \begin{pmatrix} \cos(\beta/2)e^{-i\alpha/2} \\ \sin(\beta/2)e^{+i\alpha/2} \end{pmatrix}. \tag{2.34}$$

Construct the projection operators for such a device, expressed as matrices in the z-basis. If a spin is prepared in the $S_z = +\hbar/2$ eigenstate, what is the probability of finding it in the $S_\theta = +\hbar/2$ state after the SG-θ device?

Exercise 2.5.7 Prove the operator $\hat{\Pi} = \hat{\Pi}_1 + \hat{\Pi}_2$ is also a projection operator using Properties (2.9).

Exercise 2.5.8 Consider a three-level system prepared in the normalized state

$$|\psi\rangle = a|1\rangle + b|2\rangle + c|3\rangle, \tag{2.35}$$

where a, b, c are complex coefficients and kets $|1\rangle, |2\rangle, |3\rangle$ are an orthonormal basis.

Consider the operators

$$\hat{O}_1 = \frac{1}{3}(1 + |1\rangle\langle 2| + |2\rangle\langle 1| + i|3\rangle\langle 2| - i|2\rangle\langle 3| - i|1\rangle\langle 3| + i|3\rangle\langle 1|),$$

$$\hat{O}_2 = \frac{1}{2}(1 - |2\rangle\langle 2| + i|1\rangle\langle 3| - i|3\rangle\langle 1|),$$

$$\hat{O}_3 = \frac{1}{6}(1 + 3|2\rangle\langle 2| - 2|1\rangle\langle 2| - 2|2\rangle\langle 1| + i|3\rangle\langle 1| - i|1\rangle$$
$$\langle 3| - 2i|3\rangle\langle 2| + 2i|2\rangle\langle 3|).$$

Here 1 is the identity operator.

(a) Show that the three operators form a complete set of projectors on the three-level system by verifying the fundamental properties of projection operators.

(b) If a measurement corresponding to those projection operators is implemented, what is the probability of finding result 2 (corresponding to $\hat{O}_2$) for state $|\psi\rangle$. What does the state collapse to in this case?

Exercise 2.5.9 Recall the example given in the text of a measuring device that detected total angular momentum, but not any component of the angular momentum. Now consider a detector that can only detect the z component of the orbital angular momentum, but not its total angular momentum. Suppose such a detector returned a value of $m = 0$. What values of l are consistent with this result? Write down the correct projection operator for this result, and work out the angular dependence of the postcollapse state.

3

Generalized Measurement

3.1 Measuring the Polarization of a Single Photon

We now begin our discussion of generalized measurements, where the outcome of the measurement apparatus is not perfectly correlated with the quantum state of the system being measured. For this purpose, it is instructive to introduce another motivating experiment. We turn to the subject of optics, which has been a workhorse for the field of quantum science. It is fitting to discuss one of the oldest known optical effects, first noted to occur in calcite crystals. "Iceland Spar," naturally occurring calcite ($CaCO_3$), was described by Erasmus Bartholinus at the University of Copenhagen in 1669. Although legend tells of Vikings using the "sunstone" (thought to be Iceland Spar) to navigate the seas by the partial polarization of the sky, we have no definitive evidence for this much earlier discovery of optical polarization.

Bartholinus performed experiments on the curious properties of this crystal. In gazing through it, a double image would appear. We now understand that the crystal splits the light entering it into two distinct images, corresponding to two different polarizations of the light. While we can describe the physics of optical polarization perfectly well with electromagnetic wave theory, we transition to the quantum level by considering a single photon entering such a crystal. It is a curious fact of nature that the polarization properties of a single photon correspond nearly exactly to those of a classical beam of light from the sun or a flashlight, the main difference being that while the intensity of a light beam passing through a polarizer is continuous in its intensity as the polarizer is turned, a single photon, when detected afterwards, has only two choices: it is found to be a single photon in one of the two possible polarizations. Consequently, the physics of a single photon split by a calcite crystal into one of two polarizations is very similar to the physics of the Stern–Gerlach device deflecting a silver atom up or down, depending on the quantum state of its valence electron's spin. Indeed the quantum mechanics of a

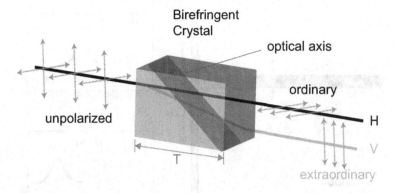

Figure 3.1 In a birefringent crystal, such as calcite, incoming radiation is transmitted unaltered if the polarization is parallel to the plane containing the optical axis and is called the ordinary. The orthogonal polarization is deflected and is called the extraordinary ray.

spin-1/2 object can be mapped on the quantum mechanics of the polarization of a single photon.

The basic physics of this effect comes from the crystalline structure of the material defining an optical axis, coming from the anisotropy of the material (see Fig. 3.1). Light is typically split into two rays: an ordinary ray, and an extraordinary ray. These correspond to two different indices of refraction associated with the two polarizations of light (n_o, n_e) resulting in different amounts of deflection of the optical ray. For a crystal of a given thickness, the optical path difference d between the two rays is the optical birefringence (defined as the difference between the two indices of refraction) times the thickness, T. That is, $d = T(n_e - n_0)$.

Just as for a spin-1/2 system we can define the quantum state of the single photon we have in mind by $|H\rangle$ and $|V\rangle$, corresponding to the horizontal and vertical polarization of light, relative to a reference, such as the optical axis of the crystal. Thus, any polarization state may be written as

$$|\psi\rangle = a|H\rangle + b|V\rangle, \tag{3.1}$$

where a, b are complex coefficients. If the coefficients are real, this state corresponds to linear polarization, but if one of the coefficients is imaginary, the state corresponds to circular or elliptical polarization.

We now consider several experiments where the thickness of the crystal is becoming thinner and thinner. At some point, we notice that the beams each have some finite transverse width that comes into play (see Fig. 3.2). This width cannot be made arbitrarily thin for a beam of light, or else we run afoul of the uncertainty

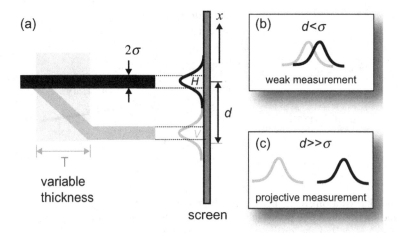

Figure 3.2 (a) After traversing a birefringent crystal, the ordinary and extraordinary rays are separated by a distance d relative to their beam waist σ, and the crystal thus performs a measurement of the polarization of the incoming light. (b) In the limit $d < \sigma$, the histograms corresponding to the different polarization states overlap, and the measurement is said to be "weak" since a single measurement outcome cannot unambiguously distinguish these states. (c) Conversely, when $d \gg \sigma$, the histograms are well separated and the measurement is said to be "projective."

principle. We consider a transverse wavefunction of the photon to be $\phi(x)$. It is usual to consider a Gaussian profile, as this is commonly found for optical states,

$$\phi(x) = \left(\frac{1}{2\pi\sigma^2} \right)^{1/4} \exp\left(-\frac{x^2}{4\sigma^2} \right). \tag{3.2}$$

Here σ is the standard deviation of the position of the photon detected in the beam, with an average of $x = 0$. It quantifies the expected scatter we expect on the final detection screen for an ensemble of measurements. Defining the origin of the x axis as the center of the ordinary beam, we can use the superposition principle to predict the state of the photon before the detection screen. Other details of the physics, such as the frequency of the photon, its temporal mode form, its speed, and so on, are suppressed for simplicity. Only the transverse degree of freedom and polarization are important to discuss here. We see that starting in the initial state

$$|\Psi\rangle = |\psi\rangle \otimes |\phi\rangle, \tag{3.3}$$

where $|\phi\rangle$ is the ket corresponding to the transverse (orbital) degree of freedom, we can find the final state before the screen by shifting the transverse state $|\phi\rangle$ to the left by an amount d. Let us then distinguish the transverse state corresponding to the horizontal polarization, $|\phi_H\rangle$, from the transverse state corresponding to the vertical polarization, $|\phi_V\rangle$, the latter being shifted to the left by an amount d. That is, $\langle x|\phi_H\rangle = \phi(x)$, while $\langle x|\phi_V\rangle = \phi(x + d)$. The superposition principle then dictates that the final state of the photon before the screen is the coherent superposition of the two possibilities,

$$|\Psi'\rangle = a|H\rangle|\phi_H\rangle + b|V\rangle|\phi_V\rangle. \tag{3.4}$$

Notice that if the separation d is much larger than the width of the profile σ, then the two distributions have no region of overlap, resulting in a nearly ideal measurement of optical polarization. That is, by simply detecting where on the screen the photon arrived, we can state confidently what the polarization of the photon is. They are perfectly correlated, so a position measurement stands in for a polarization measurement, made, for example, with a piece of Polaroid. However, for $d \leq \sigma$, the two profiles overlap, so a position measurement becomes imperfectly correlated with polarization. For example, if a photon lands to the far right of the screen, then we are pretty confident it is an H-polarized photon. However, suppose the photon lands halfway between the peaks of the two distributions. Then from position alone, we have no way of inferring whether the photon is H or V polarized.

3.2 Measuring Polarization with Position

We will now use the results of the previous chapter to work out the details of the example system introduced earlier. We wish to predict where the single photon will land on the screen. Furthermore, suppose we poke a small hole in the screen and examine only those photons that make it through the hole. What is their polarization, and does it depend on where we poke the hole in the screen?

The photon will only make it though the hole at position x if we are certain it is there. Consequently, this corresponds to a position measurement. From Section 2.4, we should then describe this situation with a position projection operator, $\hat{\Pi}_x = |x\rangle\langle x|$. Since the measurement is of position only, the action on the polarization degree of freedom is only the identity operator. The probability density in this situation is the continuous variable generalization of Eq. (2.11),

$$p(x) = \langle\Psi|\hat{\Pi}_x|\Psi\rangle, \tag{3.5}$$

and the actual probability of traversing the slit is related to this width. Taking a slit size Δx, the probability to traverse the slit is given by $P \approx p(x)\Delta x$, for suitably small Δx. Noting that $\langle x|\phi_j\rangle = \phi_j(x)$ for $j = H, V$, we can expand this expression to

$$p(x) = [\langle H|a^*\phi_H^*(x) + \langle V|b^*\phi_V^*(x)][a\phi_H(x)|H\rangle + b\phi_V(x)|V\rangle], \tag{3.6}$$
$$= |a|^2|\phi_H(x)|^2 + |b|^2|\phi_V(x)|^2. \tag{3.7}$$

This last result has a simple interpretation: we can view the probability of passing the slit as being a sum of independent events where, given the photon is in either H of V, the probability of passing the slit is either $|\phi_H(x)|^2 \Delta x$ or $|\phi_V(x)|^2 \Delta x$. The net probability is then simply the weighted probability of passing the slit, where the weighting coefficients are the probability for the photon to be in either polarization H or V, that is, either $|a|^2$ or $|b|^2$. We will return to this interpretation later on.

3.3 Polarization State Update

We are now in the position to ask our follow-up question: given that the photon made it through the slit centered at position x, what is its polarization?

We can find this because, according to the discussion around Eq. (2.14), the post-measurement state is simply $|x\rangle$ for the meter. However, what about the polarization state? This can be found from

$$|\Psi''\rangle = \hat{\Pi}_x |\Psi'\rangle, \tag{3.8}$$

$$= (a\phi_H(x)|H\rangle + b\phi_V(x)|V\rangle)|x\rangle. \tag{3.9}$$

The state of the polarization is not normalized, since we have selected on this specific event. Consequently, it must be renormalized to get a physical state, accounting for the fact the photon made it through. Substituting the original and shifted functions $\phi_{H,V}(x)$, we find

$$|\psi''\rangle = \frac{a\phi(x)|H\rangle + b\phi(x+d)|V\rangle}{\sqrt{|a|^2|\phi_H(x)|^2 + |b|^2|\phi_V(x)|^2}}. \tag{3.10}$$

As you will show in Exercise 3.7.1, the postmeasurement state of the photon can be written as $|\psi''\rangle = c_H(x)|H\rangle + c_V(x)|V\rangle$, with new coefficients that are functions of x and the other parameters of the system. This effect is identical to a rotation of the polarization of the photon, conditioned on a photon traversing the slit located at position x.

It is of interest to plot these results to get a better intuition of the result of measurement in this generalized situation. We see in Figs. 3.3 and 3.4 the coefficients of the H and V polarization state in panels (a) and (b) respectively, plotted versus the position of the slit x. Panel (c) shows the two different distributions made by vertically polarized photons (the one shifted left) corresponding to the extraordinary ray and horizontal photons (the one centered at $x = 0$), corresponding to the ordinary ray. The difference between the figures corresponds to choosing the parameter $d = 5, \sigma = 1$ for Fig. 3.3, whereas we choose $d = 0.5, \sigma = 1$ for Fig. 3.4. In both cases, position is measured in units of σ.

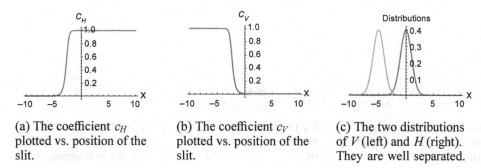

(a) The coefficient c_H plotted vs. position of the slit.

(b) The coefficient c_V plotted vs. position of the slit.

(c) The two distributions of V (left) and H (right). They are well separated.

Figure 3.3 Plots for the case $d = 5$, $\sigma = 1$, where the distributions are well separated. Position is measured in units of σ. $a = b = 1/\sqrt{2}$.

In the case of Fig. 3.3 we can more readily understand what is going on: the distributions are well separated, so when a photon is detected on the left side, it must correspond to a vertically polarized photon, whereas if it is detected on the right side, it must correspond to a horizontally polarized photon. We see there is a continuous (but rather abrupt) transition at the point between the peaks ($x_t = -2.5$), but we also notice that from the distributions shown, these values almost never occur. Consequently we have a good polarization measurement simply from looking at the position of the photon. Indeed, in such a situation, it is common to put a threshold at x_t and call anything to the right to be horizontal, and anything to the left vertical. This is not foolproof, however, because of the very slight overlap between the distributions – these lead to false positives and false negatives. However, in this case, such events are less than 1 percent.

The case of Fig. 3.4 is much more challenging to understand at first glance. We also see that there is a transition between the coefficients of the vertical and horizontal states as the position of the slit is moved across the screen. However, the transition is much more smooth. The reason for this can be seen from the mostly overlapping distributions. In this case, the position of the photon brings very little information about which polarization state the photon is in. Only in the far limits of the position being very negative or very positive can we make a confident statement about which polarization state the system is in. However, these events are very rare. The common events are those where we have very little confidence of which state the polarization is in. This is the "grey region" where we have no certainty. Consequently, the state of the photon's polarization is quite similar to the initial polarization state. Indeed, we notice at the point where we have absolutely no knowledge at all $-x = x_t = -0.25$, the postmeasurement state is exactly the same as the premeasurement state!

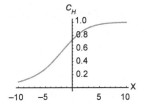 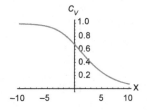 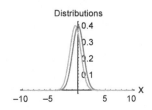

(a) The coefficient c_H plotted vs. position of the slit.

(b) The coefficient c_V plotted vs. position of the slit.

(c) The two distributions of V (left) and H (right). They are overlapping.

Figure 3.4 Plots of the state-coefficients and measurement distributions versus position for the case $d = 0.5$, $\sigma = 1$, where the distributions are well separated. Position is measured in units of σ. $a = b = 1/\sqrt{2}$.

Abstracting Away the Position of the Photon

In the previous analysis, we can simplify these considerations tremendously if we would be able to abstract away the position degree of freedom, and focus only on the polarization of the photon. Notice we can reexpress the probability density of getting the photon position (now simply called a "measurement result"), Eq. (3.7), as the following:

$$p(x) = [\langle H|a^* + \langle V|b^*]\hat{\Omega}_x^\dagger \hat{\Omega}_x[a|H\rangle + b|V\rangle], \tag{3.11}$$

where we have defined a new set of operators $\hat{\Omega}_x$ which operates in the Hilbert space of the photon's polarization. It is indexed by the result of position measurement, x. We can express this operator as a matrix in the basis defined by $|H\rangle, |V\rangle$ as

$$\hat{\Omega}_x = \begin{pmatrix} \phi_H(x) & 0 \\ 0 & \phi_V(x) \end{pmatrix}, \qquad \hat{\Omega}_x^\dagger \hat{\Omega}_x = \begin{pmatrix} |\phi_H(x)|^2 & 0 \\ 0 & |\phi_V(x)|^2 \end{pmatrix}. \tag{3.12}$$

Consequently, our probability rule for finding the measurement result x takes on a very simple form,

$$p(x) = \langle \psi|\hat{\Omega}_x^\dagger \hat{\Omega}_x|\psi\rangle. \tag{3.13}$$

Similarly, we can find the postmeasurement polarization state (3.10) as

$$|\psi'\rangle = \frac{\hat{\Omega}_x|\psi\rangle}{||\hat{\Omega}_x|\psi\rangle||}, \tag{3.14}$$

where $||\ldots||$ denotes the norm of the state. It is also important to note the probability density of result x is the square of the denominator of (3.14).

3.4 Mathematics of Generalized Measurement

The mathematical description concerning generalized measurements is an extension of the above physical example. We introduce a set of operators, called Kraus operators, in honor of Karl Kraus (1938–88), also known simply as measurement operators. Let us call this set $\{\hat{\Omega}_j\}$, where $j = 1, \ldots, M$. The operators act on the system Hilbert space alone, but there is no necessary relationship between the number of these operators and the dimension of the Hilbert space we are considering. Let us also define the operator $\hat{E}_j = \hat{\Omega}_j^\dagger \hat{\Omega}_j$. $\hat{E}_j$ is a completely positive operator; its Hermitian nature guarantees its eigenvalues are real, and they must be nonnegative. The set of $\{\hat{E}_j\}$ obeys a further constraint; they must sum to the identity operator,

$$\sum_j \hat{E}_j = 1. \tag{3.15}$$

This constraint is analogous to the condition on projection operators, Eq. (2.10). For this reason, the set of the E operators is sometimes called a positive operator-valued measure (POVM) as opposed to a projector-value measure (PVM), although we will strive to avoid this poor choice of name.

As we found in the previous section, the label j denotes different measurement outcomes, so the probability of the outcome j is given by

$$P_j = \langle \psi | \hat{E}_j | \psi \rangle = \langle \psi | \hat{\Omega}_j^\dagger \hat{\Omega}_j | \psi \rangle = ||\hat{\Omega}_j | \psi \rangle||^2. \tag{3.16}$$

The action of these operators on the quantum state may be expressed as the conditional state, given that result j has occurred,

$$|\psi_j\rangle = \frac{\hat{\Omega}_j |\psi\rangle}{||\hat{\Omega}_j |\psi\rangle||} = \frac{\hat{\Omega}_j |\psi\rangle}{\sqrt{P_j}}. \tag{3.17}$$

Therefore, every outcome has (in general) a different quantum state to which the initial state is mapped. It is important to note that neither the operators $\hat{\Omega}_j$ nor $\hat{E}_j$ are in general projection operators. That is,

$$\hat{\Omega}_j \hat{\Omega}_k \neq \hat{\Omega}_j \delta_{jk}, \quad \hat{E}_j \hat{E}_k \neq \hat{E}_j \delta_{jk}, \tag{3.18}$$

and therefore they do not describe the usual textbook projections. This will be quite important in what is to come. We stress that from the point of view of the system, the change of state (3.17) is a coherent but nonunitary process. It may be interpreted as partial collapse of the wavefunction.

The operators $\hat{\Omega}_j$ can be represented in another way in terms of the $\hat{E}_j$ operator as

$$\hat{\Omega}_j = \hat{U}_j \sqrt{\hat{E}_j}, \tag{3.19}$$

where $\hat{U}_j$ is some unitary operator. In Exercise 3.7.4, you will show this decomposition is consistent with the definition of $\hat{E}_j$.

The motivation for this decomposition is that the outcome probabilities only depend on the $\hat{E}_j$ operator. Consequently, we may view the state disturbance (3.17) as a "minimal" amount of quantum disturbance demanded by quantum mechanics, together with a unitary rotation on the state.

Constructing the Measurement Operators

While the general principles of quantum measurement turn out to be fairly simple in the end, there is no general way to construct the measurement operators. One can reformulate this conundrum as the following: what physical procedure should I implement in order to realize a set of measurement operators $\{\hat{\Omega}_j\}$? There is no unique answer to this question because we have abstracted away the measurement device. Nevertheless, we can follow the path of our motivating example and give some general procedure to follow. We consider a system together with an auxiliary system we call a meter. Let us describe them with the quantum state $|\Psi\rangle = |\psi\rangle_S \otimes |\phi\rangle_M$, where the system is in the first register, and the meter is in the second. Now allow them to interact with each other, which we model with an entangling unitary operation $\hat{U}_{SM}$, acting on both system and meter. Now project the meter with a complete set of meter projectors described by $\hat{\Pi}_j = |j\rangle_M {}_M\langle j|$, so the probability of each result is

$$P_j = \langle\Psi|\hat{U}_{SM}^\dagger|j\rangle_M {}_M\langle j|\hat{U}_{SM}|\Psi\rangle. \tag{3.20}$$

The (unnormalized) system state, postmeasurement, is described by

$$|\psi_j\rangle_S = {}_M\langle j|\hat{U}_{SM}|\psi\rangle_S \otimes |\phi\rangle_M = ({}_M\langle j|\hat{U}_{SM}|\phi\rangle_M)|\psi\rangle_S. \tag{3.21}$$

Therefore, we can associate $\hat{\Omega}_j = {}_M\langle j|\hat{U}_{SM}|\phi\rangle_M$. Although this looks like a matrix element, recall that $\hat{U}_{SM}$ acts on both system and meter, and consequently $\hat{\Omega}_j$ is still an operator acting on the system. From this result, we can then reverse engineer the appropriate meter, interaction unitary, and choice of measurements on the meter to generate the desired set of measurement operators.

3.5   von Neumann's Model: An Example with a Qubit and Free Particle Meter

While the preceding formulation looks rather abstract, let us illustrate the general result with a classic example: von Neumann's mode of measurement (sometimes called "premeasurement") [von Neumann (1955)]. We consider a qubit (two-level system such as a spin-1/2 object) as the system, and a free particle in one dimension as the meter. This is slightly generalized from the preceding section since the meter is a continuous variable, but we can use the results of Section 2.4 to fill in any gaps.

Define the initial system and meter state as $|\psi\rangle_S$ and $|\phi\rangle_M$ as in the previous section. We give the free particle its own Hamiltonian that will determine its dynamics, $\hat{H}_M = \hat{p}^2/(2m)$, and the same for the system. Here, $\hat{p}$ is the momentum operator of the meter. We take $\hat{H}_S = \epsilon \hat{\sigma}_z/2$, where σ_z is a Pauli operator for the z coordinate of the spin, and ϵ is the energy splitting between the ground and excited states. Now we allow them to physically interact via an interaction Hamiltonian. In order to illustrate the physics in a very simple way, we let the interaction Hamiltonian be time dependent, so we can clearly separate the pre- and postinteraction phase of the problem. While generally the dynamics of time-dependent Hamiltonians involves time-ordered exponentials and Dyson series, we can further simplify the problem by making the interaction impulsive in time, where a strong, but very short interaction is allowed to exist. We take the extreme of this limit and consider a Dirac delta function in time,

$$\hat{H}_{SM} = g\delta(t)\hat{\sigma}_z \otimes \hat{p}, \tag{3.22}$$

as the interaction Hamiltonian, where the hats stress the operator nature of these observables, involving both system and meter degrees of freedom.

Dividing the problem into time intervals – before the interaction, the interaction at $t = 0$, and after the interaction – allows for the construction of a piecewise solution. From the initial time t_0 to the time of the interaction, both systems have separable, independent dynamics. We can find the state just before the interaction with standard methods to give

$$|\tilde{\psi}\rangle_S = e^{-i\hat{H}_S(t-t_0)/\hbar}|\psi\rangle_S, \quad |\tilde{\phi}\rangle_M = e^{-i\hat{H}_M(t-t_0)/\hbar}|\phi\rangle_M, \tag{3.23}$$

At time $t = 0$, the interaction yields a joint unitary operator

$$\hat{U}_{SM} = \mathcal{T}\exp\left[-i\int^t dt'\hat{H}_{SM}(t')/\hbar\right] = \exp[-ig\hat{\sigma}_z \otimes \hat{p}/\hbar]. \tag{3.24}$$

The time-ordering symbol $\mathcal{T}$ does not play a role, because everything happens at $t = 0$. The time integral in the exponential removes the Dirac-delta function, giving only a unitary interaction that entangles the two degrees of freedom. The strength of the interaction is entirely encoded into the parameter g. Therefore, the state immediately following the interaction is given by

$$|\tilde{\Psi}\rangle = \exp[-ig\hat{\sigma}_z \otimes \hat{p}/\hbar]|\tilde{\psi}\rangle_S \otimes |\tilde{\phi}\rangle_M. \tag{3.25}$$

To proceed further, we recall that the displacement (or translation) operator is defined as

$$\hat{D}(\delta x) = e^{-i\delta x \hat{p}/\hbar}, \tag{3.26}$$

where δx is a position. The action of this operator on position eigenkets is its defining action,

$$\hat{D}(\delta x)|x\rangle = |x + \delta x\rangle. \tag{3.27}$$

Consequently, we can interpret the entangling unitary operator (3.24) as a conditional displacement of the state of the meter, depending on the qubit state. Let us take the qubit system state to be $|\tilde{\psi}\rangle_S = a|+\rangle_S + b|-\rangle_S$, where we choose the basis $|\pm\rangle$ that diagonalizes the system Hamiltonian, $\hat{\sigma}_z|\pm\rangle_S = \pm|\pm\rangle_S$.

Any function of an operator commutes with that operator, so the action of the exponential in Eq. (3.25) is diagonalized in the $|\pm\rangle$ basis on the system. Consequently, the state (3.25) simplifies to

$$|\tilde{\Psi}\rangle = a\exp[-ig\hat{p}/\hbar]|+\rangle_S \otimes |\tilde{\phi}\rangle_M + b\exp[ig\hat{p}/\hbar]|-\rangle_S \otimes |\tilde{\phi}\rangle_M. \tag{3.28}$$

Let us now consider a measurement to see where the meter particle is located. We describe this physics following the prescriptions of Chapter 2 and implement the meter projector $\hat{\Pi}_x = |x\rangle_{MM}\langle x|$, while the system is left untouched. Thus the operation on the system degrees of freedom is simply the identity. The projection on the system reveals some result, x, for the position. Therefore, the meter degree of freedom collapses to state $|x\rangle_M$. The postmeasurement system state, as well as the probability of finding result x, can be found by following the prescription of Subsection 3.4 to find the measurement operators

$$\hat{\Omega}_x = |+\rangle_{SS}\langle+| \, \tilde{\phi}(x - g) + |-\rangle_{SS}\langle-| \, \tilde{\phi}(x + g), \tag{3.29}$$

where the position-space wavefunction of the meter is defined as $_M\langle x|\tilde{\phi}\rangle_M = \tilde{\phi}(x)$.

We notice that similar to our motivating physical experiment – the polarization change by the calcite crystal – there is a shift of the meter. Rather than before, where the extraordinary ray was shifted while the normal ray was not, in this example, both meter states are moved in a symmetric way: the $+$ system state moves the meter to the right, while the $-$ system state moves the meter to the left. We have to add these effects coherently, so the final (unnormalized) system state is

$$|\psi'\rangle_S = a\tilde{\phi}(x - g)|+\rangle_S + b\tilde{\phi}(x + g)|-\rangle_S. \tag{3.30}$$

Therefore, we see that, conditioned on the value of x, the postmeasurement state varies smoothly between $|+\rangle_S$ and $|-\rangle_S$. We can consider this a partial wavefunction collapse, depending on how strongly correlated the system and meter are. We will explore this connection in greater depth in the next chapter.

It is interesting to consider variations on this scenario. For example, let us allow the system and meter to change in time further after the interaction for a time t. How will this change the preceding story? Similar to the beginning of this section, we must apply the separable unitary to the now entangled state,

$$|\tilde{\Psi}'\rangle = e^{-i\hat{H}_S t/\hbar} \otimes e^{-i\hat{H}_M t/\hbar}|\tilde{\Psi}\rangle. \tag{3.31}$$

Suppose now a measurement of the meter's position is made. How will the measurement operators change? As you will prove in Exercise 3.7.7, the resulting operators are given by

$$\hat{\Omega}'_x = |+\rangle_{SS}\langle+| \, e^{-i\epsilon t/(2\hbar)} \langle x|e^{-i\hat{p}^2 t/(2m\hbar)-ig\hat{p}/\hbar}|\tilde{\phi}\rangle$$
$$+ |-\rangle_{SS}\langle-| \, e^{i\epsilon t/(2\hbar)} \langle x|e^{-i\hat{p}^2 t/(2m\hbar)+ig\hat{p}/\hbar}|\tilde{\phi}\rangle. \tag{3.32}$$

Remarkably, the preceding result is still diagonal in the $\hat{\sigma}_z$ eigenbasis of the system. This property comes from the fact that the system Hamiltonian and the interaction Hamiltonian commute,

$$[\hat{H}_S, \hat{H}_{SM}] = 0. \tag{3.33}$$

In the literature, such a measurement is sometimes called a *quantum nondemolition measurement*, for historical reasons. The intuition why such a measurement is desirable is that, if one's goal is to measure the system in a textbook way in the energy eigenbasis (by making g much larger than the spread in the meter wavefunction, for example), then the measurement should not rotate the state during the measurement process. Otherwise, this would spoil what one is trying to do.

3.6 Generalization to Mixed States

At this point in the text, it is helpful to generalize our results to the situation when one has incomplete quantum information. This situation can come from ignorance of certain events in the world and is described with mixed states, corresponding to a density matrix or density operator. Readers unfamiliar with this description of quantum systems, or desiring a refresher, should stop here and read Appendix B, and then return.

If we have a description of the state as a mixed state $\hat{\rho}$, how do we generalize the statistical predictions of quantum mechanics? For projective measurements, in Appendix B, we already showed that the outcome probability for result j of a projection operator set $\hat{\Pi}_j$ is given by

$$P_j = \text{Tr}[\hat{\rho}\hat{\Pi}_j], \tag{3.34}$$

while the postmeasurement state $\hat{\rho}'$ is simply $\hat{\Pi}_j$. To extend these results to generalized measurements, we recall we can decompose any mixed state as

$$\hat{\rho} = \sum_j p_j |\psi_j\rangle\langle\psi_j|, \tag{3.35}$$

for some states $|\psi_j\rangle$ and weights $0 \leq p_j \leq 1$. Recalling results (3.16) and (3.17), we can rewrite the probability expression for some result k as

$$P_k = \text{Tr}[\hat{\Omega}_k |\psi\rangle\langle\psi| \hat{\Omega}_k^\dagger], \tag{3.36}$$

where Tr is the trace of the operator. Replacing $|\psi\rangle \rightarrow |\psi_j\rangle$, we can generalize the result to any mixed state, using the rule of total probability,

$$P_k = \sum_j p_j \text{Tr}[\hat{\Omega}_k |\psi_j\rangle\langle\psi_j| \hat{\Omega}_k^\dagger] = \text{Tr}[\hat{\Omega}_k \hat{\rho} \hat{\Omega}_k^\dagger] = \text{Tr}[\hat{\Omega}_k^\dagger \hat{\Omega}_k \hat{\rho}], \tag{3.37}$$

where we have used the cyclic property of the trace in the last equality. The important difference to Eq. (3.34) is that, while $\hat{\Pi}_j$ is a projection operator, the operator $\hat{E}_k = \hat{\Omega}_k^\dagger \hat{\Omega}_k$ is only a positive, Hermitian operator.

The measurement disturbance can be found in a similar way. Equation (3.17) can be invoked, as applied to unnormalized states; and also, using its adjoint form, we have

$$|\psi_j\rangle\langle\psi_j| \rightarrow \hat{\Omega}_k |\psi_j\rangle\langle\psi_j| \hat{\Omega}_k^\dagger. \tag{3.38}$$

We can now apply the same rule to an incoherent superposition of many such states to find

$$\hat{\rho} \rightarrow \hat{\rho}' = \sum_j p_j (\hat{\Omega}_k |\psi_j\rangle\langle\psi_j| \hat{\Omega}_k^\dagger) = \hat{\Omega}_k \hat{\rho} \hat{\Omega}_k^\dagger. \tag{3.39}$$

As in the pure state case, we must renormalize the state such that $\text{Tr}[\hat{\rho}'] = 1$ (see Appendix B). This is accomplished simply by dividing by the trace of $\hat{\rho}'$. We thus have a map of normalized states to normalized states, conditioned on result k,

$$\hat{\rho} \rightarrow \hat{\rho}' = \frac{\hat{\Omega}_k \hat{\rho} \hat{\Omega}_k^\dagger}{\text{Tr}[\hat{\Omega}_k^\dagger \hat{\Omega}_k \hat{\rho}]}, \tag{3.40}$$

where we used the cyclic property of the trace in the denominator. Notice the denominator is exactly the same as the outcome probability (3.37).

While we have used the generalized measurement formalism to derive these results, it is instructive to rederive them from the point of view of expanding the system to include a meter as we did when deriving an explicit form of the measurement operators. Consider a mixed system state and a pure meter state $|\phi\rangle_M$ for simplicity, initially separable. Now let them interact with a joint interaction unitary operator $\hat{U}_{SM}$ resulting in the state

$$\hat{\rho}_{SM} = \hat{\rho}_S \otimes |\phi\rangle_{MM}\langle\phi| \rightarrow \hat{\rho}'_{SM} = \hat{U}_{SM}\hat{\rho}_S \otimes |\phi\rangle_{MM}\langle\phi|\hat{U}^\dagger_{SM}. \tag{3.41}$$

Applying a projection operator $\hat{\Pi}_k$ on the meter degrees of freedom (and identity on the system) gives the prediction of the probability of result k (3.34),

$$P_k = \text{Tr}[\hat{\rho}'_{SM}\hat{\Pi}_k] = \text{Tr}_S\text{Tr}_M[\hat{U}_{SM}\hat{\rho}_S \otimes |\phi\rangle_{MM}\langle\phi|\hat{U}^\dagger_{SM}\hat{\Pi}_k], \tag{3.42}$$

where $\text{Tr}_{S,M}$ refers to tracing over either the system or the meter part of the Hilbert space. The trace of the meter is carried out by choosing the basis corresponding to the projection operators on the meter to find

$$P_k = \text{Tr}_S\left[\sum_l {}_M\langle l|\hat{U}_{SM}\hat{\rho}_S \otimes |\phi\rangle_{MM}\langle\phi|\hat{U}^\dagger_{SM}\hat{\Pi}_k|l\rangle_M\right], \tag{3.43}$$

$$= \text{Tr}_S\left[{}_M\langle k|\hat{U}_{SM}\hat{\rho}_S \otimes |\phi\rangle_{MM}\langle\phi|\hat{U}^\dagger_{SM}|k\rangle_M\right], \tag{3.44}$$

$$= \text{Tr}_S\left[({}_M\langle k|\hat{U}_{SM}|\phi\rangle_M)\hat{\rho}_S({}_M\langle\phi|\hat{U}^\dagger_{SM}|k\rangle_M)\right]. \tag{3.45}$$

Recalling the result (3.21), from the discussion in Section 3.4, we see that the result (3.37) is recovered with the identification $\hat{\Omega}_k = {}_M\langle k|\hat{U}_{SM}|\phi\rangle_M$. In Exercise 3.7.10, you will find the postmeasurement state of the system and meter, completing the preceding discussion.

The Mixed State Case for the von Neumann Measurement Model

We now illustrate the previous section by applying the results to the von Neumann measurement model of Section 3.5. Recall that we found, given the detection of the free particle at position x on the final screen, the measurement operators are given by Eq. (3.29), so they are diagonal in the $|+\rangle, |-\rangle$ basis.

For a qubit, the density matrix can be written in the $|+\rangle, |-\rangle$ basis as

$$\hat{\rho} = \begin{pmatrix} \rho_{++} & \rho_{+-} \\ \rho_{-+} & \rho_{--} \end{pmatrix}. \tag{3.46}$$

The positivity of the density tells us that $0 \leq \rho_{++}, \rho_{--} \leq 1$, while its normalization indicates that $\text{Tr}\hat{\rho} = \rho_{++} + \rho_{--} = 1$. The fact that $\hat{\rho}$ is Hermitian further constrains $\rho_{+-} = \rho^*_{-+}$. Generalizing the results of the previous section to continuous variables, using the results of Section 2.4, the outcome probability, Eq. (3.37), gives us the probability density of finding the free particle at position x,

$$p(x) = \text{Tr}[\hat{\Omega}^\dagger_x\hat{\Omega}_x\hat{\rho}] = \rho_{++}|\tilde{\phi}(x - g)|^2 + \rho_{--}|\tilde{\phi}(x + g)|^2. \tag{3.47}$$

Thus, only the diagonal density matrix elements enter this expression. Similarly, the unnormalized state of the qubit can be found from Eq. (3.39),

$$\hat{\rho}' \propto \begin{pmatrix} |\tilde{\phi}(x-g)|^2 \rho_{++} & \tilde{\phi}(x-g)^* \tilde{\phi}(x+g) \rho_{+-} \\ \tilde{\phi}(x+g)^* \tilde{\phi}(x-g) \rho_{-+} & |\tilde{\phi}(x+g)|^2 \rho_{--} \end{pmatrix}. \tag{3.48}$$

These results generalize those from Section 3.5. To normalize this density operator, we should divide by the sum of the diagonal elements. Thus we see that the diagonal matrix elements are weighted by the absolute square of the meter wave-functions, shifted by $\pm g$, while the off-diagonal matrix elements are weighted by $\tilde{\phi}(x+g)^* \tilde{\phi}(x-g)$, or its complex conjugate.

3.7 Quantum Bayesian Point of View

Let us return to the main results of Section 3.6, given in Eqs. (3.37) and (3.40). Further insight into the meaning of these equations can be had by using the decomposition (3.19), recalling that $\hat{E}_k = \hat{\Omega}_k^\dagger \hat{\Omega}_k$. We can then write the probability of result k and the conditional state change assignment $\hat{\rho}^{(k)}$ as

$$P_k = \mathrm{Tr}(\hat{E}_k \hat{\rho}), \quad \hat{\rho} \to \hat{\rho}^{(k)} = \frac{\hat{U}_k \sqrt{\hat{E}_k} \hat{\rho} \sqrt{\hat{E}_k} \hat{U}_k^\dagger}{P_k}. \tag{3.49}$$

We will now consider the simplest case of "plain" measurement, where no additional effective unitary operation is taking place, $\hat{U}_k = \mathbf{1}$ for all k. We consider the situation when the operators $\hat{E}_k$ are diagonalized all in the same basis so

$$\hat{E}_k |j\rangle = p(k|j)|j\rangle, \tag{3.50}$$

for all k. Here, the eigenvalues $p(k|j)$ have the natural interpretation as the conditional probability of finding result k, given that the system is in state $|j\rangle$. We can see this explicitly by finding the probability of result k after preparing the system in each one of its N states labeled by j. The total probability of result k occurring is then given by

$$P_k = \sum_j p(k|j)\rho_{jj}. \tag{3.51}$$

This is nothing more than the law of total probability described in Appendix A for classical probability theory, where we interpret the diagonal density matrix elements to be the probability of finding the system in state $|j\rangle$.

The diagonal matrix elements of the new density matrix also take on a simple form:

$$\langle j|\hat{\rho}^{(k)}|j\rangle = \rho_{jj}^{(k)} = \frac{p(k|j)\rho_{jj}}{P_k}. \tag{3.52}$$

This part of the state assignment may also be interpreted with classical probability theory as Bayes' rule (see Appendix A). The diagonal density matrix elements, interpreted as the probability to occupy state $|j\rangle$, are reassigned based on new information coming in the form of the result k, with the aid of the conditional probabilities $p(k|j)$.

The "quantum" part of the update rule can therefore be found in how the coherences change, given by

$$\rho_{ij}^{(k)} = \frac{\langle i|\sqrt{\hat{E}_k}\hat{\rho}\sqrt{\hat{E}_k}|j\rangle}{P_k} = \frac{\sqrt{p(k|i)p(k|j)}\,\rho_{ij}}{P_k}, \tag{3.53}$$

for $i \neq j$. This connection to Bayes' rule was noted in the book of Gardiner et al. (2004), and in the papers of Alexander Korotkov and collaborators (Korotkov, 1999, 2001; Jordan and Korotkov, 2006). Further elucidation of these results will be given in the next chapter.

Unselective Measurements and Decoherence

While we have so far focused on *selective measurements* (i.e. when we are interested in the conditional state disturbance corresponding to a single outcome), it is quite common to discuss *unselective* measurements, where either the meter state cannot be read, or it is read and forgotten, or averaged over. Such a situation becomes unavoidable in the case where the "meter" consists of many degrees of freedom, such as in a macroscopic environment. In this case, we can investigate the averaged quantum dynamics over all possible outcomes. We define the ensemble-averaged density matrix $\bar{\rho}$ as

$$\bar{\rho} = \sum_k P_k \hat{\rho}^{(k)} = \sum_k \hat{\Omega}_k \hat{\rho} \hat{\Omega}_k^\dagger. \tag{3.54}$$

Sometimes this mapping between $\hat{\rho} \rightarrow \bar{\rho}$ is called a *quantum channel*. In the simple case discussed previously, where $\hat{U}_k = \mathbf{1}$, and $\hat{E}_k$ are all diagonalized in the same basis $\{|i\rangle\}$, we have the diagonal density matrix elements given by

$$\bar{\rho}_{ii} = \sum_k p(k|i)\rho_{ii} = \rho_{ii}. \tag{3.55}$$

The last equality follows from the conditional probability distributions $p(k|i)$ being normalized for every i. Therefore, the diagonal matrix elements are left unchanged by the averaging over results. On the other hand, the off-diagonal elements behave as

$$\bar{\rho}_{ij} = \sum_k \sqrt{p(k|i)p(k|j)}\,\rho_{ij}. \tag{3.56}$$

Here, we see that the off-diagonal matrix elements are multiplied by an overall factor

$$C_{ij} = \sum_k \sqrt{p(k|i)p(k|j)}. \tag{3.57}$$

Recalling the interpretation of $p(k|i)$ as the conditional probability of result k, given state i, the coefficient C_{ij} is the Bhattacharyya coefficient between distributions $p(k|i)$ and $p(k|j)$ (Bhattacharyya, 1943), which is closely related to one of the Rényi divergences (Van Erven and Harremos, 2014; Averin and Sukhorukov, 2005). The Bhattacharyya distance D_{ij} is related to the Bhattacharyya coefficient as $D_{ij} = -\ln C_{ij}$. This coefficient is a measure of the similarity of the two distributions. We note that $0 \leq C_{ij} \leq 1$, with the larger distance between the distributions corresponding to smaller values of C_{ij}. Thus, for $p(k|i) \neq p(k|j)$, the coherence of the density matrix element is suppressed. This process is called *decoherence*. It is important to stress that the loss of coherence comes from the loss of information and incoherently averaging together different density matrices corresponding to different realizations of the meter or environment.

Exercises

Exercise 3.7.1   Show that the postmeasurement state (3.10) of the photon can be written as $|\psi''\rangle = c_H(x)|H\rangle + c_V(x)|V\rangle$, where the new coefficients may be written as

$$c_H(x) = \frac{a}{\sqrt{|a|^2 + |b|^2 \exp(-xd/\sigma^2 - d^2/(2\sigma^2))}}, \tag{3.58}$$

$$c_V(x) = \frac{b\exp(-xd/(2\sigma^2) - d^2/(4\sigma^2))}{\sqrt{|a|^2 + |b|^2 \exp(-xd/\sigma^2 - d^2/(2\sigma^2))}}. \tag{3.59}$$

Exercise 3.7.2   Consider a meter that is another qubit (2-state system). The system is prepared in a general state, $|\psi\rangle_S = a|H\rangle_S + b|V\rangle_S$. The meter is prepared in the state $|\phi\rangle_M = \gamma|H\rangle_M + \bar{\gamma}|V\rangle_M$. Here we have used the symbols H, V instead of 0, 1 to suggest polarization states of single photons. The physical interaction between system and meter maps the initially separable state to $|\psi\rangle_S \otimes |\phi\rangle_M$ to the state

$$|\Psi\rangle = (a\gamma|H\rangle_S + b\bar{\gamma}|V\rangle_S) \otimes |H\rangle_M + (a\bar{\gamma}|H\rangle_S + b\gamma|V\rangle_S) \otimes |V\rangle_M. \tag{3.60}$$

The meter photon is measured in the $H - V$ basis. Follow the procedure outlined in Section 3.4 to find the measurement operators appropriate for the system. This measurement was carried out by Pryde et al. (2005).

Exercise 3.7.3   Verify the result (3.14) is correct by inserting the explicit form (3.12) for the preceding equation and check to see that you recover Eq. (3.10).

Exercise 3.7.4   Show that the decomposition (3.19) is consistent with $\hat{E}_j = \hat{\Omega}_j^\dagger \hat{\Omega}_j$.

Exercise 3.7.5   Confirm for yourself that the measurement operator given in Eq. (3.29) is correct.

Exercise 3.7.6   Consider the measurement operators (3.29). Consider the case where the initial meter wavefunction is Gaussian with width σ. Work out the simplified form of the postmeasurement system state that is normalized.

Exercise 3.7.7   Show that the measurement operators (3.32) for adding dynamics to the meter and system are correct.

Exercise 3.7.8   Consider the measurement operators (3.32). Consider the case where the initial meter wavefunction is Gaussian with width σ. Work out the simplified form of the measurement operator coefficients explicitly. Hint: compute the meter matrix elements by entering the momentum basis, and then Fourier transforming back to the position basis at the end.

Exercise 3.7.9   What would happen if the initial system Hamiltonian for Section 3.5 was changed to $\hat{H}_S = \Delta \hat{\sigma}_x / 2$ instead? Would the measurement still be quantum nondemolition? Work out the new measurement operators after waiting some time t.

Exercise 3.7.10   Following the discussion in Section 3.6, find the postmeasurement state of the system and meter and show the system state is of the form (3.40), where the meter state is a projector on state $|k\rangle_M$.

Exercise 3.7.11   What would happen if the coupling Hamiltonian for Section 3.5 coupled to the meter's position operator, instead of its momentum operator? That should correspond to a splitting of the meter's state in momentum space, so subsequent free evolution in time would separate the wavepackets corresponding to different eigenstates of the system. Thus, the amount of time elapsed should also control the strength of the measurement. Confirm this is true quantitatively.

Exercise 3.7.12   Consider a fully mixed polarization state of light that enters a calcite crystal. Following a slit at position x, for $d = 0.5$ and $\sigma = 1$, find the purity of the quantum polarization state of a photon that passes the slit. Plot it as a function of x. Does this make sense?

Exercise 3.7.13   Consider the von Neumann model for when the meter is prepared in a Gaussian state of width σ. For an interaction that shifts the meter wavefunction by a distance of $\pm g$ depending on the qubit state, followed by immediate measurement, calculate the Bhattacharyya distance as a function of g and the meter width σ. For what value of g/σ will the ensemble-averaged coherence of the quantum state be reduced by a factor of 100? Of 1000?

4

Weak Measurement

The physics of weak measurement is a special case of the previous chapter where the measurement strength is taken to zero as a limit. The limit is useful because the measurement back-action goes to zero; but as we will see later in this chapter, the amount of information also limits to zero. By taking many realizations of the weak measurement process from identically prepared initial conditions, information about the system of interest may be gradually extracted in such a way that the quantum state disturbance per measurement is negligible.

4.1 The Limit of a Very Weak Stern–Gerlach Magnet

Let us consider a motivating experiment, where we take a Stern–Gerlach device with a very weak magnetic field gradient, or a piece of calcite crystal that is just a sliver, as was done in the experiment of Ritchie et al. (1991b) (see Fig. 4.1).

In this limit, the splitting of the atomic or optical beam corresponding to the different system state is very small, so the ability to distinguish the system is nearly nonexistent. In the limit $d \ll \sigma$, we Taylor expand the measurement operators of the calcite crystal case (3.12) to first order in d to find

$$\hat{\Omega}_x = \phi(x)\hat{\Pi}_H + \phi(x+d)\hat{\Pi}_V \approx \phi(x)\mathbf{1} + d\phi'(x)\hat{\Pi}_V, \tag{4.1}$$

where $\phi'(x)$ indicates the spatial derivative, and we have used the projection operator property $\hat{\Pi}_H + \hat{\Pi}_V = \mathbf{1}$.

In this limit, we can determine the probability density to find the photon at position x, given an initial polarization state $|\psi\rangle = a|H\rangle + b|V\rangle$ as

$$p(x) = \langle\psi|\hat{\Omega}_x^\dagger\hat{\Omega}_x|\psi\rangle \approx |\phi(x)|^2 + 2d\,\mathrm{Re}[\phi'(x)\phi(x)^*]|b|^2 + \hat{\mathcal{O}}(d^2). \tag{4.2}$$

As you will show in Exercise 4.5.2, this distribution is normalized to first order in d. Notice that if $d = 0$, the polarization has no role to play, but for small d, there

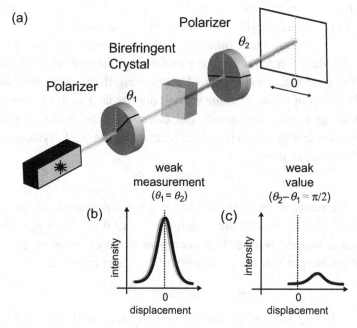

Figure 4.1 (a) In the experiment of Ritchie, Story, and Hulet (Ritchie et al., 1991b), a laser source is used to produce a collimated beam that is passed through a pair of polarizers with transmission angles θ_1 and θ_2. In between these polarizers is a birefringent crystal, which generates a small deflection in one polarization direction. (b) Without the final polarizer, a weak measurement is performed and the output intensity is the combination of two, very slightly shifted Gaussian profiles, depending on the initial polarization. (c) If the two polarizers are tuned to near orthogonality, the two Gaussian profiles destructively interfere, reducing the overall intensity but causing a sizable shift of the pattern, readily revealing the weak value associated with this choice of input and output states.

is a slight shift of the probability distribution to find the photon at position x, that involves the population of the vertical polarization, $|b|^2$.

The new polarization state, assuming the photon passes a thin slit at position x, is given by

$$|\tilde{\psi}\rangle = \frac{\hat{\Omega}_x|\psi\rangle}{||\hat{\Omega}_x|\psi\rangle||} \propto \phi(x)|\psi\rangle + d\phi'(x)b|V\rangle \propto |\psi\rangle + d(\ln\phi(x))'b|V\rangle, \quad (4.3)$$

where, in the last equality, we have replaced $\phi'(x)/\phi(x) = (\ln\phi(x))'$. In Eq. (4.3), we have kept the state unnormalized to show the main effect – the state is still almost the same as $|\psi\rangle$, but with a small correction in the direction of the vertical polarization. One interesting feature of this result is that the amount of shift depends on the position x. For example, if we take a Gaussian meter state,

$\phi(x) \propto \exp(-x^2/(4\sigma^2))$, then $(\ln\phi(x))' = -x/(2\sigma^2)$. This gives a prefactor of $-dxb/(2\sigma^2)$ in front of the $|V\rangle$ state of Eq. (4.3). On one hand, we notice that if for some reason, x were to take on a large value, this correction could be, in fact, quite large. On the other hand, from the probability distribution (4.2), we see that such values of x are highly improbable. Nevertheless, these values are possible and underpin the effect of the *weak value* we will discuss later in this chapter.

Now let us generalize this result, but considering an interaction strength g between a system and meter. In this case, we can express the $\hat{E}_j$ system operator in the limit of very small g as

$$\hat{E}_j = p_j \mathbf{1} + g\Gamma_j \hat{O}, \tag{4.4}$$

where p_j is the meter probability of finding result j in absence of the system, Γ_j is a coefficient associated with the detector, while $\hat{O}$ is a system operator. Such a measurement is defined as a *weak measurement* of system operator $\hat{O}$. For such a decomposition, we can find the probability of the meter result j as

$$P_j = \mathrm{Tr}(\hat{\rho}\hat{E}_j) = p_j + g\Gamma_j\langle\hat{O}\rangle, \tag{4.5}$$

where $\langle\hat{O}\rangle$ is the expectation value of the system operator $\hat{O}$ in the state $\hat{\rho}$. Notice that, in the previous chapter, the general theory of measurement lost connection with the concept of measuring a system observable. Here, in the weak measurement limit, we regain it. Knowledge of the detector parameters p_j, Γ_j and the coupling strength g allows us to infer the expected system observable $\langle\mathcal{O}\rangle$ given sufficient trials in identically prepared initial conditions.

4.2 Information–Disturbance Trade-off

We have seen in the previous section that, in the weak measurement limit, there is typically little disturbance to the quantum state, but also that typically only a little information can be extracted per measurement. Let us now quantify that relationship. Returning to the simple case of plain (or QND) measurement of Section 3.7, where the operators E_k are all diagonalized in the same basis, (3.50), then the probability of result k (3.51) contains the accessible information about the quantum state. Let us view this situation as the probability of distinguishing which of the j states the system resides in. The information acquired from the measurement can be quantified by the statistical overlap of the distributions

$$C_{ij} = \sum_k \sqrt{p(k|i)p(k|j)}. \tag{4.6}$$

We have already encountered this quantity as the Bhattacharyya coefficient. A *quantum limited detector* corresponds to when the information acquisition equals the ensemble-averaged degree of decoherence, defined as $d_{ij} = -\ln(\sum_k P_k \rho_{ij}^{(k)} / \rho_{ij})$ of the coherence of states i and j. More generally, we have the inequality

$$D_{ij} \leq d_{ij}, \tag{4.7}$$

where again $D_{ij} = -\ln C_{ij}$ is the Bhattacharyya distance. This inequality implies that not all possible information about the quantum state is usually revealed to the detector, and information can be lost in a variety of ways. These include not measuring in the optimal basis, not using the optimal meter states, and more practically, loss, inefficiency of the detector, and other sources of technical noise, such as detector dark counts, thermal noise, and the like. Such effects either increase decoherence or decrease distinguishability of the different quantum states. A more general situation can occur with even quantum limited detectors, where the decomposition (3.19) is considered with the unitary term U_j reintroduced. In this case, an ensemble average will still result in decoherence, but only because of the different unitaries applied for different members of the ensemble. On any given run, perfect coherence is retained in the quantum system. In fact, the effect of the outcome-dependent unitary can be eliminated with the use of suitable coherent feedback, a subject we will return to in Chapter 10.

Case Study: Selective Tunneling and a Null-Result Measurement

Better intuition about these results can be obtained by studying some physical examples. In this section, we take the case of selective tunneling, which gives a natural strength parameter for the measurement (see Fig. 4.2). This experiment was carried out in the group of John Martinis (Katz et al., 2006). We give a slightly simplified description of the experiment below.

The phase qubit is defined as a superconducting loop, interrupted by a Josephson junction, and controlled by an external magnetic flux Φ_e. The phase across the junction ϕ obeys an equation of motion

$$C\Phi_0^2 \ddot{\phi} + E_J \sin \phi = \Phi_0 I, \tag{4.8}$$

where C is the capacitance of the junction, E_J is the Josephson energy, and $\Phi_0 = h/(2e)$ is the magnetic flux quantum. The external current controlling the junction I is here replaced by the external flux control via the loop inductance $I = \Phi_e/L$. This dynamical equation may be interpreted as describing a fictitious particle of mass $C\Phi_0^2$ moving in a potential well given by

$$V(\phi) = -\frac{\Phi_0 \Phi_e}{L} \phi - E_J \cos \phi. \tag{4.9}$$

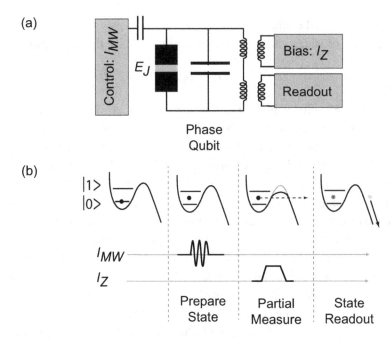

Figure 4.2 (a) A phase qubit generally consists of a capacitively shunted Joseph-son junction that is biased with a static current I_Z to reduce the number of quantized levels in the potential energy surface to a few. In this particular real-ization, the current bias is applied through a so-called flux transformer; a similar structure is also used to sense a voltage that develops when the junction switches to the dissipative state. A microwave signal I_{MW} is used to drive transitions between the quantum levels. The basic pulse sequence utilized in the experiment carried out by Katz et al. (2006) is shown in (b): a microwave pulse is used to create a desired quantum superposition of the ground and excited states; the tun-neling barrier is lowered for a short duration to perform a partial measurement; the state of the system is inferred by associating the presence of a voltage pulse with occupancy of the excited state, originating from state-selective quantum tunneling out of the potential well.

Such a potential that oscillates while having a slope is often called a "tilted wash-board" potential. When the tilt of the washboard gets too large, the local minima disappear, and the potential has no stable solutions.

The phase qubit is arranged so that the applied flux is just under the critical amount, so shallow minima exist that are 2π periodic. A quantum treatment of the circuit imposes the canonical commutation relation on the charge and flux degrees of freedom, $[\hat{\phi}, \hat{Q}] = i\hbar$, so the quantum physics of the circuit gives rise to quantized energy levels, coherent superpositions, and the rest of the quantum phenomena. Among those quantum phenomena is *quantum tunneling*, where a par-ticle can escape from a metastable potential well, even if its energy is less than the barrier height. This phenomenon is very important for this section. The tunneling rate for escaping the potential well takes the form

$$\Gamma = \omega_a e^{-S_c}, \tag{4.10}$$

where ω_a is the attempt frequency, while $S_c(E)$ is the classical action of the escape path in an inverted potential, given an energy E. We consider the simple case where the potential well is sufficiently shallow that there are only two energy levels, named $|g\rangle$ with energy E_g and $|e\rangle$ with energy E_e, for ground and excited state. The higher energy of the excited state (see Fig. 4.2) implies that the associated classical action of the escape path is smaller than the same for the ground state. The exponential dependence of the tunneling rate on this action leads us to the approximation that the ground state escape rate can be set to 0 for the timescale of the experiment we will consider, so only the excited state $|e\rangle$ will have a nonzero escape rate Γ.

We are interested in the tunneling events as a state-selective event – the upper level can tunnel out, while the lower level cannot. The tunneling event can be detected in the following way: if the particle tunnels out of the well, it has sufficient energy to run down the washboard without getting trapped in any metastable states. This results in the phase ϕ increasing rapidly in time, which gives rise to an additional magnetic flux through the ring. This can be detected with a sensitive magnetometer, which will register an event when the tunneling occurs. In the experiment of Katz et al. (2006), this was a "DC SQUID," also a ring, but with two Josephson junctions and connected to leads with a current source that gives rise to interference.

Let us analyze the measurement physics of this situation. Consider an arbitrary initial state of the phase qubit – it can be prepared by applying an electromagnetic signal, oscillating at the frequency of $(E_e - E_g)/\hbar$, to drive a "Rabi oscillation" of variable time to make any state on demand. Setting the clock to $t = 0$ when the initial state is prepared, let us wait some time t and see if the detector registers a tunneling event (named d) or not (named $\bar{d}$). If the system is prepared in the excited state, the event d (or $\bar{d}$) will occur with probability $p_d = 1 - \exp(-\Gamma t)$ and $p_{\bar{d}} = \exp(-\Gamma t)$, which can be found by solving a simple rate equation, $\dot{p}_{\bar{d}} = -\Gamma p_{\bar{d}}$. The lower level will not decay on the timescales of the experiment. That information is sufficient to construct the operators $\hat{E}_d$ and $\hat{E}_{\bar{d}}$ corresponding to the two possibilities, after waiting time t,

$$\hat{E}_d = [1 - \exp(-\Gamma t)]|e\rangle\langle e|, \quad \hat{E}_{\bar{d}} = \exp(-\Gamma t)|e\rangle\langle e| + |g\rangle\langle g|. \tag{4.11}$$

We can now find the probabilities of finding either a no-tunneling event or a tunneling event if the system is prepared in the state $\hat{\rho}$, with matrix elements ρ_{ij}, where $i,j = e, g$. Applying the formalism of the previous chapter, we find

$$P_d = \text{Tr}(\hat{\rho}\hat{E}_d) = \rho_{ee}(1 - e^{-\Gamma t}), \quad P_{\bar{d}} = \text{Tr}(\hat{\rho}\hat{E}_{\bar{d}}) = \rho_{gg} + \rho_{ee}e^{-\Gamma t}. \quad (4.12)$$

As is clear from the preceding formulas, the distribution has positive probabilities and is normalized. At time $t = 0$, the detector brings no information, so we always register a null result. However, after waiting a time $t > \Gamma^{-1}$, then we begin to have better distinguishability between the states $|e\rangle$ and $|g\rangle$: the detection event is associated with the system being in the excited state, while the null event is associated with the system being in the ground state. We stress an important observation: the no-result measurement (when the detector sits quietly, doing nothing) also brings information. This is reminiscent of the incident of the "dog that did not bark," told by Sir Arthur Conan Doyle in one of his short stories, "The Adventure of Silver Blaze" (1894). The following exchange occurs between Gregory, the Scotland Yard detective, and Sherlock Holmes:

- **Gregory:** Is there any other point to which you would wish to draw my attention?
- **Holmes:** To the curious incident of the dog in the night-time.
- **Gregory:** The dog did nothing in the night-time.
- **Holmes:** That was the curious incident.

Just like the dog *not* barking brought the new information needed to solve the mystery, the detector *not* clicking brings new information and indeed, as we shall see, results in a change of system state.

To find the quantum state dynamics of the system, we must account for the measurement dynamics, as well as the Hamiltonian. Fortunately, both of these effects are diagonal in the $|g\rangle, |e\rangle$ basis (a QND measurement), so we can combine them together. Later on the book, we will encounter examples of the unitary dynamics and the measurement dynamics that interfere with each other. Let us first consider the case where the system tunnels, resulting in a detection event. This corresponds to a *destructive* event, because once this occurs, the state no longer occupies the metastable well, and the system no longer exists. If such an event occurs, the system must be reset in order to start over. This phenomenon is similar to photo-detection. In that case the detection of the photon also goes together with destroying it, so further experiments can no longer take place. We can say, though, that before the phase qubit tunneled and was destroyed, it was in its excited state. However, the no-tunneling event is more interesting. One might be tempted to simply project the phase qubit onto the ground state, but this would be incorrect. We can apply the formalism of the last chapter to find the measurement operator,

$$\hat{\Omega}_{\bar{d}} = e^{i\omega t/2}|g\rangle\langle g| + e^{-i\omega t/2 - \Gamma t/2}|e\rangle\langle e|, \quad (4.13)$$

where the frequency $\omega = (E_e - E_g)/\hbar$ is introduced to account for the system Hamiltonian. The post–null measurement state can now be calculated,

$$\hat{\rho}' = \frac{\hat{\Omega}_{\bar{d}}\hat{\rho}\hat{\Omega}_{\bar{d}}^{\dagger}}{P_{\bar{d}}} = \frac{1}{\rho_{gg} + e^{-\Gamma t}\rho_{ee}} \begin{pmatrix} \rho_{gg} & \rho_{ge}e^{i\omega t - \Gamma t/2} \\ \rho_{eg}e^{-i\omega t - \Gamma t/2} & \rho_{ee}e^{-\Gamma t} \end{pmatrix}. \qquad (4.14)$$

In the last expression, the density matrix is expressed in the $|g\rangle, |e\rangle$ basis. It is important to stress that, given the no-tunneling event, the prediction for the conditional state (4.14) is deterministic. We note that the new conditional density matrix is properly normalized. We can now associate the state disturbance with the information gain. For time $t = 0$, we have gained no information, but we also see that $\hat{\rho}'(t = 0) = \hat{\rho}$. In the opposite limit, for $\Gamma t \gg 1$, the postmeasurement state (corresponding to no-tunneling) is simply a projection operator on the ground state, corresponding to statistical certainty that the qubit must be in its ground state. We see there is a continuous transition between these two extremes controlled by the amount of time waited for an event that never comes.

These state dynamics may be interpreted as a partial collapse of the wavefunction, controlled by the measurement strength Γt. The fact that the system does not collapse to the ground state immediately comes from the fact that there are two possible explanations for the absence of the detection event: either the qubit is in its ground state the entire time, or it is in its excited state, but we simply have not given it enough time to tunnel out of the barrier. It is important to note that if the system begins in a coherent superposition of the ground and excited states that is pure, it stays in a pure state throughout the gradual collapse process.

4.3 Weak Value

Now that we understand weak measurements, we can combine them together with more weak measurements, or with projective (or strong) measurements. An interesting application of weak measurements is the *weak value*, a concept introduced by Aharonov et al. (1988). We motivate the effect by considering the experiment of Ritchie et al. (1991a) in the group of R. Hulet, where a thin birefingent crystal (quartz, in this case) is followed by a polarizer, to select a given polarization of one's choosing. We continue the treatment of Section 4.1, and rather than place a screen with a pinhole, the experiment is terminated with a polarizer followed by a detecting screen.

Let us first revisit the case with no final polarizer. The thin quartz crystal weakly measures $\hat{\Pi}_V$ as before. Recalling the probability of finding the photon at position (4.2), the probability density for a photon to land at position x is given by

$$p(x) = \frac{1}{\sqrt{2\pi\sigma^2}}e^{-\frac{x^2}{2\sigma^2}}\left(1 - \frac{xd|b|^2}{\sigma^2}\right), \qquad (4.15)$$

where we have assumed a Gaussian meter wavefunction with width σ. We note the distribution is normalized, with a mean of $\langle x \rangle = -d|b|^2$, so the mean is shifted by anywhere between $[-d, 0]$, which we can associate with d times the eigenvalues of the $-\hat{\Pi}_V$ projection operator.

Now, insert the polarizer. Suppose the initial polarization state is $|\psi\rangle = a|H\rangle + b|V\rangle$, as before, and the polarizer is tilted to allow only the polarization $|\tilde{\psi}\rangle = a'|H\rangle + b'|V\rangle$ to pass through, where a', b' are different coefficients. Suppose the measurement is weak ($d \ll \sigma$), so the probability of passing the polarization is approximately given by Malus' law, $P_{pass} = |\langle\tilde{\psi}|\psi\rangle|^2$, given by the squared cosine of the relative angle of the photon polarization and the polarizer. To find where on the screen the photon is likely to land, we can return to our initial discussion of this effect, (3.8), and project the meter state onto position $|x\rangle$, while we project the system state with the projection operator $\hat{\Pi}_{\psi'} = |\tilde{\psi}\rangle\langle\tilde{\psi}|$. This procedure leaves the meter state as

$$\phi(x) = \phi_0(x)\langle\tilde{\psi}|\psi\rangle + d\phi_0'(x)\langle\tilde{\psi}|V\rangle\langle V|\psi\rangle, \tag{4.16}$$

where we denote $\phi_0(x)$ as the original Gaussian meter state. Taking all states and wavefunctions to have real values for simplicity, we can simplify the probability density for where the photon will land as

$$p_{pol}(x) = |\langle\tilde{\psi}|\psi\rangle|^2 \phi_0^2 \left(1 - \frac{xd}{\sigma^2}\frac{\langle\tilde{\psi}|V\rangle\langle V|\psi\rangle}{\langle\tilde{\psi}|\psi\rangle}\right). \tag{4.17}$$

The first term, $|\langle\tilde{\psi}|\psi\rangle|^2$, we have already interpreted as the probability for the photon to pass the polarizer. Accordingly, the probability for the photon to land anywhere on the screen is given by

$$\int_{-\infty}^{\infty} dx\, p_{pol}(x) = |\langle\tilde{\psi}|\psi\rangle|^2, \tag{4.18}$$

the same as if there were no spatial mode. However, we can ask a more probing question: given that the photon passes the polarizer, where will it land? We can find the mean position as

$$\langle x\rangle_{ps} = \frac{\int_{-\infty}^{\infty} dx\, x p_{pol}(x)}{\int_{-\infty}^{\infty} dx\, p_{pol}(x)} = -d\frac{\langle\tilde{\psi}|\hat{\Pi}_V|\psi\rangle}{\langle\tilde{\psi}|\psi\rangle}, \tag{4.19}$$

where the variance is unchanged: σ^2. We see the presence of the polarizer has shifted the center of the distribution from $-d|b|^2 = -d|\langle V|\psi\rangle|^2$ to a new position, $-d(\hat{\Pi}_V)_w$, where we define the weak value of the operator $\hat{\Pi}_V$ as

$$(\Pi_V)_w = \frac{\langle\tilde{\psi}|\hat{\Pi}_V|\psi\rangle}{\langle\tilde{\psi}|\psi\rangle}. \tag{4.20}$$

This new position shift controlled by the weak value involves all the ingredients: the initial state $|\psi\rangle$, the final state $|\tilde{\psi}\rangle$, and the operator that is being weakly measured, $\hat{\Pi}_V$.

Notice that some strange things are happening with this new shift. When the initial polarization state of the photon and the final state of the polarizer become orthogonal, $\langle\tilde{\psi}|\psi\rangle \to 0$, corresponding to $aa' + bb' \to 0$, the weak value can diverge since we can arrange that bb' remains finite. Therefore, while without the polarizer, the beam shift is bounded by the eigenvalues of the operator that is being weakly measured (scaled by the weak measurement strength), with the polarizer, the shift can be made arbitrarily larger than the eigenvalue bounds! When applied to a similar effect in the Stern–Gerlach device, Aharonov, Albert, and Vaidman called this effect "How the result of a measurement of a component of the spin of a spin-1/2 particle can turn out to be 100."

Let us understand in greater detail how this interesting effect can arise. Recall that the conditional shift in the polarization shift of the photon, provided it passed through a slit at position x, is given by Eq. (4.3). Notice that, while usually the polarization of the photon is only weakly rotated by the weak measurement, occasionally very large values of position occur (although they are very unlikely). When $|x| > \sigma^2/d$, this leads to the rotation of the polarization by a large amount. We can interpret that to mean that although the coupling is weak, a rare event can make a large change in the system state. The weak value is similar to the reverse of this effect. When we demand that the photon be in a nearly orthogonal state to the polarization it began in, only very large values of x are capable of doing this, so their average must be anomalously large. Notice also that the divergence of the weak value coincides with the probability of the photon passing through the polarizer going to zero. This implies an important trade-off between how large the weak value can be, and how frequently the anomalous shift can occur. We will return to this point later in the chapter.

Weak Value Derivation within the von Neumann Model

Consider the von Neumann model with $H_{int} = g\delta(t)\hat{p} \otimes \hat{A}$, where $\hat{p}$ is the meter momentum, and $\hat{A}$ is a general system operator. Here we consider a finite-dimensional Hilbert space. Consequently, the unitary generating the entanglement is given by

$$\hat{U}_{SM} = e^{-ig\hat{p}\hat{A}/\hbar}. \tag{4.21}$$

Let the system and meter state be $|\psi\rangle \otimes |\phi\rangle$, and we take the weak measurement limit so that g is much less than the width of the meter states. If the eigenstates and eigenvalues of A are defined as

$$\hat{A}|a_i\rangle = a_i|a_i\rangle, \tag{4.22}$$

then expressing the meter in the position basis $|x\rangle$, we find the system/meter wavefunction takes the form

$$\langle x|\hat{U}_{SM}|\psi\rangle\phi\rangle = \sum_i c_i|a_i\rangle\phi(x - ga_i). \tag{4.23}$$

Here, we used the fact that the unitary is a displacement operator in the position basis, and that the entire state is a coherent superposition of varying degrees of spatial shift, depending on the eigenvalues of A. For large values of g such that there is no longer any overlap of the states of the meter, an x measurement is sufficient to determine the eigenvalues of A. This is because the eigenvalues can be associated to the well-separated peaks of the meter, for which the coefficients c_i control the relative weights of the states.

We now consider when the system is measured projectively with the operator $\hat{\Pi}_{\tilde{\psi}} = |\tilde{\psi}\rangle\langle\tilde{\psi}|$. We focus on the outcome corresponding to this state. Operationally, this is called *postselection* – we simply wait until this event, of all the possible events, occurs, and examine what happens in that case. We then have a meter state that corresponds to

$$\phi(x) = \langle\tilde{\psi}|\langle x|e^{-ig\hat{p}\hat{A}}|\psi\rangle|\phi\rangle \tag{4.24}$$

$$\approx \langle\tilde{\psi}|\langle x|(1 - ig\hat{p}\hat{A})|\psi\rangle|\phi\rangle \tag{4.25}$$

$$= \langle\tilde{\psi}|\psi\rangle\phi(x) - ig\langle\tilde{\psi}|\hat{A}|\psi\rangle\langle x|\hat{p}|\phi\rangle \tag{4.26}$$

$$\approx \langle\tilde{\psi}|\psi\rangle\langle x|e^{-ig\hat{p}A_w}|\phi\rangle \tag{4.27}$$

$$= \langle\tilde{\psi}|\psi\rangle\phi(x - gA_w). \tag{4.28}$$

In line (4.25), we have Taylor expanded to first order in g. In line (4.27), we have factored out the quantity $\langle\tilde{\psi}|\psi\rangle$ and re-exponentiated. In doing so the weak value of the operator A is defined as

$$A_w = \frac{\langle\tilde{\psi}|\hat{A}|\psi\rangle}{\langle\tilde{\psi}|\psi\rangle}. \tag{4.29}$$

The interpretation of the result (4.28) is that the meter wavefunction is shifted by the quantity gA_w, while the amplitude is reduced by the overlap between initial and final states.

The weak value (4.29) has several interesting properties. One must "preselect" and "postselect" the initial and final states. We have already noted that it can exceed its eigenvalue range. The weak value is symmetric with respect to exchange of these two states, which is related to time-reversal symmetry of the weak value in the weak measurement limit. The weak value can be a complex number. Physically, this is related to shifting the meter in the position basis (for the real weak value) and/or the momentum basis (for the imaginary weak value).

Different interpretations of the weak value exist in the literature. The most common one is that it represents the value the operator takes in a pre- and postselected ensemble, acting as a generalization of an eigenvalue to this situation. There are a variety of interesting phenomena to support this view, such as the influence of the system on any weakly coupled system acting like the system operator takes the weak value – however, this remains an interpretation of the physics.

4.4 Weak Value Amplification

In this section, we discuss an application of the weak value to an interesting problem in the science of making precise measurements. The basic idea is to note that we can turn our perspective around to take the physics of the weak value for granted and focus on the estimation of the measurement parameter g, taken as an unknown parameter, by a known weak value. Applied to the physics of the Stern–Gerlach device, Aharonov, Albert, and Vaidman point out: "Another striking aspect of this experiment becomes evident when we consider it as a device for measuring a small gradient of the magnetic field" (Aharonov et al., 1988). The basic idea is that without the postselection, the average of the meter's position is bounded between the eigenvalues of the measured operator, while with the postselection, the shift of the meter, by the weak value, can greatly exceed the eigenvalue range. Consequently, because the meter reading is much larger with the postselection, by dividing the meter reading by the weak value, a more precise determination of the parameter g can be made. However, there is a trade-off for the weak value, mentioned earlier: a larger weak value comes at the expense of a smaller postselection fraction, reducing the amount of data. As the weak value diverges, the amount of data goes to zero!

This idea was put to use in two influential experiments to make high-precision measurements of optical parameters. While the experiments were done with coherent states of light, and also can be explained with electromagnetic wave interference, we adopt the quantum formalism that inspired them, which carries over to the single-photon experiments. Hosten and Kwiat were trying to measure a tiny optical effect predicted to exist: the optical spin Hall effect, an effect analogous to the spin Hall effect in condensed matter physics, when an electric field is applied to a semiconductor, a dissipationless spin-dependent current perpendicular to the field can be generated (Hosten and Kwiat, 2008). In the optical case, a modified form of spin-orbit interaction can exist, resulting in a polarization-dependent spatial shift of the photon, where the role of the electric field is played by a change in the index of refraction. This effect occurs even in the most mundane case of light

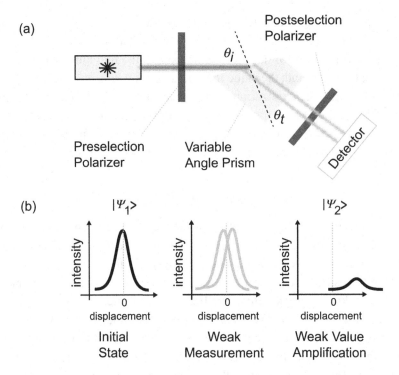

Figure 4.3 (a) In the weak value amplification experiment by Hosten and Kwiat, a light beam splits into two displaced beams when incident on a glass prism on account of the spin Hall effect for light (Hosten and Kwiat, 2008). The variable angle prism consists of two round wedge BL7 prisms joined with index matching fluid, allowing the two surfaces to be oriented such that light beams exit the optic at normal incidence without any further perturbation. Polarizers before and after the prisms allow for pre- and post-selection of the quantum state being measured, resulting in a weak value amplification of the tiny displacement between the two spin Hall shifted beams, as illustrated in (b).

passing from air to glass. In order to detect this effect, weak value amplification methods were used. As shown in Fig. 4.3, initially polarized light was incident on an air–glass interface, resulting in a slight displacement of the two polarization components. This effect is different from birefringence, which corresponds to different amounts of bending. By placing a polarizer after the interface, with the orientation nearly orthogonal to the initial polarization of the light, the deflection was greatly magnified. Using an optical detector sensitive to the location of the optical beam resulted in an accurate measurement of the size of the effect, with the precision of an angstrom.

In a second experiment by Dixon et al. (2009), a method was devised to make a very sensitive measurement of optical beam deflection, as shown in Fig. 4.4.

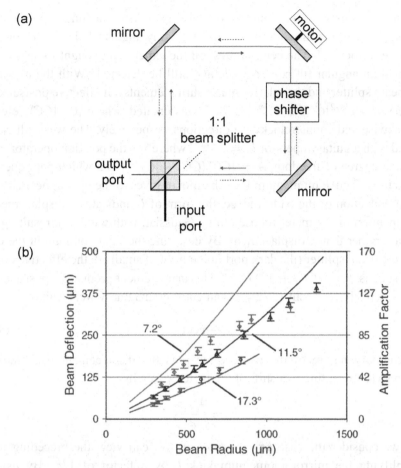

Figure 4.4 (a) Dixon, Starling, Jordan, and Howell conceived of and implemented a general scheme for weak value amplification in the optical domain using a Sagnac interferometer (Dixon et al., 2009). In this device, incident light is split at the input port with a 1:1 beam slitter, with the two components traversing a square in opposite directions. A motorized mirror can induce a deflection, changing the effective path length for the two light beams and resulting in a signal at the output port that is perfectly dark for identical paths. A tunable phase shift can be imposed using a continuously variable wave-plate such as a Soleil–Babinet compensator and other elements to postselect the quantum state being measured. (b) Data showing the measured beam deflection versus beam width for different settings of the variable wave-plate, and thus choice of quantum states. The curves were all taken for an unamplified deflection of 2.95 µm. Adapted with permission from the American Physical Society.

A Sagnac interferometer was constructed by causing two optical paths to make a cycle in a square geometry and then interfere on a beam splitter. By inserting polarization controlling optical elements, a relative phase shift ϕ was induced between the clockwise and counterclockwise path. By attaching a piezoelectric actuator to one of the interferometer mirrors, a tiny angular deflection of the mirror can be introduced. The central question the experiment sought to answer was how small an angular tilt can be made and still be detected? With the use of the 50/50 beam splitter and the relative phase shift elements, an effective postselection state $|\psi\rangle = (ie^{i\phi/2}|CC\rangle + e^{-i\phi/2}|C\rangle)/\sqrt{2}$ was created, where $|C\rangle, |CC\rangle$ refers to the clockwise and counterclockwise directions, respectively. The weak tilt can be modeled with a unitary operator $\hat{U} = e^{-ik\hat{x}\hat{A}}$, where $\hat{x}$ is the position operator of the transverse degree of freedom, $\hat{A} = |CC\rangle\langle CC| - |C\rangle\langle C|$ is a which-path operator, and k is the effective momentum kick the mirror gives to the optical beam, resulting in a deflection of the path. Indeed, the form of $\hat{U}$ indicates a displacement in the momentum of the meter (chosen to be Gaussian with width σ), resulting in a deflection rather than a displacement. By detecting the light coming out the other port of the beam splitter (the dark port in the limit of small ϕ), the effective postselection state is $\tilde{\psi} = (|CC\rangle + i|C\rangle)/\sqrt{2}$. This arrangement produces a postselection probability $|\langle\tilde{\psi}|\psi\rangle|^2 = \sin^2(\phi/2)$, and an entirely imaginary weak value,

$$A_w = -i \cot\frac{\phi}{2}. \tag{4.30}$$

The imaginary nature of the weak value puts the displacement back into the position space, resulting in a shift of the meter, given by

$$\langle\hat{x}\rangle = \frac{4k\sigma^2}{\phi}, \tag{4.31}$$

where we consider the case $k\sigma \ll \phi \ll 1$. We can view the preceding result as amplifying the mirror momentum kick k by a factor of $1/\phi$. By using a position-sensitive detector for the displacement of this weak optical beam, a very precise measurement of parameter k can be measured. Indeed, this first experiment obtained a precision of the mirror angle down to around 500 frad, with a precision of travel of the piezoelectric actuator to 20 fm. This angular precision is equivalent to a hair's width at the moon's distance (Steinberg, 2010).

Signal-to-Noise Ratio

We can make a more quantitative analysis of this estimation procedure by introducing the concept of signal-to-noise ratio, a very useful concept that experimentalists spend a great deal of time with. In this way, we can discuss the estimation precision without entering into a detailed discussion of the field of quantum metrology, which would take us too far away from the subject of our focus. The idea is that the measured signal, r, is proportional to the desired quantity that is to be measured, g, but is masked by some noise that could be from a variety of sources: thermal fluctuations, instrumental noise, shot noise of electrons or photons, $1/f$ noise from the electronics of the detector, and so on. We will consider the simplest type of noise for this discussion, "white noise," which is additive and uncorrelated with itself from moment to moment. Thus, the measurement result can be written as

$$r = cg + \xi, \tag{4.32}$$

where c is a constant of proportionality, related to the experimental setup, and ξ is a random variable that models the noise. The value of the random variable is different from run to run and stands in for some unpredictability in the measurement. We assume the noise term has zero mean and has a finite variance, $\langle \xi^2 \rangle = \Sigma^2$, where the angle brackets here represent a statistical average over many realizations of the random process. Consequently, there is statistical uncertainty in the estimate of g, given by $g = r/c \pm \Sigma/c$. The uncertainty fixes how accurately we can estimate the parameter g. The signal-to-noise ratio (SNR) is defined as the ratio of the mean to the standard deviation of the measured signal,

$$\mathcal{R} = \frac{cg}{\Sigma}, \tag{4.33}$$

which indicates the relative contribution of the signal, versus the noise in the measurement result. The larger $\mathcal{R}$ is, the more confident we are about the value of g. A common way to increase the SNR, assuming the value of g does not change, or at least changes slowly, is to take many measurements and average the results together. Let a time series of measurement results be $\{r_1, r_2, \ldots r_N\}$. We can then find the average as

$$\bar{r} = \frac{1}{N} \sum_{i=1}^{N} r_i = cg + \frac{1}{N} \sum_{i=1}^{N} \xi_i. \tag{4.34}$$

We calculate the variance of $\bar{r}$ under the assumption of uncorrelated noise, so $\langle \xi_i \xi_j \rangle = \sigma^2 \delta_{ij}$. Thus,

$$\langle \bar{r}^2 \rangle - \langle \bar{r} \rangle^2 = \frac{1}{N^2} \sum_{i=1, j=1}^{N} \langle \xi_i \xi_j \rangle = \frac{1}{N^2} \sum_{i=1, j=1}^{N} \sigma^2 \delta_{ij} = \frac{\sigma^2}{N}. \tag{4.35}$$

Averaging the signal increases the signal-to-noise ratio by a factor of $\sqrt{N}$. The central limit theorem indicates this result is more general, because adding together N iterations of the same random variable gives an average whose distribution becomes Gaussian under a wide range of conditions, so only the variance needs to be specified.

We can now apply these considerations to the weak value estimation of the parameter g. Under the weak value amplification process, where the operator $\hat{A}$

is weakly measured, the signal is boosted as $c = a_{max} \to A_w$, so the largest eigenvalue (under the best case conditions) is replaced by the weak value, $A_w \gg a_{max}$. The standard deviation σ is interpreted as the spatial width of the meter, which is unchanged after the postselection. However, the postselection only occurs for a small fraction of all the events when the weak value is large, so the number of events is reduced as $N \to N|\langle \tilde{\psi}|\psi \rangle|^2$. Consequently, the SNR is given by

$$\mathcal{R} \to \frac{gA_w}{\sigma}\sqrt{N|\langle \tilde{\psi}|\psi \rangle|^2} = \frac{g}{\sigma}\sqrt{N}\langle \tilde{\psi}|A|\psi \rangle. \qquad (4.36)$$

Here, we take $\langle \tilde{\psi}|A|\psi \rangle$ to be real for simplicity. The important point in the preceding formula is that the divergence of the weak value exactly compensates the postselection loss, so the SNR is a wash for a quantum-limited system (Starling et al., 2009), up to the difference between a_{max} and $\langle \tilde{\psi}|A|\psi \rangle$, which are typically comparable. There is, however, an important difference: about the same SNR is obtained by using a very small fraction of the data that would have been obtained with all the data in the standard measurement. We can thus interpret this method as a kind of filtering procedure. This immediately gives a number of advantages to this technique:

- In an optical context, a small portion of the laser light beam can be shaved off to measure optical properties with the same precision as if the entire beam had been measured, but then using the rest of the beam in a subsequent experiment (Starling et al., 2010).
- In an optical context, it is typically easy to use more photons by increasing the power of a laser, but most laboratory detectors can accept only a finite amount of power before they begin to burn or saturate. Consequently, weak value amplification allows experiments to use much more optical power than can be accepted by the detector and still retain the full advantage of all the photons (Starling et al., 2009).
- The non-postselected events can sometimes be recycled, reinjecting the rejected events back into the experiment to further utilize finite resources to increase precision (Dressel et al., 2013; Lyons et al., 2015; Krafczyk et al., 2021).
- Very often there are situations where other noise sources present, or the experimental setup implementing the measurements, are important (Jordan et al., 2014), where "less is more." It is sometimes experimentally advantageous to have more signal, but fewer events (Magaña-Loaiza et al., 2014), such as in the presence of systematic noise (Pang et al., 2016), or to put more time between colored noise events (Feizpour et al., 2011; Sinclair et al., 2017).

Let us now give a simple illustration of this physics with the von Neumann measurement model. Consider a preselected state $|\psi \rangle = (|+\rangle + |-\rangle)/\sqrt{2}$, and a postselected state $|\tilde{\psi}\rangle = \sin(\pi/4 + \phi)|+\rangle - \cos(\pi/4 + \phi)|-\rangle$. Let us weakly measure the operator $\hat{\sigma}_z$. In Exercise 4.5.17, you will show $(\hat{\sigma}_z)_w = \cot\phi$, with postselection probability $\sin^2\phi$. In this case, an initial Gaussian meter with standard deviation σ is shifted in position by the amount g if the system is prepared in an eigenvalue $\hat{\sigma}_z = 1$, but by the amount $g(\sigma_z)_w$ if it is pre- and postselected. Consequently, a measurement of the meter's position will give a much larger reading, improving the signal of an unknown g measurement. Applying our result (4.36) indicates that the SNR of this measurement is given by

$$\mathcal{R} = \frac{g(\hat{\sigma}_z)_w\sqrt{N|\langle\tilde{\psi}|\psi\rangle|^2}}{\sigma} = \frac{g\cos\phi\sqrt{N}}{\sigma}. \tag{4.37}$$

Consequently, as $\phi \to 0$ (while keeping g even smaller), the SNR recovers the SNR with all the data from the eigenstate measurement. The observant student may have a bright idea at this point: perhaps we can do *even better* if we also use the non-postselected data to further improve our precision. In Exercise 4.5.18 you will show that in fact this is not the case, and that any well-designed weak-value amplification experiment has negligible information in the rejected events, effectively concentrating all information into the rare postselected events.

The previous section only scratches the surface about weak value amplification. There are now hundreds of experiments applying variants of this method to all manner of estimation problems. An early review can be found in Dressel et al. (2014). We mention here a few highlights of exciting recent work in this area of research. The group of Aephraim Steinberg demonstrated weak-value amplification of the nonlinear effect of a single photon (Hallaji et al., 2017), where one photon effectively acted like many photons. Weak value amplification has also been demonstrated in integrated optical platforms in the group of Jaime Cardenas (Song et al., 2021). In that work, miniaturized multimode interferometers were designed on-chip to sensitively measure phase shifts induced by a heater as well as measure a change in optical frequency (with the help of a ring resonator) down to a precision of 2 kHz. Finally, recent work has considered the advantages of postselected metrology in quantum physics more generally (Lupu-Gladstein et al., 2022), beyond the weak value paradigm. The parameter estimation protocol has come to be the primary application of weak values.

4.5 Generalized Eigenvalues for Any Measurement Type

In the treatment of generalized measurements of Chapter 3, we learned about new types of measurement, how to calculate their probabilities, and the resulting change of the system state. However, we lost an important notion of what we were measuring. That is, the link to the system's observable property became obscured. In the case of weak measurement, as discussed in the previous section, there is a clear interpretation of weakly measuring a system observable. We now discuss how to generalize this notion to arbitrary types of measurement by introducing the concept of *generalized eigenvalues* of an observable, labeled $\{\kappa_j\}$, where $j = 1, \ldots, M$, associated with the possible outcomes of a generalized measurement. The assignment of these generalized eigenvalues depends on the kind of generalized measurement to be made. These thus depend on the measurement context, so these values are sometimes called "contextual values" (Dressel et al., 2010).

In order to assign these values, given a set of measurement operators $\{\hat{\Omega}_j\}$, we can express the expectation value of the operator $\hat{O}$ (2.19) as

$$\langle \hat{O} \rangle = \sum_{i=1}^{N} \lambda_i P_i = \sum_{j=1}^{M} \kappa_j \mathcal{P}_j, \tag{4.38}$$

where we replace $\lambda_i \to \kappa_j$ and the probability of the projective measurement result $P_i = \text{Tr}[\hat{\rho} \hat{\Pi}_i]$ with the probability of the generalized measurement, $\mathcal{P}_j = \text{Tr}[\hat{\rho} \hat{E}_j]$. Note that $M \neq N$ in general. This assignment should be true for all possible states $\hat{\rho}$, so we have then an operator equation,

$$\hat{O} = \sum_{i=1}^{N} \lambda_i \hat{\Pi}_i = \sum_{j=1}^{M} \kappa_j \hat{E}_j. \tag{4.39}$$

It is clear from this expression that the generalized eigenvalues must be real numbers. Once a valid assignment of the generalized eigenvalues is made, we note that the expectation of any moment of the operator $\hat{O}$ (2.20) can be found as

$$\langle \hat{O}^n \rangle = \sum_{i=1}^{N} \lambda_i^n P_i = \sum_{j_1, j_2, \ldots, j_n = 1}^{M} \kappa_{j_1} \kappa_{j_2} \ldots \kappa_{j_n} \mathcal{P}_{j_1, j_2, \ldots, j_n}, \tag{4.40}$$

where we have defined

$$\mathcal{P}_{j_1, j_2, \ldots, j_n} = \text{Tr}[\hat{E}_{j_1} \hat{E}_{j_2} \ldots \hat{E}_{j_n} \hat{\rho}], \tag{4.41}$$

which may be interpreted as the probability of finding results $j_1, j_2, \ldots, j_n$ in a sequence of n generalized measurements. We have made the simplifying assumption that the observable $\hat{O}$ and all operators $\hat{E}_j$ commute.

Before a general discussion of how to satisfy our main equation of this section, Eq. (4.39), we first give a simple example, based on Exercise 3.7.2. This problem concerns coupling a two-level system to a two-level meter, realized as the polarization of a single photon. If you solved the problem correctly, you found that the measurement operators satisfy $\hat{E}_{1,2} = (1/2)\mathbf{1} \pm g\hat{\sigma}_z$, where we label 1 and 2 as when the meter photon registers the results H and V, respectively. In this case, we have the operator equation (4.39) given by

$$\hat{O} = \sum_{i=1}^{N} \lambda_i \hat{\Pi}_i = \frac{\kappa_1 + \kappa_2}{2} \mathbf{1} + g \frac{\kappa_1 - \kappa_2}{2} \hat{\sigma}_z. \tag{4.42}$$

From this expression, it is clear that we can reconstruct operators in the span of $\{\mathbf{1}, \hat{\sigma}_z\}$. Let us choose the operator $\hat{O} = \hat{\sigma}_z = |H\rangle\langle H| - |V\rangle\langle V|$, to be definite, with eigenvalues $\lambda_{1,2} = \pm 1$ in the $H - V$ basis. The solution to Equation (4.42) is then uniquely given by

$$\kappa_1 = \frac{1}{g}, \quad \kappa_2 = -\frac{1}{g}. \tag{4.43}$$

It is interesting to note that as the measurement strength g limits to zero, $\kappa_{1,2}$ diverge to plus or minus infinity. For any value of $g < 1$, the generalized eigenvalues exceed the eigenvalue bounds ± 1, reminiscent of the weak value from the previous section. Nevertheless, despite this behavior, we can confirm that the desired behavior for the expectation value of $\hat{O}$ is satisfied for a general pure state $|\psi\rangle = a|H\rangle + b|V\rangle$:

$$\langle \hat{O} \rangle = \langle \psi | \hat{\sigma}_z | \psi \rangle = |a|^2 - |b|^2 = \sum_{j=1,2} \kappa_j \langle \psi | \hat{E}_j | \psi \rangle = \sum_{j=1,2} \kappa_j \mathcal{P}_j. \tag{4.44}$$

Indeed, we find that either way, we get the correct expression. To have a deeper understanding, we write the probabilities of the generalized measurement as $\mathcal{P}_j = 1/2 \pm g \langle \psi | \hat{\sigma}_z | \psi \rangle = 1/2 \pm g(|a|^2 - |b|^2)$. The fact that these probabilities have only a very weak dependence on the system properties $|a|^2 - |b|^2$ when g is small, implies that in order to extract the system dependence, we must effectively amplify it with the generalized eigenvalues. This is why they must diverge as g becomes small, and be opposite signed: The constant term must be canceled off, and the asymmetric terms must be amplified.

Recalling the defining equation for the generalized eigenvalues, Eq. (4.39), we see that, if $N = M$, the solution is unique, while if $N > M$ the system is overdetermined, while, if $N < M$, then the system is underdetermined. Physically, this dictates our ability to construct averages of system operators from the type of measurements and information at hand. For example, if we carry out a two-outcome generalized measurement on a position of a particle, then it is not too surprising that with that data, it is not possible to accurately calculate the particle's expected position, $\langle x \rangle$, in general. On the other hand, if you have access to an infinite number of possible outcomes weakly measuring a two-state system, there can be many ways to extract the desired information about that system. Nevertheless, in both the over- and underdetermined cases, there is a principled way to make a choice of the assigned values, either to make the best available approximation in the overdetermined case or to minimize the norm of the solution in the underdetermined case. To illustrate why the later condition is desirable here, consider the following example.

Let us return to the von Neumann measurement model (3.5), where the measurement operators, indexed by the position of the meter registration, is given by

$$\hat{\Omega}(x) = \phi(x - g\hat{\sigma}_z), \tag{4.45}$$

where $\phi(x)$ is the meter's position wavefunction. It is shifted either left or right by g, depending on the state of the system. Let us suppose we are interested to find the expectation value of $\hat{\sigma}_z$ of the two-level system in whatever state it is in. The operator $\hat{E}(x)$ is given by $\hat{E}(x) = P_m(x - g\hat{\sigma}_z)$, where $P_m(x)$ is the meter's probability density. Let us consider two possible choices (of the many) to assign to $\kappa(x)$ to estimate $\hat{\sigma}_z$:

$$\kappa_A(x) = \frac{x}{g}, \quad \kappa_B(x) = \frac{P_m(x - g) - P_m(x + g)}{a - b}, \tag{4.46}$$

where a, b are constants, given by $a = \int dx P_m(x)^2, b = \int dx P_m(x - g)P_m(x + g)$. In Exercise 4.5.13, you will show that both choices satisfy the condition (4.39).

In terms of pros and cons of which one of these functions to use, notice that while the choice $\kappa_A(x)$ is simplest, and independent of even the type of meter wavefunction, its value continues to increase linearly as x grows, while the choice $\kappa_B(x)$ rapidly decays to zero once x extends beyond the typical width of the meter wavefunction. This difference plays an important role in experiments. When collecting data, many rounds of the experiment are done from the same initial conditions. In terms of the operational procedure, the experimentalist would register the events x_k where k indexes the round number, and the x axis would usually be binned to create statistics. The average would then be constructed by weighting the generalized eigenvalue assigned to that bin by the frequency of events in a given bin, and then summing. Once this procedure is understood, the advantage of choice κ_B over κ_A becomes clear: given finite statistics, the convergence of the weighted average will be much faster. Therefore, we can make an optimal choice for $\kappa(x)$, that is, independent of the state of the system, by choosing the assignment that minimizes the norm of κ_j, that is, $\sum_j \kappa_j^2$. This choice can be found by assuming that the operator $\hat{O}$ and the operator $\hat{E}_j$ are all diagonal in the same basis. We can then express the operator equation (4.39) as a matrix equation,

$$\lambda = \bar{\bar{F}}\kappa, \quad F_{ij} = \text{Tr}[\hat{\Pi}_i \hat{E}_j], \tag{4.47}$$

where the vectors κ and λ are column vectors of length M and N, respectively, with elements κ_j and λ_i. $\bar{\bar{F}}$ is a real $N \times M$ matrix with the elements defined earlier.

We recognize the constitutive equation (4.47) as a linear system of equations for κ. Solving this linear system in the underdetermined or overdetermined cases is a well-studied problem in linear algebra. If $M = N$ we can usually simply find the inverse of the matrix F^{-1} to solve the problem, $\kappa = \overline{\overline{F}}^{-1} \lambda$. In the overdetermined case, a solution is generally impossible; however, we can find the "least-squares" solution – that is, the solution that minimizes the error, $||\lambda - \overline{\overline{F}}\kappa||$, where $|| \ldots ||$ is the Euclidean norm. This solution is given by

$$\kappa_{ls} = (\overline{\overline{F^T}} \, \overline{\overline{F}})^{-1} \, \overline{\overline{F^T}} \lambda, \tag{4.48}$$

also known as the pseudo-inverse of a left-invertible matrix. A similar pseudo-inverse can be constructed in the undetermined case. We quote the most general type of pseudo-inverse, also known as the Moore–Penrose pseudo-inverse, which uses the singular value decomposition. Any real matrix has a singular value decomposition, $\overline{\overline{F}} = \overline{\overline{U}} \, \overline{\overline{\Sigma}} \, \overline{\overline{V^T}}$, where $\overline{\overline{U}}$ is an $N \times N$ orthogonal matrix, $\overline{\overline{\Sigma}}$ is an $N \times M$ real matrix with nonnegative elements on its diagonal, and $\overline{\overline{V}}$ is an $M \times M$ orthogonal matrix. The diagonal elements of the rectangular matrix $\overline{\overline{\Sigma}}$ are called the singular values. With this construction, the Moore–Penrose pseudo-inverse is given by

$$\overline{\overline{F^+}} = \overline{\overline{V}} \, \overline{\overline{\Sigma^+}} \, \overline{\overline{U^T}}, \tag{4.49}$$

where the rectangular $M \times N$ matrix $\overline{\overline{\Sigma^+}}$ is defined by inverting all nonzero elements of $\overline{\overline{\Sigma^T}}$. The benefit of the solution $\kappa_{mp} = \overline{\overline{F^+}} \lambda$ is that it satisfies our earlier desire of having the minimum norm. All other solutions to Eq. (4.47) can be written as $\kappa = \kappa_{mp} + \mathbf{n}$, where $\mathbf{n}$ is another vector in the null-space of F. This additional vector can only lengthen the norm of the solution. Further discussion of these solutions and examples can be found in Dressel et al. (2010) and Dressel and Jordan (2012a).

Conditioned Averages and the Recovery of the Weak Value

As an illustration of the application of the generalized eigenvalues, we now discuss how to recover the weak value formula in the weak measurement limit. We begin by considering a generalized measurement, followed by a projective measurement. Consider an initial state $|\psi\rangle$. We treat the pure state for simplicity, but our treatment is straightforward to generalize to the mixed state case. We consider the measurement of the system operator $\hat{O}$. The generalized eigenvalues of the first measurement are taken to be $\{\kappa_j\}$, along with measurement operators $\{\hat{\Omega}_j\}$, and the final projection is described with projectors $\{\hat{\Pi}_k\}$. The effect of the two measurements together is then $\hat{M}_{jk} = \hat{\Pi}_k \hat{\Omega}_j$. The probability of finding both results j and k in a measurement sequence is given by

$$P(j,k) = \langle \psi | \hat{\Omega}_j^\dagger \hat{\Pi}_k \hat{\Omega}_j | \psi \rangle. \tag{4.50}$$

Recall that the average of operator $\hat{O}$ is given in Eq. (4.38) as a sum of the generalized eigenvalues weighted by the probability of the result. It is then natural in the measurement sequence to define the *conditioned average* of operator $\hat{O}$, when postselecting on a particular final state, $\hat{\Pi}_f = |\tilde{\psi}\rangle\langle\tilde{\psi}|$, as follows,

$$_{\tilde{\psi}}\langle \hat{O} \rangle_\psi = \sum_j \kappa_j P(j|k=f). \tag{4.51}$$

Here, the probability $P(j|k=f)$ is the conditional probability of result j, given a final postselection on $k=f$. We can calculate this probability using Bayes' rule (see Appendix A):

$$P(j|k=f) = \frac{P(j,k=f)}{P_f}, \qquad P(f) = \sum_j P(j,k=f). \tag{4.52}$$

Before proceeding further, we notice a natural explanation for why it is now possible for the conditional average to exceed the eigenvalue bounds: the generalized eigenvalues can exceed the normal eigenvalue bounds, so if we weight them with a different distribution, it is possible for the result to also exceed the eigenvalue bounds. This is related to the effect that if we take the variance of the generalized eigenvalues, it is not given by the variance of the operator; see (4.40).

It is now fairly easy to see how the weak value emerges in the weak measurement limit. Recall that if we introduce a measurement strength g, then the form (4.4) holds in the weak measurement limit. We consider the simplest case of a QND measurement, where in the decomposition (3.19), we choose $U_j = 1$, or at least a U_j that limits to the identity to higher order in g (note that the effect of U_f can be included by unitarily rotating the postselection state). We then have

$$\hat{\Omega}_j = \sqrt{\hat{E}_j} = \sqrt{p_j 1 + g\Gamma_j \hat{O}} \approx \sqrt{p_j} 1 + \frac{g\Gamma_j}{2\sqrt{p_j}} \hat{O}, \tag{4.53}$$

where we exclude $p_j = 0$ for any j, for simplicity, and we drop higher-order terms in g. Let us then consider the denominator in Eqs. (4.51, 4.52):

$$P_f = \sum_j \langle \psi | (\sqrt{p_j} \hat{\Pi}_f \sqrt{p_j} | \psi \rangle + O(g) \approx |\langle \tilde{\psi} | \psi \rangle|^2. \tag{4.54}$$

We see that so long as the overlap between the initial and final states is not zero, the leading-order term is independent of g. Let us now turn to the numerator N of Eq. (4.51):

$$N = \sum_j \kappa_j \left| \langle \tilde{\psi} | \sqrt{p_j} 1 + \frac{g\Gamma_j}{2\sqrt{p_j}} \hat{O} | \psi \rangle \right|^2, \tag{4.55}$$

$$= \sum_j \kappa_j \left| \langle \tilde{\psi} | \psi \rangle \sqrt{p_j} + \frac{g\Gamma_j}{2\sqrt{p_j}} \langle \tilde{\psi} | \hat{O} | \psi \rangle \right|^2, \tag{4.56}$$

$$\approx \sum_j \kappa_j \left[p_j |\langle \tilde{\psi} | \psi \rangle|^2 + \frac{g\Gamma_j}{2} (\langle \tilde{\psi} | \psi \rangle^* \langle \tilde{\psi} | \hat{O} | \psi \rangle + \langle \tilde{\psi} | \psi \rangle \langle \tilde{\psi} | \hat{O} | \psi \rangle^*) \right], \tag{4.57}$$

where we drop higher-order terms in g. To further simplify this result, note that since $\sum_j \kappa_j \hat{E}_j = \hat{O}$ by construction, and since $\hat{E}_j \approx p_j 1 + g\Gamma_j \hat{O}$ in the small g limit, we must have the relations

$$\sum_j \kappa_j p_j = 0, \qquad g \sum_j \kappa_j \Gamma_j = 1. \tag{4.58}$$

These sum rules then simplify the numerator (4.57) by eliminating the first term, and eliminating the $\Gamma_j \kappa_j$ factors in the second term. We then have a final result,

$$\tilde{\psi}\langle \hat{O} \rangle_\psi = \frac{\langle \tilde{\psi} | \psi \rangle^* \langle \tilde{\psi} | \hat{O} | \psi \rangle + \langle \tilde{\psi} | \psi \rangle \langle \tilde{\psi} | \hat{O} | \psi \rangle^*}{2|\langle \tilde{\psi} | \psi \rangle|^2} = \text{Re} \frac{\langle \tilde{\psi} | \hat{O} | \psi \rangle}{\langle \tilde{\psi} | \psi \rangle}. \tag{4.59}$$

It is important to this derivation that the choice of generalized eigenvalues isolates the desired operator to be measured, and amplifies it, in order to get a finite result depending on the system operator as g limits to zero. We point out that very few assumptions have been made about the type of measurement to obtain the weak value in the small g limit, indicating a robustness, if not universality, of this result. A final point about our result is that it is the real part of the weak value that appears. In hindsight, this comes from our starting point (4.51) which stresses an operational interpretation of dealing with data from an experiment. Such data is always real, and since our generalized eigenvalues are also defined to be real, it is guaranteed to obtain a real number for any type of average of such numbers. The imaginary part of the weak value also has a physical interpretation as discussed earlier in this chapter. See Steinberg (1995), Aharonov and Botero (2005), Jozsa (2007), and Dressel and Jordan (2012b) for further reading.

Exercises

Exercise 4.5.1 Consider the von Neumann measurement model in the weak measurement limit, where the parameter g is very small. What system observable is being weakly measured? What function plays the role of Γ_j in Eq. (4.4)?

Exercise 4.5.2 Demonstrate that the distribution (4.2) remains normalized to $\mathcal{O}(d)$.

Exercise 4.5.3 A common model for "pure dephasing" is to add a random phase θ to a coherent superposition, and to then average the phase over a distribution we can take to be Gaussian. For a qubit, work out what the quantum channel for this process is. That is, find the appropriate operators $\hat{\Omega}_j$.

Exercise 4.5.4 For Exercise 3.7.2, study the weak measurement limit. What system operator is being weakly measured by the meter?

Exercise 4.5.5 For the von Neumann measurement model, calculate the Bhattacharyya coefficient for the measurement probabilities. Using a Gaussian meter wavefunction, check to see if this is a quantum limited detector by investigating the ensemble averaged decoherence. What happens if the meter wavefunction is a complex function?

Exercise 4.5.6 For Exercise 3.7.2, calculate the Bhattacharyya coefficient for the measurement probabilities. Check to see if this is a quantum limited detector by investigating the ensemble averaged decoherence.

Exercise 4.5.7 Consider a particle of mass m in a harmonic trap of frequency ω. Suppose you want to weakly measure its position with an impulsive interaction using a free particle meter. What interaction Hamiltonian should be designed? What about a weak momentum measurement? A weak measurement of the mechanical energy?

Exercise 4.5.8 Consider a metastable potential approximated as a cubic function. Use the WKB approximation to calculate the tunneling rate out of the well. Verify that the rate depends exponentially on the energy of the particle inside the well.

Exercise 4.5.9 Verify that another way of expressing the result (4.14) for the no-tunneling conditional evolution of the phase qubit is

$$\frac{\rho_{ee}(t)}{\rho_{gg}(t)} = e^{-\Gamma t}\frac{\rho_{ee}(0)}{\rho_{gg}(0)}, \qquad \frac{\rho_{eg}(t)}{\sqrt{\rho_{gg}(t)\rho_{ee}(t)}} = e^{i\omega t}\frac{\rho_{eg}(0)}{\sqrt{\rho_{gg}(0)\rho_{ee}(0)}}, \qquad (4.60)$$

while maintaining the norm 1 and Hermitian nature of the density matrix.

Exercise 4.5.10 Verify that in Eq. (4.14), if the initial density matrix is pure, that is, $\hat{\rho} = |\psi\rangle\langle\psi|$ for some state $|\psi\rangle$, that the state remains pure during the continuous collapse.

Exercise 4.5.11 For the previous problem, write the initial pure state as $|\psi\rangle = \cos\theta_i|g\rangle + \sin\theta_i|e\rangle$. Find the continuous collapse rule for the null measurement case after time t. That is, find $\theta(t)$ that begins at θ_i and ends at $\theta = 0$ in the long time limit.

Exercise 4.5.12 Following the discussion in Section 4.4, what are the generalized eigenvalues for the projection operators for H or V polarization, either $\hat{\Pi}_H$ or $\hat{\Pi}_V$?

Exercise 4.5.13 Verify that both choices, κ_A and κ_B in Eq. (4.46), obey the central equation (4.39), suitably generalized to continuous variables. Hint: Consider one matrix element at a time if you need to.

Exercise 4.5.14 Use the properties of the complete set of projections operators $\hat{\Pi}_i$ to show the matrix equation (4.47) is true.

Exercise 4.5.15 Suppose the two-level system is measured with a three-outcome measurement. The observable being measured is $\hat{\sigma}_z$. The measurement operators, expressed in the $\hat{\sigma}_z$ eigenbasis, are $\hat{E}_1 = \text{diag}(1/3, 1/2)$, $\hat{E}_2 = \text{diag}(1/3, 1/5)$, $\hat{E}_3 = \text{diag}(1/3, 3/10)$. Find the generalized eigenvalues of $\hat{\sigma}_z$ for this measurement that minimizes the norm of the solution.

Exercise 4.5.16 A three-level system is measured with a two-outcome measurement, described by the operators $\hat{E}_1 = \text{diag}(1/2 + g, 1/2 - g, 1/2 + w)$, $\hat{E}_2 = \text{diag}(1/2 - g, 1/2 + g, 1/2 - w)$, where $-1/2 < w, g < 1/2$. Find the generalized eigenvalue needed to measure a general observable in this basis of the form $\hat{O} = \text{diag}(a, b, c)$, where a, b, c are real numbers, that has the minimum error.

Exercise 4.5.17 Show for the von Neumann model that for the choices in Section 4.5.1, the postselection probability and weak value are given by

$$|\langle \tilde{\psi} | \psi \rangle|^2 = \sin^2 \phi, \qquad (\sigma_z)_w = \cot \phi. \qquad (4.61)$$

Exercise 4.5.18 See if there is any information in the rejected events from a weak value amplification model of Section 4.5.1 by finding the SNR using the deflection signal from the other postselection state, $|\psi_\perp\rangle = \cos(\pi/4 + \phi)|+\rangle + \sin(\pi/4 + \phi)|-\rangle$ in the limit as $\phi \to 0$. This experiment was carried out in Viza et al. (2015).

5

Continuous Measurement: Diffusive Case

In the previous chapters, we have learned about projective measurements, generalized measurements, weak measurements, and sequences of the same. The weak value involved a weak measurement followed by a strong one. We now introduce the concept of a continuous measurement by considering a sequence of many weak measurements, one after the other, not bothering to reset the system back to the same initial condition. This weak measurement sequence has results $r_1, r_2, \ldots r_N$, where r_j can be either continuous or discrete. We consider a timeline, discretized into equally spaced time chunks of size Δt, so that after N measurements, a total time $T = N\Delta t$ has elapsed. We consider the limit where the measurement progressively weakens as $\Delta t \to 0$, to get a well-defined mathematical limit.

Continuous quantum measurements were developed by a number of scientists coming at the problem from different perspectives, including Balavkin, Barchielli, Carmichael, Milburn, Wiseman, Diósi, and Gisin, starting in the mid 1980s. Depending on the type of detectors involved, one can realize different types of continuous quantum measurements. In this chapter we focus on the diffusive quantum trajectories, while in the next chapter we focus on quantum jump continuous measurements.

Each measurement j is associated with a measurement operator $\hat{\Omega}_j$, so starting with a well-defined state $|\psi_0\rangle$, we define a quantum map from each time to the next,

$$|\psi_j\rangle = \frac{\hat{\Omega}_j|\psi_{j-1}\rangle}{||\hat{\Omega}_j|\psi_{j-1}\rangle||}, \tag{5.1}$$

where $j = 1, \ldots, N$. This creates a chain of states, dependent on the initial state and all subsequent measurements. The strength of the measurement can be controlled by Δt, so as $\Delta t \to 0$, the sequence of states can become a diffusive process in the Hilbert space. Importantly, this is not the only option: another class of quantum trajectories is called quantum jumps and will be discussed in the next chapter.

The sequence of measurement results $\{r_j\}$ is called a *measurement record*, and the sequence of quantum states $\{|\psi_j\rangle\}$ is called a *quantum trajectory*.

Rather than launch into the abstract, mathematical description of this physics, we begin with motivating physical experiments: an electron in a double-quantum-dot being measured with a quantum point contact, and a superconducting circuit being dispersively measured by microwave frequency electromagnetic radiation.

5.1 Measuring the Location of an Electron on a Double Quantum Dot with a Quantum Point Contact

We consider a first experiment involving condensed matter physics. In a two-dimensional electron gas, formed by a layered semiconductor material that confines electron motion to a plane, metallic top-gates further constrict electron motion with electrical potential control. These metallic gates can cause islands of electrons to form in small regions, called *quantum dots*. If they are small enough, they become a kind of artificial atom because the energy of the electrons is quantized, as a result of being in a confining potential.

We can complicate this picture by considering a double quantum dot, where we imagine two potential well bowls close together, making a kind of dumbbell (see Fig. 5.1). In this geometry, we consider the simplest case of a single electron, which can be localized on the left side of the double quantum dot (DQD), or on the right side. Taken in isolation, these dots have their own energy levels, but the possibility of the electron tunneling from left or right or vice versa results in a hybridization of the energy levels. If we denote $|L\rangle$ and $|R\rangle$ as the states of an electron confined to the ground state of either the left or the right well, and the wells are symmetric, then $|\psi_{\pm}\rangle = (|L\rangle \pm |R\rangle)/\sqrt{2}$ describes the ground and first excited states, respectively. In Exercise 5.6.1 you will prove this statement.

If we suppose the electron is localized in the left well (for example), then the tunneling process will cause flopping back and forth between wells, described by the state

$$|\psi(t)\rangle = \cos(\Delta t/\hbar)|L\rangle - i\sin(\Delta t/\hbar)|R\rangle. \tag{5.2}$$

More generally, if we suppose the potential double well is slightly tipped, so the difference between the left and right wells' energy is ϵ, this situation can be modeled within the two-state approximation by adding a term $\epsilon\hat{\sigma}_z/2$ to the Hamiltonian, so the more general two-state system Hamiltonian is given by

$$H = \frac{\epsilon}{2}\hat{\sigma}_z + \frac{\Delta'}{2}\hat{\sigma}_x. \tag{5.3}$$

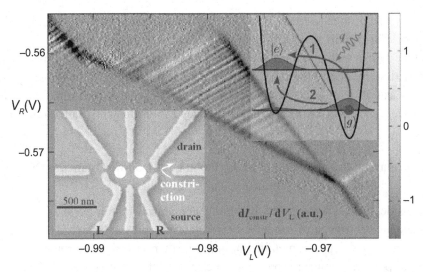

Figure 5.1 A double quantum dot with a quantum point contact charge sensor is shown in the scanning electron micrograph in the lower left. The light grey structures are metal electrodes used to gate an AlGaAs/GaAs heterostructure with a two-dimensional electron system 80 nm beneath the surface. The main plot shows the transconductance, measured at 30 mK, of the quantum point contact sensor in response to a small modulation of the left electrode voltage. This response is then plotted over a range of bias voltages applied to the left and right gates. The bright/dark lines correspond to single-electron transitions (Granger et al., 2012). The stripes are attributed to an interference pattern between two different paths for single-phonon absorption which create a nonequilibrium population of the excited state, illustrated in the upper right. Adapted with permission from Springer Nature and Stefan Ludwig.

Here we have substituted $\Delta = \Delta'/2$ for a symmetric Hamiltonian. More generally, a $\hat{\sigma}_y$ could be added, but a basis transformation can always align the operator along one transformed transverse axis.

How can such a system be measured? Note that the electron has an electric charge on it. This physical property can be used to determine where it is. We must therefore design a charge detector that can rapidly and efficiently determine where the electron is located. We will employ a nearby electrical conductor to sense the presence of the nearby charge. An optimal choice turns out to be a *quantum point contact*. This device is formed using metallic top gates to further constrict electron flow to a narrow channel, similar to a river channeled through a narrow constriction. The quantum point contact (QPC) is a mesoscopic version of a resistor; we can think of the constriction as an electron-scattering barrier that the electrons must traverse if they are to cross from one side to the other. The contact is connected on both sides to electron reservoirs, so an applied electrical bias V results in the flow of electrical current I. The current obeys Ohm's law, $I = GV$, where G is the conductance of the contact. While a detailed discussion of the mesoscopic transport physics is beyond the scope of this book, we quote the Landauer–Büttiker formula for transport (Büttiker, 1986), that relates conductance to transmission,

$$G = \frac{2e^2}{h}\mathcal{T}. \tag{5.4}$$

Here, e^2/h is the conductance quantum, the factor of 2 corresponds to two spin components, while $\mathcal{T}$ is the transmission coefficient of the scattering region. The qualitative physics behind this formula is that when an electrical bias is applied, the Fermi energy of the electrons on one side is elevated by the chemical potential. This allows fermions in filled states (at zero temperature) to find empty states on the other side of the barrier. The constriction only allows a discrete number of quantum channels to open up (corresponding to quantizing the transverse wave-function according to the boundary condition), quantizing conductance in units of $2e^2/h$. We focus here on the case where the contact is operating between the 0 and 1 channels. The scattering transmission probability is controlled by the shape of the scattering potential, and its height in particular. If the peak of the barrier exceeds $E_F + eV$, then only tunneling processes can transport electrons across the barrier, so we have very small current, $\mathcal{T} \ll 1$. On the other hand, as the energy of the electrons is increased (or the barrier height is decreased), the transmission gradually increases, up to a fully open channel.

Now that we have described both the DQD and the QPC, we can combine them, where the DQD is viewed as the system and the QPC is viewed as the detector (see Fig. 5.2). By fabricating these (semi)conductors nearby, an effective geometric capacitance exists between them. Therefore, a single electron being either L or R on the DQD causes the QPC potential to be V_L or V_R, causing a slight difference in the constriction of the contact. This results in the electrical conductance being either G_L or G_R, which in turn causes the electrical current to be I_L or I_R, a little like Schrödinger's cat. Using this deductive chain, by measuring electrical current with an ammeter, we can measure the position of the charge on the DQD. While very sensitive, the small difference in position of the electron, together with weak coupling, results in a *weakly responding detector*, so we typically have the situation where $I_L - I_R \ll (I_L + I_R)/2$. Importantly, the detector has intrinsic and extrinsic sources of noise that must be accounted for to obtain a complete picture. This goes hand in hand with the sensitivity of the device, a topic we now examine.

The potential landscape around the QPC can be modeled as a saddle-point: there the potential rises and falls as the electrons traverse the hill, while the walls of the pass rise up on both sides. This potential takes the form $V(x, y) =$

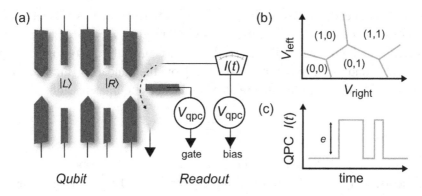

Figure 5.2 (a) A double quantum dot can be formed by arranging metallic electrodes on the surface of a semiconductor substrate doped with a 2D electron gas
slightly below the surface. Applying voltages on these gates can pinch off charge-
rich regions, forming dots of a variable number of electrons. High-frequency
signals are coupled with select electrodes to induce quantum state transitions. (b)
A typical stability diagram is shown where voltages applied via the left and right
gates tune the number of electrons in each dot, expressed as a pair (left, right). A
common measurement scheme involves charge sensing one or both dots though a
quantum point contact placed in proximity to the quantum dot. (c) A current traveling through the narrow conductive channel of the point contact is very sensitive
to single electron changes in the dot.

$V_0 - (1/2)m\omega_x^2 x^2 + (1/2)\omega_y^2 y^2$, where the electron passes in the x direction, and
the walls rise up in the y direction. For this potential, a scattering problem with
particles coming from the negative x direction with energy E can be solved exactly
(Büttiker, 1990) to give the transmission coefficient

$$\mathcal{T} = \frac{1}{1 + e^{-\epsilon}}, \qquad \epsilon = \frac{2\pi}{\hbar\omega_x}\left(E - V_0 - \frac{\hbar\omega_y}{2}\right). \qquad (5.5)$$

Here we see that if the electron's energy is below the top of the energy barrier V_0,
the transmission becomes very small, while if the electron's energy is well above
V_0, then the transmission limits to 1, indicating an open channel. The conductance
takes into account electrons traversing the barrier over a range of energies, but
if the applied voltage is small compared to the Fermi energy, $eV \ll E_F$, we can
approximate all transport as taking place at the Fermi level, so the conductance is
given by (5.4), where the electron's energy is taken at the Fermi energy, $E = E_F$.

The electron position on the DQD controls the potential energy V_0 of the QPC
potential. Thus, the place of maximum sensitivity for the detector corresponds to
the point of maximum slope,

$$\frac{\partial I}{\partial V_0} = \frac{2e^2 V}{h}\frac{\partial \mathcal{T}}{\partial \epsilon}\frac{\partial \epsilon}{\partial V_0} = -\frac{2e^2 V}{h}\frac{2\pi}{\hbar\omega_x}\frac{e^{-\epsilon}}{(1 + e^{-\epsilon})^2}. \qquad (5.6)$$

From this expression, we can see the slope is maximized in magnitude when $\epsilon = 0$, or $\mathcal{T} = 1/2$.

To fully understand the detector physics, we must also understand the noise. Detector noise can come from many places: there is thermal noise (also called Johnson–Nyquist noise) that is present in all dissipative devices, there is $1/f$ noise at low frequency, electronic flicker and telegraph noise, and so on. Here we focus on a fundamental source of noise intrinsic to the detection process – shot noise of the electrons. So long as $eV \gg k_B\Theta$, the thermal energy, we can neglect the thermal noise contribution, where Θ is the temperature. Electronic shot noise originates from the granularity of the electron charge. Each electron can transmit or reflect from the scattering region, giving rise to partition noise. Analogous to hailstones falling on a tin roof, the electronic current will experience fluctuations. Writing the electrical current as $I = GV + \delta I(t)$, the current fluctuates on a very fast timescale, $\tau_0 = h/eV$, that is much shorter than any other timescale in the problem. Consequently, we can treat the fluctuations as uncorrelated on the timescale of the detection process, also called the "white noise" approximation. The strength of the shot noise is characterized by

$$S_I = \int dt \langle \delta I(t + \tau)\delta I(\tau)\rangle, \tag{5.7}$$

where S_I is independent of τ by stationarity of the process. The quantity S_I may be interpreted as the zero frequency noise power. The fermionic nature of the particles gives rise to an exclusion process, so the noise power is given by (Blanter and Büttiker, 2000).

$$S_I = (eV)\frac{4e^2}{h}\mathcal{T}(1 - \mathcal{T}). \tag{5.8}$$

In the limit where $k_B\Theta \ll eV \ll E_F$, we consider for simplicity. We note that if either $\mathcal{T}$ is 0 or 1, there is no noise because all electrons are either reflected or transmitted; otherwise, the partitioning of the electron current is noisy. Note that $S_I = 2e\langle I \rangle (1 - \mathcal{T})$, where $\langle I \rangle$ is the average electrical current. The expression $e\langle I \rangle$ is the Schottky expression for the electronic noise of vacuum tubes, resulting from Poissonian statistics of well-spaced electrons. The multiplicative factor of $1 - \mathcal{T}$ for mesoscopic circuits indicates that the statistics are sub-Poissonian for this detector, as is generically the case for coherent electronic conductors.

Returning to the problem of quantum detection, we seek to distinguish currents I_L from I_R. Let us introduce the measured quantity,

$$\mathcal{Q} = \int_0^T dt' I(t'), \tag{5.9}$$

which may be interpreted as the accumulated charge from a measurement lasting a time T. Supposing no tunneling occurs for the DQD ($\Delta = 0$) in the simplest case. Then the signal accumulated is $Q_L - Q_R = T(I_L - I_R)$, while the variance of Q is given by

$$\langle \delta Q^2 \rangle = \int_0^T dt_1 dt_2 \langle \delta I(t_1) \delta I(t_2) \rangle. \tag{5.10}$$

Using the stationary nature of the process and changing variables to $t_s = (t_1 + t_2)/2, t_d = t_1 - t_2$, we have in the limit where T is long compared to τ_0 the result $\langle \delta Q^2 \rangle \approx T S_I$. Regardless of the statistics of $\delta I(t)$ (under certain assumptions about how quickly the tails of the distribution fall off), the central limit theorem dictates that the distribution of Q/T limits to a Gaussian with the mean of I_L or I_R (depending on the initial state of the DQD), and a variance as shown earlier.

We now appeal to the concept of the signal-to-noise ratio, introduced in the previous chapter in Section 4.4. The detector's signal-to-noise ratio, distinguishing the electron being on the left or right dot, is given by

$$\mathcal{R} = \frac{Q_L - Q_R}{\sqrt{\langle \delta Q^2 \rangle}}. \tag{5.11}$$

In principle, one can have different noise powers for the two different configurations. However, in the weakly responding detector limit, there is little difference between the two noise powers, so we can use the formula (5.8) for the noise power, where the transmission is taken to be the average, $\mathcal{T} = (\mathcal{T}_L + \mathcal{T}_R)/2$. Thus, we find a simplified expression for the SNR,

$$\mathcal{R} = \sqrt{\frac{T}{\tau_0}} \frac{\mathcal{T}_L - \mathcal{T}_R}{\sqrt{\mathcal{T}(1 - \mathcal{T})}}, \tag{5.12}$$

which grows as the square root of the time of the duration of the experiment, divided by the fast correlation time of the transport, and also depends on the dimensionless ratio of combinations of the transmission coefficient.

The preceding SNR gives a new timescale to the problem: a characteristic time needed to distinguish the signal from the background shot noise. It is defined by letting $\mathcal{R} = k$, where k is a constant of order 1. Shortly, we will use $k = 2$. Solving (5.12) for the associated time gives

$$\tau_m = \frac{k^2 S_I}{(I_L - I_R)^2}. \tag{5.13}$$

We call this timescale the *characteristic measurement time*, not to be confused with the duration of the measurement. We notice that, if $I_L \rightarrow I_R$, the characteristic measurement time diverges, so it takes longer and longer to distinguish the small signal from the noise. The characteristic measurement time is an important timescale in describing the quantum state dynamics of the DQD during the measurement process.

Recalling the form of the transmission of the QPC for a saddle-point contact potential (5.5), notice that the noise power S_I depends on the combination

$$\mathcal{T}(1 - \mathcal{T}) = \frac{e^{-\epsilon}}{(1 + e^{-\epsilon})^2}. \tag{5.14}$$

This function has its maximum at the point $\epsilon = 0$, corresponding to $\mathcal{T} = 1/2$, which is also the point of highest sensitivity! In order to find at what point the SNR is highest, we consider the linear response limit, so we can approximate

$$I_L - I_R = \frac{\partial I}{\partial \epsilon} \delta\epsilon, \tag{5.15}$$

where $\delta\epsilon$ accounts for the small shift in the potential energy V_0 for the two positions the DQD electron induces, $\delta\epsilon = -2\pi \delta V_0 / \hbar\omega_x$. Once the device structure is made, the geometric capacitances and potential landscape of the sample are fixed, but the transmission of the contact can be controlled with a top gate, shifting the overall value of V_0 with an applied gate voltage. Recalling the result (5.6), the measurement time is given by

$$\tau_m = \frac{4k^2 \tau_0}{\delta\epsilon^2} \cosh^2(\epsilon/2). \tag{5.16}$$

The assumption that $\delta\epsilon \ll 1$ guarantees that $\tau_m \gg \tau_0$. For $\delta\epsilon$ fixed by the geometry of the sample, this quantity is minimized at $\epsilon = 0$, corresponding to $\mathcal{T} = 1/2$. Consequently, the fastest measurement one can make with this detector is for a half-open channel, despite the fact the noise is largest there. This fact comes from the sensitivity of the detector rapidly degrading as one approaches either the open or closed contact limit.

5.2 Measuring the State of a Superconducting Quantum Circuit with Electromagnetic Radiation

Another physical system that emerged in the early 2000s that is both a flexible laboratory for quantum measurement science and a promising candidate for quantum computing architectures is the circuit quantum electrodynamics architecture (cQED), as realized by superconducting circuits (Blais et al., 2004). While bearing many similarities to QED systems constructed from ultra-cold atoms, the electrical circuit version has a life of its own, allowing access to many parameter regimes difficult to implement in atomic systems and thus enabling new types of experiments. The basic architecture consists of a superconducting circuit embedded in a

(a) (b)

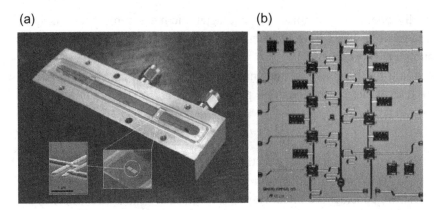

Figure 5.3 (a) A typical transmon qubit in a 3D geometry. A submicron Josephson tunnel junction is connected with two large paddles, which form the shunting capacitor and also an antenna to control the coupling to the electromagnetic field in the cavity. Shown here is half of a rectangular aluminum cavity, which is sandwiched together with the chip placed on a ledge in the center. (b) The transmon qubit can also be interfaced with a 2D cavity formed by a transmission line resonator on a planar chip. Shown here are eight transmon qubits on a 1-cm-square silicon chip connected via individual coupling capacitors to a single resonator bus.

microwave frequency cavity, either a 3D realization or a planar resonator colocated with the circuit on a semiconductor substrate.

As a canonical example, we consider a superconducting transmon qubit in a microwave resonator. In the previous chapter, we treated the case of a phase qubit, which, as its name implies, is well described by the dynamics of a fictitious particle in a tilted sinusoidal potential with a coordinate given by the Josephson junction phase. Qubits can also be constructed where the charge n on an isolated superconducting grain is a well-defined quantum number rather than the phase φ. The charge basis effectively reflects the eigenstates of a quantum circuit when the charging energy E_C, the energy associated with transferring a charge on/off the grain, is much larger than the Josephson energy E_J, which represents the energy associated with a Cooper pair tunneling across the junction. Conversely, states of definite phase are appropriate in the opposite limit. The ground state of a Cooper box (Vion, 2004) with $E_J \sim E_C$ was first probed in 1996, a superposition of ground and excited states was demonstrated in 1999, an improved version already fitted with single-shot readout and protection against dephasing was demonstrated in 2002, and the now ubiquitous transmon version with $E_J/E_C \sim 10 - 100$ appeared in 2006. For a description of the transmon, let us start with an isolated Josephson junction shunted by a capacitor C_S and charged biased via a gate capacitor C_g. The junction itself has a self-capacitance C_J and inductive energy E_J. The shunt capacitance allows control of the qubit frequency with the added benefit of shifting a

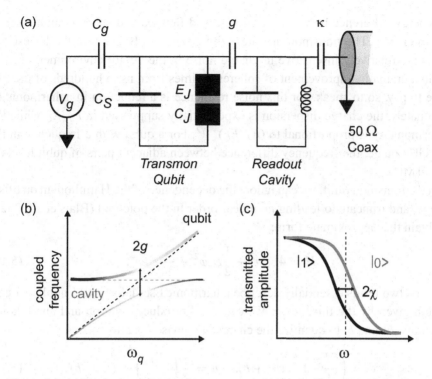

Figure 5.4 (a) Equivalent circuit of a transmon qubit formed by a Josephson junction shunted by a capacitor C_S and connected to a readout cavity. Qubit transitions are driven by an external voltage V_g, coupled via a capacitor C_g. The qubit-cavity coupling is parameterized by g, and the cavity decay rate to an external feedline is given by κ. (b) The frequencies of the coupled qubit-cavity system are plotted as a function of the qubit frequency for fixed cavity frequency. On resonance, an avoided crossing with splitting $2g$ is present. (c) In the dispersive limit, the qubit state pulls the bare cavity frequency, resulting in a quantum-state-dependent phase shift when the cavity is probed in transmission.

large fraction of the stored electromagnetic energy into a lower-loss capacitance, say one with a vacuum dielectric rather than an amorphous film. The Hamiltonian of the circuit can be written as

$$H = 4E_C(n - n_g)^2 - E_J \cos(\varphi), \qquad (5.17)$$

where n is the Cooper pair charge on the circuit, $n_g = C_g V_g / 2e$ is the charge bias expressed in Cooper pairs applied by an external voltage source V_g, and $E_C = e^2/2(C_S + C_J)$ is the charging energy which has contributions from both the junction capacitance and the external shunt.

This Hamiltonian can solved analytically in the phase basis using Mathieu functions (Koch et al., 2007). When $E_J/E_C < 1$, the circuit realizes a charge qubit and the energy difference between the ground and first excited states is highly sensitive to charge. The transmon operates with $E_J/E_C \sim 100$, where the lowest two levels are relatively insensitive to charge noise in the circuit environment, resulting in a dramatic improvement of coherence times from ns to hundreds of μs. The price to pay, so to speak, for this noise resilience is a reduction in anharmonicity. Fortunately, the charge dispersion is exponentially suppressed in E_J/E_C while the anharmonicity is proportional to $(E_J/E_C)^{-1/2}$. For a qubit with a frequency in the few GHz range, the frequency difference between adjacent pairs of qubit levels is $\sim 100\,\text{MHz}$.

In the transmon limit, we can ignore the dependence of the Hamiltonian on offset charge, and truncate to leading nonlinear order in the potential (Blais et al., 2021) to obtain this approximate form:

$$\hat{H} = 4E_C n^2 + \frac{1}{2}E_J\varphi^2 - \frac{1}{24}E_J\varphi^4. \tag{5.18}$$

The first two terms essentially describe a harmonic oscillator with a nonlinear correction given by the third term. We can thus introduce creation and annihilation operators, $\hat{b}^\dagger$ and $\hat{b}$, to quantize the charge and phase operators using

$$\hat{\varphi} = \left(\frac{2E_C}{E_J}\right)^{1/4}(\hat{b}^\dagger + \hat{b}), \quad \hat{n} = \frac{i}{2}\left(\frac{E_J}{2E_C}\right)^{1/4}(\hat{b}^\dagger - \hat{b}). \tag{5.19}$$

The creation and annihilation operators obey the commutation relation $[\hat{b}, \hat{b}^\dagger] = 1$, so the number and phase operators obey the canonical commutation relation $[\hat{\varphi}, \hat{n}] = i$. Substituting these expressions into (5.18) and discarding any terms that do not conserve excitation number and would be oscillating in a rotating frame of reference, we obtain

$$\hat{H} = \hbar\omega_q \hat{b}^\dagger\hat{b} - \frac{E_C}{2}\hat{b}^\dagger\hat{b}^\dagger\hat{b}\hat{b}. \tag{5.20}$$

Here the qubit frequency associated with the lowest two levels

$$\hbar\omega_q = \sqrt{8E_JE_C} - E_C$$

is effectively the frequency of plasma oscillations in the Josephson potential offset by a Lamb shift of E_C. Notably, transitions to higher states are shifted by the second term that represents a Kerr-type nonlinearity.

To read out the state of the qubit, we couple it to a cavity. For simplicity, we can approximate the system as being described by the Jaynes–Cummings model for a spin 1/2 particle, thus only considering the two lowest states of the transmon coupled to a single mode of a bosonic field. In this limit, we can write the state of the qubit by a Pauli operator, say $\hat{\sigma}_z$. The simplified Hamiltonian for the coupled system is given by

$$\hat{H} = \hbar\omega_c \hat{a}^\dagger \hat{a} + \hbar\omega_q \frac{\hat{\sigma}_z}{2} + \hbar g(\hat{a}^\dagger + \hat{a})(\hat{\sigma}_+ + \hat{\sigma}_-). \tag{5.21}$$

Here, the creation and annihilation operators $\hat{a}^\dagger$ and $\hat{a}$ refer to the excitations of the electromagnetic field, and $\hat{\sigma}_+$ and $\hat{\sigma}_-$ excite and relax the state of the two-level qubit described by $\hat{\sigma}_z$. The last term parameterizes the interaction between the qubit and the cavity and is of the form $\hat{\mathbf{p}} \cdot \hat{\mathbf{E}}$, which describes the typical interaction of a dipole with quantum operator $\hat{p} = \hat{\sigma}_+ + \hat{\sigma}_-$ with an electric field $\hat{E} = E_0(\hat{a}^\dagger + \hat{a})$ with zero-point-fluctuation amplitude E_0. Here we align the field in one direction and drop the vector nature of the field and dipole moment. The energy scales in the problem are set by the cavity frequency ω_c, the qubit frequency ω_q, and the light–matter coupling strength g. To further simplify (5.21), we can expand the product in the last term and drop products that do not conserve excitation number. These would correspond to transitions with energies much larger than the coupling strength. The Hamiltonian written in this interaction picture with respect to the first two terms of the Hamiltonian, or rotating frame, where we have dropped counterrotating terms is

$$\hat{H} = \hbar\omega_c \hat{a}^\dagger \hat{a} + \hbar\omega_q \frac{\hat{\sigma}_z}{2} + \hbar g(\hat{a}^\dagger \hat{\sigma}_- + \hat{a}\hat{\sigma}_+). \tag{5.22}$$

There are two basic decay processes in this system, parameterized by Γ, the spontaneous emission rate of the qubit, and κ, the cavity decay rate due to coupling to the external environment. We can express the coupling constant g in terms of circuit parameters,

$$g = \omega_c \frac{C_g}{C_J + C_S} \left(\frac{E_J}{2E_C}\right)^{1/4} \sqrt{\frac{\pi Z_c}{R_Q}}, \tag{5.23}$$

where Z_c is the impedance of the cavity mode and $R_Q = h/e^2$ is the resistance quantum.

As mentioned earlier, we can view the coupling energy $\hbar g$ as arising from a dipole interaction between the transmon and the vacuum electric field. To pen a dipole moment p_0, we can assume a characteristic length l associated with the motion of a Cooper pair in the qubit, yielding

$$p_0 = 2el \left(\frac{E_J}{32E_C}\right)^{1/4}, \quad E_0 = \frac{\omega_c}{l} \frac{C_g}{C_J + C_S} \sqrt{\frac{\hbar Z_c}{2}}. \tag{5.24}$$

Both of these quantities can be made large by appropriate circuit design. The distance a charge travels in a superconducting circuit is on a mesoscopic rather than an atomic length scale, giving rise to a large dipole moment which is efficiently coupled to volume occupied by the vacuum electric field, E_0. Thus, in electrical circuits, the coupling g can be made larger than both of these decay rates to achieve "strong coupling," where $g > \Gamma, \kappa$ (Wallraff et al., 2004). This regime is considerably more difficult to achieve in naturally occurring atomic systems which have a much smaller electric dipole moment, with the atom only occupying a small volume of the electric field. We note that while the two-level approximation captures many of the basic features of a cQED system, many important effects trace their origin to the additional quantized levels typically present in a superconducting circuit. Furthermore, cQED systems can also reach "ultrastrong" coupling where the hybridized states of the system involve multiple cavity levels. In addition to enabling more sophisticated tests of light–matter interaction, the presence of additional levels can also add functionality for both encoding and measuring quantum information.

We consider two basic limits of the Jaynes–Cummings model as referenced by the detuning $\Delta = \omega_q - \omega_c$. When $\Delta = 0$, the qubit and cavity are maximally entangled and the collective excitations are polariton-like. One cannot distinguish the individual character of either the cavity or the qubit. This regime has been used to study the radiative relaxation of an atom in squeezed vacuum. Gardiner calculated the variation of the spontaneous emission decay rate of an atom when coupled to an environment with intrinsic structure (Gardiner, 1986). For the case of quadrature squeezing, as demonstrated by Murch and coworkers, the transverse decay rate of a superconducting qubit varies depending on whether it is coupled to the quadrature with enhanced or suppressed fluctuations (Murch et al., 2013a). This experiment is very difficult to conduct using atomic systems interacting with free-space radiation as the vacuum in many spatial modes has to be squeezed and efficiency coupled. In the case of a superconducting qubit strongly interacting on resonance with a cavity, the system is essentially an artificial atom with a large electric dipole moment in a one-dimensional waveguide and can be readily coupled with a superconducting parametric amplifier generating a squeezed-state output. In this case, no signal is fed to the amplifier and it is only used to squeeze vacuum at its input. Superconducting parametric amplifiers are discussed later in Chapter 8 of this text.

More commonly, cQED systems operate in the dispersive limit with finite detuning Δ. When the detuning is large compared to the coupling strength g, the Jaynes–Cummings model can be approximated as

$$\hat{H} \approx \frac{1}{2}\hbar\left(\omega_q + \frac{g^2}{\Delta}\right)\hat{\sigma}_z + \hbar\omega_c(\hat{a}^\dagger\hat{a} + \frac{1}{2}) + \hbar\frac{g^2}{\Delta}\hat{\sigma}_z\hat{a}^\dagger\hat{a}, \tag{5.25}$$

as you will show in Exercise 5.6.4. The mutual interaction between the electromagnetic field and the qubit results in a modification of the frequencies of both systems. The terms in the Hamiltonian can be regrouped in different ways to make this apparent. From the point of view of the cavity, we can write

$$\hat{H} \approx \frac{1}{2}\hbar \left(\omega_q + \frac{2g^2}{\Delta}\hat{a}^\dagger\hat{a} + \frac{g^2}{\Delta} \right) \hat{\sigma}_z + \hbar\omega_c \left(\hat{a}^\dagger\hat{a} + \frac{1}{2} \right), \qquad (5.26)$$

where the qubit frequency has been shifted by g^2/Δ – this is a Lamb shift (Lamb and Retherford, 1947), caused by the interaction of the qubit with the vacuum fluctuations of the cavity. The presence of photons in the cavity further shifts the qubit frequency via the ac Stark effect (Autler and Townes, 1955). We can also view the interaction from the viewpoint of the qubit,

$$\hat{H} \approx \frac{1}{2}\hbar\omega_q\hat{\sigma}_z + \hbar \left(\omega_c + \frac{g^2}{\Delta}\hat{\sigma}_z \right) \left(\hat{a}^\dagger\hat{a} + \frac{1}{2} \right). \qquad (5.27)$$

Here the resonant frequency is pulled by a dispersive shift $\pm\chi$ whose magnitude per photon is given by g^2/Δ. The sign of the effect depends on the state of the qubit, as reflected by the value of $\hat{\sigma}_z$.

This frequency shift forms the basis for a convenient measurement of the qubit state when the cavity is irradiated by a microwave tone. The magnitude of this shift also depends on average number of photons $\bar{n} = \langle\hat{a}^\dagger\hat{a}\rangle$ in the cavity. When an incident microwave signal is scattered off the qubit-cavity system, the outgoing wave is shifted in phase in a qubit-state-dependent fashion. For a small displacement in cavity frequency, we can estimate the expected angular response when the cavity is probed in a transmission measurement. We can express the phase response θ of the cavity as a function of the probe frequency ω and the decay rate to an external waveguide κ using

$$\theta = \arctan\left(\frac{2}{\kappa}(\omega_0 - \omega) \right), \qquad (5.28)$$

where the bare cavity resonant frequency is ω_0. If the cavity is probed at ω_0, which is now the average of the dressed cavity frequencies with the qubit in its ground and excited states, we should ideally observe a total phase shift $4\chi/\kappa$.

Much like in the Stern–Gerlach experiment, repeating the measurement results in a histogram of values for microwave signal phase. The width of these histograms are a result of the measurement apparatus, which has fluctuations. In an idealized measurement, there would be no technical noise due to imperfections of the detector and only unavoidable quantum fluctuations would be present. If the histograms corresponding to the two qubit states are separated by greater than their width, a projective measurement is realized. Conversely, overlapping histograms realize a partial or weak measurement of the system.

A number of experimental parameters can be varied to smoothly go from one regime to the other. For a given coupling g, a larger frequency detuning Δ reduces the qubit-induced dispersive shift. Reducing the number of photons in the cavity also reduces the displacement of the histograms. Finally, the measurement time can be varied to narrow the histogram width, provided the measurement is not dominated by technical noise. The power signal-to-noise ratio $\mathcal{R}^2$ is a good proxy for the measurement strength and is obtained by taking the square of the ratio of the histogram separation to histogram standard deviation, as measured in amplitude units of the detected microwave field, and can be compactly expressed as

$$\mathcal{R}^2 = 64\chi^2 \left(\frac{\bar{n}T}{\kappa} \right). \tag{5.29}$$

Here, T is the measurement duration, and this expression assumes perfect measurement efficiency. Indeed, the use of near-quantum-noise-limited superconducting amplifiers has proven critical for implementing weak measurements where information is not lost into unmonitored degrees of freedom. In contrast, for situations with extra added noise or lost information, the measurement histograms overlap, but this is not a faithful weak measurement of the system, but rather the hallmark of an inefficient setup. To paraphrase Schrödinger, there is a difference between a shaky or out-of-focus photograph and a snapshot of clouds and fog banks (Trimmer, 1980).

5.3 Stochastic Schrödinger and Master Equations

Now that the physical systems have been outlined and discussed in detail, we will discuss the quantum state dynamics arising from these detection processes. Despite their clear differences, we will be able to describe both the QPC/DQD system and the cQED system with microwave readout in a unified way. We can consider both detection processes of the quantum systems as a kind of scattering process, where the scattering matrix elements depend on the qubit state. In the QPC/DD process, we will detect a change in the transmission coefficient, while in the cQED case, we will detect a phase shift of the optical readout.

Quantum State Change for the QPC/DQD Case in the Single Electron Limit

In the QPC/DQD case, the elementary process is an individual electron incident from the source electrode in a state with energy near the Fermi energy, described by state $|\text{in}\rangle$, with the qubit in state $|\psi\rangle = \alpha|L\rangle + \beta|R\rangle$. The scattering process maps this input state into an output scattering state,

$$|\text{in}\rangle(\alpha|L\rangle + \beta|R\rangle) \rightarrow \alpha(r_L|S\rangle + t_L|D\rangle)|L\rangle + \beta(r_R|S\rangle + t_R|D\rangle)|R\rangle, \tag{5.30}$$

where states $|S\rangle, |D\rangle$ are asymptotic scattering states corresponding to electrons ending in the source (S, reflected) or drain (D, transmitted) of the conductor. Here also, t_j, r_j where $j = L, R$ are the transmission or reflection amplitudes that depend on the qubit state and can be seen as elements of a scattering matrix $\bar{\bar{\mathbf{S}}}_j$,

$$\begin{pmatrix} b_S \\ b_D \end{pmatrix} = \bar{\bar{\mathbf{S}}}_j \begin{pmatrix} a_S \\ a_D \end{pmatrix}, \qquad \bar{\bar{\mathbf{S}}}_j = \begin{pmatrix} r_j & \bar{t}_j \\ t_j & \bar{r}_j \end{pmatrix}. \tag{5.31}$$

The scattering matrix maps the incoming scattering state $\mathbf{a}$ to the outgoing scattering states $\mathbf{b}$. While the elements t_j, r_j correspond to transmission and reflection amplitudes for an electron incident from the source, the elements $\bar{t}_j, \bar{r}_j$ correspond to transmission and reflection amplitudes for an electron incident from the drain. The scattering matrix is unitary, which relates some of the matrix elements to each other.

The resulting state from the scattering process (5.30) results in entanglement between the position of the scattering electron and the left/right orientation of the electron in the DQD. The electron's transport state is detected as contributing to the electrical current in the conductor, which can be viewed as making a projection of the electron onto the source (S) or drain (D). It follows from our discussion in Chapter 3 that the measurement operators, expressed in the $|L\rangle, |R\rangle$ basis, are given by the matrices

$$\bar{\bar{\Omega}}_S = \begin{pmatrix} r_L & 0 \\ 0 & r_R \end{pmatrix}, \qquad \bar{\bar{\Omega}}_D = \begin{pmatrix} t_L & 0 \\ 0 & t_R \end{pmatrix}. \tag{5.32}$$

Consequently, we can write the E_j matrices as follows:

$$\bar{\bar{\mathbf{E}}}_S = \bar{\bar{\Omega}}_S^\dagger \bar{\bar{\Omega}}_S = \begin{pmatrix} R_L & 0 \\ 0 & R_R \end{pmatrix}, \qquad \bar{\bar{\mathbf{E}}}_D = \bar{\bar{\Omega}}_D^\dagger \bar{\bar{\Omega}}_D = \begin{pmatrix} T_L & 0 \\ 0 & T_R \end{pmatrix}, \tag{5.33}$$

where $T_j = |t_j|^2, R_j = |r_j|^2$. We notice that the coarse-grained position of the electron serves as an effective qubit meter in the preceding analysis, similar to Exercise 3.7.2. Next, the new quantum state, given the electron is found in the drain, is

$$|\psi'_D\rangle = \frac{\hat{\Omega}_D|\psi\rangle}{||\hat{\Omega}_D|\psi\rangle||} = \frac{t_L\alpha|L\rangle + t_R\beta|R\rangle}{\sqrt{T_L|\alpha|^2 + T_R|\beta|^2}}, \tag{5.34}$$

while the new quantum state, given the electron is found in the source, is

$$|\psi'_S\rangle = \frac{\hat{\Omega}_S|\psi\rangle}{||\hat{\Omega}_S|\psi\rangle||} = \frac{r_L\alpha|L\rangle + r_R\beta|R\rangle}{\sqrt{R_L|\alpha|^2 + R_R|\beta|^2}}. \tag{5.35}$$

To complete the analysis, the probability P of finding the electron in the drain D (or source S) is given by

$$P(D) = T_L|\alpha|^2 + T_R|\beta|^2, \qquad P(S) = R_L|\alpha|^2 + R_R|\beta|^2. \tag{5.36}$$

As before, we can view this as an expression of the law of total probability. It turns out from symmetry considerations that, if the QPC is spatially symmetric, the scattering matrix elements are real, which gives the optimal efficiency, as we will see later on.

Quantum State Change for the cQED Dispersive Limit Case

A similar analysis can be done for the cQED setup in the dispersive limit. We consider a coherent state of light at microwave frequencies that is sent from a microwave source to the cavity via a transmission line. If the frequency of the light is near the resonant frequency of the cavity, it will enter it for a time, interact with the qubit, and then leave the cavity. The altered light can then be amplified and examined. In the dispersive limit, no energy is exchanged between qubit and cavity, but the interaction can give a conditional phase shift to the coherent state. We begin the analysis with the qubit in state $|\psi\rangle = a|0\rangle + b|1\rangle$, and the cavity coherent state (taken as the meter) in state $|\alpha\rangle$, where $|\alpha|^2$ characterizes the average number of photons in the state. The resulting interaction takes the form

$$(a|0\rangle + b|1\rangle)|\alpha\rangle \rightarrow a|0\rangle|\alpha_0\rangle + b|1\rangle|\alpha_1\rangle, \tag{5.37}$$

where the modified coherent states are given by

$$|\alpha_j\rangle = |\alpha e^{i\theta_j}\rangle. \tag{5.38}$$

Here, θ_j are the phase shifts acquired by the interaction, which are related to the product of the dispersive shift χ and the interaction time of the system and meter. The states of the cavity and the qubit are now entangled, and we assume the cavity is sufficiently lossy that the light field quickly leaks out without further interaction with the qubit. This will be satisfied if $\kappa \gg \chi$. The light state is now measured by an amplification process, which will be described in detail in Chapters 7 and 8. For now, we will describe the result of the process as a projection of the coherent state onto one of its quadrature variables. Defining a position-like operator (in the context of the quantum harmonic oscillator) $\hat{X} = (\hat{a}+\hat{a}^\dagger)/\sqrt{2}$ for the light field, we will project on $\hat{\Pi}_x = |x\rangle\langle x|$, where $|x\rangle$ are the eigenstates of operator $\hat{X}$ with eigenvalues x, a continuous variable. The amplifier has the freedom to amplify along any quadrature that is a linear combination of $\hat{a}$ and $\hat{a}^\dagger$, but that is also equivalent to changing the phase of the input coherent state, so we choose to keep the amplifier quadrature fixed and vary the input phase. It is important to note that the phase is relative to an external clock which is an important part of the amplification process. Taking the phase shift from the qubit to be $\theta_j = \pm\theta_0$ (typically a very small angle), the choice of phase for α can result in varying information revealed about the qubit state, ranging from 0 (when α is real) to a maximal value (when α is imaginary).

We note the total phase shift is $2\theta_0$. This is because the projection onto the real axis results in partially distinguishable distributions, depending on the phase of α. Thus, we can view the meter as a continuous degree of freedom, like the position of the Stern–Gerlach device, or the optical beam position for the birefringence case. From the methods of Chapter 3, the measurement operator is therefore given by

$$\hat{\Omega}_x = \langle x|\alpha_0\rangle|0\rangle\langle 0| + \langle x|\alpha_1\rangle|1\rangle\langle 1|. \tag{5.39}$$

The coefficients $\langle x|\alpha_j\rangle$ are simply the wavefunctions of the two coherent states. These are given by

$$\langle x|\alpha_j\rangle = \frac{1}{\pi^{1/4}} \exp\left[-\frac{1}{2}\left(x - \sqrt{2}\mathrm{Re}(\alpha_j)\right)^2 + i\,\mathrm{Im}(\alpha_j)\left(\sqrt{2}x - \mathrm{Re}(\alpha_j)\right)\right]. \tag{5.40}$$

From this wavefunction, it can be verified that $\langle x\rangle_j = \sqrt{2}\mathrm{Re}(\alpha_j)$, and $\mathrm{Var}[x] = 1/2$, corresponding to the quantum vacuum fluctuations of the coherent state of light. This corresponds to the $\overline{\overline{E}}_x$ operator given in matrix form by

$$\overline{\overline{E}}_x = \begin{pmatrix} |\langle x|\alpha_0\rangle|^2 & 0 \\ 0 & |\langle x|\alpha_1\rangle|^2 \end{pmatrix}. \tag{5.41}$$

The measurement operator allows us to find the probability distribution P for x, when the qubit is in a coherent superposition,

$$P(x) = |a\langle x|\alpha_0\rangle|^2 + |b\langle x|\alpha_1\rangle|^2, \tag{5.42}$$

so we have $\langle x\rangle = \sqrt{2}[|a|^2\mathrm{Re}(\alpha_0) + |b|^2\mathrm{Re}(\alpha_1)]$, and $\mathrm{Var}[x] = 1/2$. Let us choose $\alpha = i|\alpha|$ to get the best signal, corresponding to

$$\mathrm{Re}(\alpha_j) = \pm|\alpha|\sin\theta_0 \approx \pm\sqrt{\bar{n}}\theta_0, \tag{5.43}$$

for small values of θ_0, where $\bar{n}$ is the average number of photons in the cavity. Appealing to the concept of signal-to-noise ratio introduced in Section 4.4, we can quantify our ability to discriminate the two qubit states by taking the difference of the average detected signals, divided by their standard deviation,

$$\mathcal{R} = \frac{|\langle x\rangle_1 - \langle x\rangle_0|}{\sqrt{\mathrm{Var}[x]}} = 4\sqrt{\bar{n}}\theta_0. \tag{5.44}$$

This number is typically much less than 1. Given result x has been recorded for the result of operator $\hat{X}$ on the meter, the new system state is given by

$$|\psi'\rangle_x = \frac{\hat{\Omega}_x|\psi\rangle}{||\hat{\Omega}_x|\psi\rangle||} = \frac{\langle x|\alpha_0\rangle|0\rangle + \langle x|\alpha_1\rangle|1\rangle}{\sqrt{|\langle x|\alpha_0\rangle|^2 + |\langle x|\alpha_1\rangle|^2}}. \tag{5.45}$$

5.4 Continuous Measurement

In both examples of the QPC/QDQ and the cQED systems, the previous two sections are not a realistic description of the actual measurements made, but are useful idealizations. In practice, the timescale of the electron passing the barrier $\tau_0 \sim h/eV$, or the duration of light in the cavity $\sim \kappa^{-1}$, is very short. It is, in fact, the cumulative effect of many electrons interacting with the system and forming a quasi-continuous noisy current $I(t)$, or the resulting voltage signal from the optical amplifier, forming a quasi-continuous noisy quadrature signal $x(t)$. We can pass from a discrete to quasi-continuous limit by time-indexing the result s_i, where we give a unified treatment by letting s be either I or x. In both cases, let us consider the time average over a duration T,

$$\bar{s}(T) = \frac{1}{N} \sum_{i=1}^{N} s_i = \frac{1}{T} \int_0^T dt\, s(t), \tag{5.46}$$

where we go over to a continuous description, $s(t_i) = s(t)$. From the central limit theorem, the distribution of the random variable $\bar{I}$ converges to a Gaussian (so long as the tails of the distribution of s_i do not decay too slowly), so we can specify the statistics with the mean and variance of $\bar{s}$, which depend only on the first and second moments of $s(t)$, depending on the qubit state

$$\langle s \rangle = \begin{cases} I_{L,R}, & \text{QPC/DQD}, \\ \sqrt{2}\mathrm{Re}(\alpha_{0,1}), & \text{cQED}. \end{cases} \tag{5.47}$$

Here the angle brackets are taken over a statistical ensemble. We take the time step discretization to be in units of the correlation time τ_0 for both systems, so the variables are uncorrelated in different time windows. The variance of $\bar{s}$ is given by

$$\mathrm{Var}[\bar{s}] = \sum_{i,j=1}^{N} \langle \delta s_i \delta s_j \rangle = \mathrm{Var}[s]/N = \frac{1}{(T)^2} \int_0^T \langle \delta s(t) \delta s(t') \rangle dt dt'. \tag{5.48}$$

The duration is $T = N\tau_0$, so we conclude that

$$\langle \delta s(t + \tau)\delta s(t) \rangle = S_s \delta(\tau), \tag{5.49}$$

in the white noise limit (assuming we consider times longer than the correlation time), where

$$S_s = \begin{cases} S_I, & \text{QPC/DQD}, \\ \tau_0/2, & \text{cQED}. \end{cases} \tag{5.50}$$

In the cQED case, we associate the correlation time with the timescale of the cavity ringing up or down, given by $\tau_0 = \kappa^{-1}$, while the phase shift due to the qubit is given by $\theta_0 = 2\chi/\kappa$ in the dispersive approximation. We then have the SNR for $T\bar{s}$ given by

$$
\mathcal{R} = \begin{cases} \frac{(I_L - I_R)}{\sqrt{S_L}}\sqrt{T}, & \text{QPC/DQD}, \\ \frac{8\chi\sqrt{\bar{n}}}{\kappa}\sqrt{\kappa T}, & \text{cQED}. \end{cases} \tag{5.51}
$$

A complementary, more detailed derivation of this result is given in Chapter 7. It is convenient to now shift and scale the $\bar{s}$ variable to make it symmetric and dimensionless. We define the continuous measurement result

$$
r = \frac{2\bar{s} - (s_0 + s_1)}{s_0 - s_1}, \tag{5.52}
$$

so when $s = s_0$, the result corresponds to $r = 1$, while when $s = s_1$, the result corresponds to $r = -1$. Given this convention, we define the characteristic measurement time τ_m such that the probability distributions $P_j(r)$ corresponding to $j = 0, 1$ are given by

$$
P_j(r) = \sqrt{\frac{T}{2\pi\tau_m}} \exp\left[-\frac{T}{2\tau_m}(r \mp 1)^2\right]. \tag{5.53}
$$

Here $j = 0, 1$ corresponds to $-, +$, respectively. The mean of r is ± 1, while its variance is τ_m/T. With this convention (corresponding to the choice $k = 2$ in (5.13)), the characteristic measurement times are given by

$$
\tau_m = \begin{cases} \frac{4S_I}{(I_L - I_r)^2}, & \text{QPC/DQD}, \\ \frac{\kappa}{16\chi^2\bar{n}}, & \text{cQED}. \end{cases} \tag{5.54}
$$

Quantum Trajectories: a First Look

Drawing on treatment of generalized quantum measurement, we can predict how the state will be disturbed with the measurement operator $\hat{\Omega}_r$, corresponding to finding a result r lying in a region of bin size Δr, so the probability is given by

$$
P(r)\Delta r = \langle\psi|\hat{\Omega}_r^\dagger\hat{\Omega}_r|\psi\rangle = |\alpha|^2 P_0(r)\Delta r + |\beta|^2 P_1(r)\Delta r. \tag{5.55}
$$

The new state is given by $|\psi'\rangle = \hat{\Omega}_r|\psi\rangle/||\hat{\Omega}_r|\psi\rangle||$. Notice that any factor appearing in $\hat{\Omega}_r$ that is common to both states may be dropped, because it will cancel out in the denominator during the renormalization. Consequently, we can simplify

$$
\hat{\Omega}_r = \begin{pmatrix} \sqrt{P_0(r)\Delta r} & 0 \\ 0 & \sqrt{P_1(r)\Delta r} \end{pmatrix} \propto \hat{\Omega}'_r = \begin{pmatrix} e^{\frac{Tr}{2\tau_m}} & 0 \\ 0 & e^{-\frac{Tr}{2\tau_m}} \end{pmatrix}. \tag{5.56}
$$

The form of $\hat{\Omega}'$ is illuminating: we see that if a result $r > 0$ is obtained, partial collapse toward state $|0\rangle$ occurs, while a result $r < 0$ gives partial collapse toward state $|1\rangle$.

This behavior can be described with a differential equation. Let us define for simplicity the unnormalized state coefficients $\tilde{\alpha}, \tilde{\beta}$ (the true state can always be found by renormalizing). It then follows by taking a time derivative and letting $T \to 0$ that

$$\frac{d}{dt}\begin{pmatrix} \tilde{\alpha} \\ \tilde{\beta} \end{pmatrix} = \frac{r}{2\tau_m}\hat{\sigma}_z \begin{pmatrix} \tilde{\alpha} \\ \tilde{\beta} \end{pmatrix}. \tag{5.57}$$

The solution is easily found by integration,

$$\tilde{\alpha}(t) = \tilde{\alpha}(0)\exp\left(\frac{1}{2\tau_m}\int_0^t dt' r(t')\right), \quad \tilde{\beta}(t) = \tilde{\beta}(0)\exp\left(-\frac{1}{2\tau_m}\int_0^t dt' r(t')\right). \tag{5.58}$$

Consequently, only the time-integrated signal of $r(t)$ matters as time develops, that is, $\int^t dt' r(t')$. The integrated signal is also Gaussian with mean $t\langle\hat{\sigma}_z\rangle(0)$ and variance of $t\tau_m$. We will see later this property no longer holds once we add in a system Hamiltonian. Notice that we can make the prediction of the quantum state given that we are provided $r(t)$, for example by an experiment. However, we have seen that the probability distribution of r depends on the state, while the equation of motion for the state depends on the measurement result. We will learn how to deal with this full solution of the problem in the next section.

Describing Quantum Trajectories: Stochastic Differential Equations

The full solution of quantum trajectories described as a continuous stochastic process requires learning about the mathematical formalism behind stochastic processes, such as stochastic differential equations, Fokker–Planck equations, stochastic path integrals, and so on. Before launching into this, let us take a step back to the time-discretized case. We divide the timeline up into pieces of size Δt. The index $j = 1, \ldots, N$ stands for the time index, so the duration $T = N\Delta t$. The quantum trajectory is defined as an iterative map,

$$|\psi_{j+1}\rangle = \frac{\hat{\Omega}_{r_j}|\psi_j\rangle}{||\hat{\Omega}_{r_j}|\psi_j\rangle||}, \tag{5.59}$$

where the measurement result r_j is drawn from the probability (density)

$$P(r_j) = \langle\psi_j|\hat{\Omega}^\dagger_{r_j}\hat{\Omega}_{r_j}|\psi_j\rangle. \tag{5.60}$$

The set of results $\{r_j\}$, where $j = 1, \ldots, N$ is the measurement record, and the set of states $\{|\psi_j\rangle\}$ is the quantum trajectory. As $\Delta t \to 0$ this becomes a continuous process with a well-defined mathematical limit; but, of course, in experiments there is always a smallest timescale of the correlation time of the detector than which the considered time steps should always be longer. This is typical of all stochastic processes in nature. We can also add in unitary dynamics with a system Hamiltonian H_S via the unitary operator

$$\hat{\Omega}_{r_j} \rightarrow \hat{U}_j \hat{\Omega}_{r_j}, \qquad \hat{U}_j = e^{-i\Delta t H_S(t_j)/\hbar}, \tag{5.61}$$

where we assume $\Delta t H_s/\hbar$ is much smaller than the identity. The ordering of the operators does not matter at this stage, since the commutator of the two operators will be order Δt^2 and can be neglected.

Recalling Eqs. (5.53) and (5.49), we can introduce $r(t)$ as a time-dependent stochastic variable with mean $|\alpha|^2 - |\beta|^2$ that is uncorrelated with itself unless the time index is the same, with variance τ_m/T, where we consider a duration $T = \Delta t$ for the moment. Letting $\Delta t \rightarrow 0$ formally, we can write equivalently

$$r(t) = \langle \hat{\sigma}_z \rangle(t) + \sqrt{\tau_m} \xi(t). \tag{5.62}$$

Here, the variable $\xi(t)$ is a Gaussian random variable of zero mean and is delta-correlated in time,

$$\langle \xi(t) \rangle = 0, \qquad \langle \xi(t)\xi(t') \rangle = \delta(t - t'). \tag{5.63}$$

These definitions reproduce the statistics of $r(t)$. The singular nature of the noise as $\Delta t \rightarrow 0$ can lead to mathematical difficulties, so it is also convenient to introduce the Wiener increment ΔW, also a Gaussian random variable defined so that the variance is given by $\text{Var}[\Delta W] = \text{Var}[\Delta t\, r]/\tau_m = \Delta t$, which vanishes as the time interval limits to 0. We can then define

$$\xi(t) = \lim_{\Delta t \rightarrow 0} \frac{\Delta W}{\Delta t} = \frac{dW}{dt}. \tag{5.64}$$

We will return to the Wiener differential dW shortly.

We will now derive equations of motion for the normalized qubit state based on our first result (5.57). Let us generalize the treatment by going to a density matrix formulation of the problem and finding the stochastic equation of motion for the density matrix, with elements

$$\rho_{ij} = \frac{\langle i|\psi \rangle \langle \psi|j \rangle}{\langle \psi|\psi \rangle} = \frac{\psi_i \psi_j^*}{\sum_k |\psi_k|^2}, \tag{5.65}$$

expressed in the $|0\rangle, |1\rangle$ basis, where the normalization allows us to consider also the unnormalized state $|\psi\rangle$, and $\psi_i = \langle i|\psi \rangle$. We can now apply a time-derivative to ρ_{ij} and use the rules of calculus to find

$$\dot{\rho}_{ij} = \frac{\dot{\psi}_i \psi_j^* + \psi_i \dot{\psi}_j^*}{\sum_k |\psi_k|^2} - \frac{2\psi_i \psi_j^* (\sum_l (\dot{\psi}_l \psi_l^* + \psi_l \dot{\psi}_l^*))}{(\sum_k |\psi_k|^2)^2}. \tag{5.66}$$

We assume at $t = 0$, the density matrix begins as the pure state

$$\hat{\rho} = \begin{pmatrix} |\alpha|^2 & \alpha\beta^* \\ \alpha^*\beta & |\beta|^2 \end{pmatrix}. \tag{5.67}$$

Inserting our previous results (5.57), we find the following differential equation for each matrix element,

$$\dot{\rho}_{00} = \frac{r}{\tau_m}\rho_{00} - \frac{r}{\tau_m}\rho_{00}(\rho_{00} - \rho_{11}), \tag{5.68}$$

$$\dot{\rho}_{11} = -\frac{r}{\tau_m}\rho_{11} - \frac{r}{\tau_m}\rho_{11}(\rho_{00} - \rho_{11}), \tag{5.69}$$

$$\dot{\rho}_{01} = (\dot{\rho}_{10})^* = -\frac{r}{\tau_m}\rho_{01}(\rho_{00} - \rho_{11}). \tag{5.70}$$

It is interesting to check that, since this is an equation for a normalized density matrix, the time derivative of the trace, given by $\dot{\rho}_{00} + \dot{\rho}_{11}$, is indeed exactly 0. It is often useful to reparameterize ρ for a qubit with the Bloch coordinates (x, y, z) defined by

$$\hat{\rho} = \frac{1}{2}\begin{pmatrix} 1+z & x-iy \\ x+iy & 1-z \end{pmatrix}, \tag{5.71}$$

or inverting the relation, $z = \rho_{00} - \rho_{11}$, $x = 2\text{Re}(\rho_{01})$, $y = -2\text{Im}(\rho_{01})$. All possible states (with trace 1, and eigenvalues between 0 and 1) must lay within a ball defined by $x^2 + y^2 + z^2 \leq 1$, with pure states residing on the surface, and mixed states inside the ball. We may reexpress the equations of motion for the qubit in these coordinates, as well as add in the effect of the system Hamiltonian $H_S = (\epsilon/2)\hat{\sigma}_z + (\Delta/2)\hat{\sigma}_x$, which may be accounted for as additional terms, corresponding to the Schrödinger equation, $d\hat{\rho}/dt = -i[H_S, \hat{\rho}]$, to give the unified equations

$$\dot{x} = -\frac{r}{\tau_m}xz - \epsilon y, \tag{5.72}$$

$$\dot{y} = -\frac{r}{\tau_m}yz + \epsilon x - \Delta z, \tag{5.73}$$

$$\dot{z} = \frac{r}{\tau_m}(1 - z^2) + \Delta y. \tag{5.74}$$

There equations are supplemented with the stochastic readout, given as a random variable, that depends on the qubit state as

$$r = \text{Tr}[\hat{\rho}\hat{\sigma}_z] + \sqrt{\tau_m}\xi = z + \sqrt{\tau_m}\xi. \tag{5.75}$$

Thus, the readout traces the instantaneous value of qubit signal $z(t)$, but is masked by detector noise. Although the readout does not depend on x, y, knowledge of the initial state together with the system parameters allows us to predict the values of the entire state, given the noisy measurement readout $r(t)$. Inserting (5.75) into the preceding equations of motion then gives a complete description of the stochastic process of the continuously measured qubit as a set of coupled stochastic differential equations.

Solving the Stochastic Differential Equations

We must now enter a point of mathematical subtlety that has confused many a physicist. Namely, how can we construct solutions of equations of the form (5.72, 5.73, 5.74)? We must stress at the outset that the underlying physical prediction of the quantum state at a given time is perfectly well defined, as is clear from the discussion in Section 5.4. Nevertheless, when stochastic differential equations are written, they cannot be considered meaningful unless an "interpretation" is also given. The most common ones in the literature are the Itô and the Stratonovich interpretations. The interpretation refers to a problem that is encountered when trying to integrate these equations. We are faced with how to deal with expressions of the form $\int dt \xi(t) f(q(t))$, where q stands in for any system variable, and f is some function of the variable. Let us consider a generic stochastic differential equation of the form

$$\dot{q} = a(q) + b(q)\xi. \tag{5.76}$$

When $b(q)$ is nonconstant, this is sometimes referred to as multiplicative noise and causes the ambiguity mentioned before. Let us write a discretized version of this equation,

$$q_{k+1} - q_k = a(\bar{q}_k)\Delta t + b(\bar{q}_k)\Delta W_k, \tag{5.77}$$

where again, ΔW_k is the Wiener increment. Thus, the solution to $q(T)$ is given by the sum of increments,

$$q(T) = (q_N - q_{N-1}) + (q_{N-1} - q_{N-2}) + (q_{N-2} - q_{N-3}) + \ldots + q_0$$
$$= q_0 + \sum_{k=0}^{N-1} a(\bar{q}_k)\Delta t + b(\bar{q}_k)\Delta W_k. \tag{5.78}$$

The mathematical ambiguity comes into how to assign $\bar{q}_k$. We could assign it at the beginning of the interval (Itô convention, $\beta = 0$), or at the midpoint (Stratonovich convention, $\beta = 1/2$), or anywhere else we like,

$$\bar{q}_k = \beta q_{k+1} + (1 - \beta)q_k, \tag{5.79}$$

where $\beta \in [0, 1]$. Depending on how we choose this convention as $\Delta t \to 0$, the above increment sum will converge to different answers. This is due to the stochastic nature of ΔW_k, which is continuous, but not differentiable. Despite this seeming ambiguity, we must remember van Kampen's dictum, "From a physical point of view the Itô–Stratonovich controversy is moot" (Van Kampen, 1981). The point is that any stochastic differential equation that has nonlinear terms in it must also be supplemented by a convention about the choice of β. Without it, the stochastic differential equation is just a meaningless string of symbols.

It turns out that Eqs. (5.72, 5.73, 5.74) must be interpreted as Stratonovich-form stochastic differential equations, because we used the ordinary rules of calculus in the time derivative of ρ_{ij}. We can also shift the time increments by half a step and define the time derivative in a symmetrized form, $\dot{q} = [q(t+\delta t/2) - q(t-\delta t/2)]/\delta t$ in the Stratonovich interpretation. In order to use other choices of β, new forms of stochastic calculus must be applied. The most common one is Itô, which is typically favored by mathematicians, the subject of the next section, while physicists typically favor the Stratonovich interpretation because of its deeper, more physical origin. We will go into more detail about this point later in the chapter. Indeed, coming back again to van Kampen, "The final conclusion is that a physicist cannot go wrong by regarding the Itô interpretation as one of those vagaries of the mathematical mind that are of no concern to him." (Van Kampen, 1981). Nevertheless, because this interpretation is common in the quantum trajectory literature and in statistical physics more generally, we will discuss it now.

Introduction to Itô Stochastic Calculus

We begin by considering a continuous random walk, described by the variable $W(t)$, which is a random variable, with distribution

$$P(W) = \sqrt{\frac{2\pi}{t}} \exp\left(-\frac{W^2}{2t}\right).$$ (5.80)

This process can be developed by starting with $W = 0$ and adding the Wiener increment $\Delta W(t) = W(t_k + \Delta t) - W(t_k)$ repeatedly. The variance of Δt will simply add with each step, so the $t = N\Delta t$ variance of $W(t)$ is recovered. We can keep t fixed and take $N \to \infty$ while $\Delta t \to 0$. Each additional increment is statistically independent of all past values.

In the limit, Δt shrinks to the infinitesimal dt, and we define the limit of the Wiener increment to be $\Delta W \to dW$. Consequently, we can represent Eq. (5.77) as

$$dq = a(q)dt + b(q)dW$$ (5.81)

in this limit. Importantly, in the Itô interpretation, $\bar{q}$ is taken at q_k, while $\Delta W_k = W_{k+1} - W_k$, so we can say $b(q)$ is "nonanticipating." This convention then permits the nice simplification

$$\langle b(q)dW \rangle = 0,$$ (5.82)

because the average over the noise involves integrating over W_{k+1}, which has zero mean, while $b(q_k)$ does not involve the next time step.

The variance of ΔW is Δt, indicating that we should regard ΔW to be of order $\sqrt{\Delta t}$ in expansions with respect to time. This is very important because the usual chain rule does not apply in Itô calculus. The correct procedure is to expand the relevant equations with respect to dt and dW, and to keep to first order in dt, but to second order with respect to dW. As $\Delta t \to 0$, we can drop terms $dt^2, dt dW$, and higher order. To see why this is, we return to the discrete time case, so time is labeled by t_k, and the duration Δt times the number of steps N is the total duration t. We define the stochastic integral

$$\int_0^t dW = \lim_{N \to \infty} \sum_{k=1}^N \Delta W_k = W(t), \tag{5.83}$$

where $W_k = W(t_k)$. Let us now consider the stochastic integral

$$X = \int_0^t dW^2 = \lim_{N \to \infty} \sum_{k=1}^N (\Delta W_k)^2. \tag{5.84}$$

The claim we made earlier is equivalent to $X = t$ exactly. Let us check this by first taking the statistical average of X,

$$\langle X \rangle = \lim_{N \to \infty} \sum_{k=1}^N \langle (\Delta W_k)^2 \rangle = \lim_{N \to \infty} N \Delta t = t. \tag{5.85}$$

If $X = t$ exactly, then the variance of X should be zero. Let us check that this is the case:

$$\text{Var}[X] = \lim_{N \to \infty} \langle \sum_{k,l=1}^N (\Delta W_k)^2 (\Delta W_l)^2 \rangle - t^2. \tag{5.86}$$

We can simplify the first term as

$$\langle \sum_{k,l=1}^N (\Delta W_k)^2 (\Delta W_l)^2 \rangle = \sum_{k=1}^N \langle (\Delta W_k)^4 \rangle + \sum_{k \neq l=1}^N \langle (\Delta W_k)^2 \rangle \langle (\Delta W_l)^2 \rangle. \tag{5.87}$$

Here we use the fact that the Wiener increments at different times are uncorrelated, so their average product can be replaced by their product of averages. The average $\langle \Delta W_k^4 \rangle = 3 \langle \Delta W^2 \rangle^2 = 3 \Delta t^2$ is the fourth moment of a Gaussian distribution, while we can replace $\langle (\Delta W_k)^2 \rangle \langle (\Delta W_l)^2 \rangle = \Delta t^2$. We must now account for the fact that there are N of the fourth moment terms, while there are $N^2 - N$ of the $k \neq l$ terms, to find

$$\text{Var}[X] = \lim_{N \to \infty} 3N \Delta t^2 + (N^2 - N)\Delta t^2 - t^2 = \lim_{N \to \infty} \frac{2t^2}{N} = 0. \tag{5.88}$$

In the last equality, we have replaced $\Delta t = t/N$, keeping the time interval t fixed. This derivation justifies $dW^2 = dt$ as a deterministic replacement in the infinitesimal limit.

Stochastic Schrödinger Equation

We may now apply the above methods to derive the Itô form of the stochastic differential equations. In the infinitesimal limit, the measurement result $r(t)\Delta t$ is replaced by the increment dr, where

$$r(t)\Delta t \to dr = \langle \hat{\sigma}_z \rangle dt + \sqrt{\tau_m} dW. \tag{5.89}$$

The state update (5.56) can be expanded to second order in r:

$$\psi_j' = \psi \left(1 \pm \frac{dr}{2\tau_m} + \frac{1}{2} \frac{dr^2}{(2\tau_m)^2} \right) / |||\psi'\rangle||, \tag{5.90}$$

where $\psi_j = \tilde{\alpha}, \tilde{\beta}$, for $j = 0, 1$ respectively. Making the replacement $dW^2 = dt$ and dropping higher-order terms gives

$$\psi_j' = \psi_j \left(1 \pm \langle \hat{\sigma}_z \rangle \frac{dt}{2\tau_m} + \frac{dt}{8\tau_m} \pm \frac{dW}{2\sqrt{\tau_m}} \right) / |||\psi'\rangle||. \tag{5.91}$$

Taylor expanding the norm of the new state to second order in dW gives

$$\frac{1}{|||\psi'\rangle||} = 1 - \langle \hat{\sigma}_z \rangle^2 \frac{dt}{8\tau_m} - \frac{dt}{4\tau_m} - \frac{\langle \hat{\sigma}_z \rangle dW}{2\sqrt{\tau_m}}. \tag{5.92}$$

Multiplying out all terms and keeping to order $dW^2 = dt$ gives the increment $d\alpha = \alpha' - \alpha$ and $d\beta = \beta' - \beta$ as

$$d\alpha = -\frac{\alpha(1 - \langle \hat{\sigma}_z \rangle)^2}{8\tau_m} dt + \frac{\alpha(1 - \langle \hat{\sigma}_z \rangle)}{2\sqrt{\tau_m}} dW, \tag{5.93}$$

$$d\beta = -\frac{\beta(1 + \langle \hat{\sigma}_z \rangle)^2}{8\tau_m} dt - \frac{\beta(1 + \langle \hat{\sigma}_z \rangle)}{2\sqrt{\tau_m}} dW. \tag{5.94}$$

In the physics literature, it is common to divide both sides of this equation by dt and replace $dW/dt = \xi$ to obtain a stochastic differential equation in Itô interpretation. We can add in the dynamical terms from the standard Schrödinger equation, $i\hbar \partial_t |\psi\rangle = \hat{H}|\psi\rangle$, where $\hat{H} = (\epsilon/2)\hat{\sigma}_z + (\Delta/2)\hat{\sigma}_x$ for a more general analysis incorporating both dynamics from the measurement and dynamics from the Hamiltonian. This type of equation is called a stochastic Schrödinger equation.

Stochastic Master Equation

If we convert the pure state dynamics into the dynamics of the density matrix (see Appendix 2 for a review of mixed quantum states), this is called a stochastic master equation. It can be derived from $\rho_{ij} = \psi_i \psi_j^*$, where no normalization is required since we already normalized the pure state dynamics. This form is more useful for experiments, because decoherence and measurement inefficiency can be incorporated, which are unavoidable in the lab. We derive the Itô form of the stochastic master equation element by element. We can express $\rho'_{00} = \alpha'^* \alpha' = (\alpha + d\alpha)^* (\alpha + d\alpha)$ and substitute the increment (5.93). Expanding to first order in dt and second order in dW, we find

$$d\rho_{00} = \rho_{00} \frac{(1 - \langle \hat{\sigma}_z \rangle)}{\sqrt{\tau_m}} dW, \tag{5.95}$$

where we replace $|\alpha|^2$ with ρ_{00} at time $t = 0$ and note $\langle \hat{\sigma}_z \rangle = \rho_{00} - \rho_{11}$. Similarly for ρ_{11}, we express $\rho'_{11} = \beta'^* \beta' = (\beta + d\beta)^* (\beta + d\beta)$, and substitute the increment (5.94). Expanding as before, we find

$$d\rho_{11} = -\rho_{11} \frac{(1 + \langle \hat{\sigma}_z \rangle)}{\sqrt{\tau_m}} dW, \tag{5.96}$$

where $|\beta|^2$ is replaced with ρ_{11}. For the off-diagonal matrix elements $\rho_{01} = \rho_{10}^*$, we have $\rho'_{01} = \alpha' \beta'^* = (\alpha + d\alpha)(\beta^* + d\beta^*)$. Substituting the increments (5.93, 5.94) and expanding to first order in dt and second order in dW, we find

$$d\rho_{01} = -\frac{\rho_{01}}{2\tau_m} dt - \frac{\rho_{01} \langle \hat{\sigma}_z \rangle}{\sqrt{\tau_m}} dW, \tag{5.97}$$

where we have replaced $\alpha \beta^*$ at $t = 0$ with ρ_{01}. We note that the sum $d\rho_{00} + d\rho_{11} = 0$, indicating that the state remains normalized, $\rho_{00} + \rho_{11} = 1$, at the end of the increment as well. Stochastic averages over the noise dW can be conveniently done in the Itô convention using property (5.82), which does not hold in other interpretations. Writing the ensemble averaged density matrix as $\bar{\rho}$, we find the equations of motion:

$$\frac{d\bar{\rho}_{00}}{dt} = \frac{d\bar{\rho}_{11}}{dt} = 0, \qquad \frac{d\bar{\rho}_{01}}{dt} = -\frac{\rho_{01}}{2\tau_m}. \tag{5.98}$$

These results indicate that the diagonal matrix elements remain at their initial values, while the coherence decays exponentially in time with rate $1/(2\tau_m)$. Thus a continuous measurement process where the results of the measurement are discarded is simply a decoherence process.

As we did before, it is convenient to express the quantum trajectory equations in the Bloch ball coordinates and include the qubit Hamiltonian

$H = (\epsilon/2)\hat{\sigma}_z + (\Delta/2)\hat{\sigma}_x$ as well to write the final form of the Itô stochastic master equation,

$$\frac{dx}{dt} = -\frac{x}{2\tau_m} - \frac{xz}{\sqrt{\tau_m}}\xi - \epsilon y, \tag{5.99}$$

$$\frac{dy}{dt} = -\frac{y}{2\tau_m} - \frac{yz}{\sqrt{\tau_m}}\xi + \epsilon x - \Delta z, \tag{5.100}$$

$$\frac{dz}{dt} = \frac{1 - z^2}{\sqrt{\tau_m}}\xi + \Delta y. \tag{5.101}$$

We stress that although these equations look different from (5.72, 5.73, 5.74) after the readout (5.75) is substituted in for r, these equations describe exactly the same physics and have the same solution once the correct interpretation of the stochastic differential equation is taken into account.

Coming back to the beginning of the analysis, we recall that the quantum state of the system remains pure during the entire process, and only by averaging over the measurement results does decoherence come into the picture. Nevertheless, in experiments, interaction with an unmonitored environment is difficult to avoid, so it is also often helpful to add in terms $-\gamma x$ or $-\gamma y$ to the first two equations to account for environmental dephasing with the rate γ. Relaxation to the ground state by spontaneous emission can be accounted for with an extra term $-\rho_{ee}/T_1$ applied to the matrix element of the excited state $|e\rangle$. Often the *efficiency* of the detector is accounted for in a similar way as described in Section 4.2. If we multiply the ensemble average decoherence rate, $d\bar{\rho}/dt = \Gamma_\phi$, by twice the measurement time (in our convention), this results in the inverse efficiency of the detector, characterizing how much information is lost to the environment,

$$\eta = \frac{1}{2\tau_m\Gamma_\phi}. \tag{5.102}$$

Here η is bounded between $0 \le \eta \le 1$, where an efficiency of 1 indicates no lost information, while an efficiency of 0 indicates all the information is lost. This latter case recovers the dynamics (5.98) and is equivalent to an "open quantum system," a topic we will return to soon.

This theory can now be tested against experimental practice. In Fig. 5.5, data is shown from an experiment using a superconducting quantum circuit (Murch et al., 2013b; Weber et al., 2014). Quantum trajectories are represented as expectation values of Pauli matrices, and plotted on a slice of the Bloch ball. Single trajectories may be constructed from the initial state of the qubit and data from the quantum amplifier as described in this chapter, and are plotted in the figure. By averaging over many realizations of the experiment, smooth curves (the ensemble average) are also obtained and may be compared with theory. An outstanding

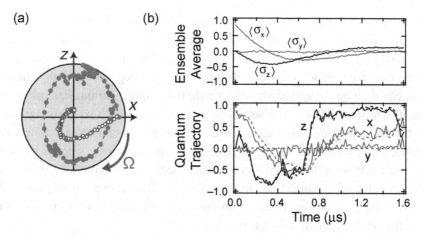

Figure 5.5 Experimental data taken of quantum trajectories with Rabi drive $\Delta = \Omega$, starting near the state $(|0\rangle + |1\rangle)/\sqrt{2}$ – repeated experimental preparation of this state is subject to slight dephasing, resulting in the experiment beginning slightly inside the Bloch ball (Murch et al., 2013b). The combination of effects coming from the Rabi drive and the continuous measurement is manifest. (a) Closed dots are data taken from a single run of the experiment. The quantum trajectory, plotted in the $x - z$ plane of the Bloch ball, is inferred from theory via the noisy output of the quantum amplifier. Open dots are the ensemble averaged quantum trajectory over many experiments, starting from the same initial state. (b) Upper panel – the ensemble averaged trajectories, represented as expectation values of the Pauli matrices are plotted versus time. Lower panel – The solid line is the quantum trajectory of a single trajectory is plotted versus time. The dashed line is the experimental tomographic reconstruction of that single trajectory, showing good agreement with theory. See the main text for a description of the tomographic procedure. Adapted with permission from Springer Nature.

question is how to experimentally validate the predictions of the theory. In panel (b) of Fig. 5.5, the following procedure was used. The experiment is allowed to proceed from the same initial state until a variable time t. At that point the continuous measurement is stopped, and the state of the qubit at that time is checked with quantum state tomography. This means a variable unitary operator is applied, followed by a conventional projective measurement, populating the readout cavity with a large number of photons, and projecting the qubit into one of its eigenstates. By repeating this procedure many times, reliable statistics can be used to validate the assigned quantum state's predictions in the different possible choices of basis. A subtlety here is that many millions of measurements must be made, because each realization of the continuous measurement process produces a different quantum state after a fixed period of time. By varying the amount of time from the beginning of the experiment until quantum state tomography measurements are performed, a full catalogue of the correct quantum state assignment is produced. It is then possible to validate a single quantum trajectory: each value of the quantum state at time t, along with the readout of the optical amplifier, predicts the correct state at the time time step, which can be looked up in the catalog. Those dots, when connected, can then be compared with the single trajectory quantum state assignment for each time step. The comparison, seen in panel (b), shows excellent agreement. From this experimental analysis, we conclude that it is possible to accurately make predictions of the quantum state all through the duration of a *single experiment*. Thus, while it is sometimes said that quantum mechanical predictions are only true for an ensemble of outcomes, we see here that this Bayesian point of view enables quantum state assignments even for single events that are physically powerful to predict the correct statistics of future events.

We end this subsection by pointing out that this experiment allows us to peer into the inner workings of the quantum state collapse process. It is no longer a mysterious black box to be discussed only in philosophy books, but an empirical process that can be characterized and tested extensively in the laboratory (Jordan, 2013).

5.5 Stochastic Path Integral

A powerful alternative to stochastic differential equation approaches to quantum trajectories and their statistical properties is the stochastic path integral. Just as the Feynman path integral gives a complementary understanding to quantum phenomena as well as an alternative way of calculating than the Schrödinger equation, the stochastic path integral approach does the same for the stochastic Schrödinger and master equations.

The basic idea introduced by Chantasri et al. (2013) (also called the CDJ formalism) is to calculate a master probability density of all possible measurement outcomes and associated quantum trajectories from which every possible statistical question can be answered. We return to a time-discretized description, shown in Fig. 5.6, with a time step of Δt, so there are a collection of continuous measurement results $\{r_k\}$, where $k = 1, 2, \ldots, N$, and associated quantum states $\{\hat{\rho}_k\}$, where we adopt a mixed-state picture for a fully general treatment. It is convenient to parameterize the density matrix with a set of parameters $\mathbf{q}$ that are the generalization of the Bloch coordinates. To describe higher-dimensional spaces, the Pauli matrices for a two-dimensional quantum system are expanded to a set of $n^2 - 1$ matrices in an n-dimensional system, also known as the generalized Gell-Mann matrices, appropriate for generating the group SU(n). In a two-dimensional Hilbert space, $\mathbf{q} = (x, y, z)$, the Bloch coordinates. The mapping (5.56) from one density matrix to the next time step may be expressed as $\mathbf{q}_{k+1} = \mathcal{E}(\mathbf{q}_k)$, where $\mathcal{E}$ describes

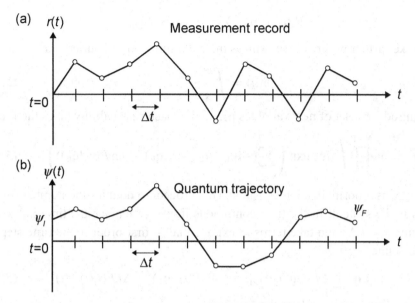

Figure 5.6 A conceptualization of the time-discredited formulation of quantum trajectory theory is shown. (a) The measurement readout $r(t)$ is plotted versus time t in time-slicing steps of Δt, taking values r_k shown in open circles. The experimental detector naturally has a finite time-resolution, given by its inverse bandwidth. (b) The corresponding quantum trajectory is schematically shown as a time-sliced function, taking values ψ_k at the open circles. Initial and final states are ψ_I and ψ_F. The progressive value of the quantum state at the next time step, ψ_{k+1}, is calculated from the current value of the state, ψ_k, and the current value of the measurement record, r_k.

the operation of the measurement operator on the parameterized density matrix (a function taking n inputs and outputting n outputs).

Let us define the master probability density of all possible measurement outcomes and associated quantum trajectories as $\mathcal{P}(\{\mathbf{q}_k\}, \{r_k\}|\mathbf{q}_i)$, starting from the initial quantum state $\mathbf{q}_i$. As is the case earlier in the chapter, we make the assumption that the time step Δt is longer than the correlation time of the noise, so the coordinates at each step depend only on the previous one (Markov approximation). In that case, we can express $\mathcal{P}$ as

$$\mathcal{P} = \prod_{k=0}^{N-1} P(\mathbf{q}_{k+1}|\mathbf{q}_k, r_k)P(r_k|\mathbf{q}_k). \tag{5.103}$$

Here, $P(r_k|\mathbf{q}_k)$ is the conditional probability distribution for r_k, given the system is in state $\mathbf{q}_k$, while $P(\mathbf{q}_{k+1}|\mathbf{q}_k, r_k)$ is the conditional probability for the new state at time step $k + 1$, given result r_k is obtained. The latter is given by the update rule mentioned before, now expressed as an n-dimensional delta function,

$$P(\mathbf{q}_{k+1}|\mathbf{q}_k, r_k) = \delta^{(n)}(\mathbf{q}_k - \mathcal{E}(\mathbf{q}_k)). \tag{5.104}$$

To make further progress, we express the delta function in Fourier form,

$$\delta(q) = \int_{-\infty}^{\infty} \frac{dp}{2\pi} e^{-iqp}, \tag{5.105}$$

and introduce a set of new variables $\mathbf{p}_k$, so the master probability takes the form

$$\mathcal{P} = \mathcal{N} \prod_{k=0}^{N-1} \int d\mathbf{p}_k \exp\left(\sum_{k=0}^{N-1} (-i\mathbf{p}_k \cdot [\mathbf{q}_k - \mathcal{E}(\mathbf{q}_k)] + \ln P(r_k|\mathbf{q}_k))\right), \tag{5.106}$$

where $\mathcal{N}$ is a normalization constant. This procedure doubles the number of variables in the system. In the time-continuous limit, we replace the time index k by the time $t = k\Delta t$ and take terms in exponential to first-order in the time-step Δt, so we define

$$\mathcal{E}(\mathbf{q}_k) \approx \mathbf{q}_k + \Delta t \mathcal{L}[\mathbf{q}(t), r(t)], \qquad \ln P(r_k, \mathbf{q}_k) \approx \Delta t \mathcal{F}[\mathbf{q}(t), r(t)]. \tag{5.107}$$

Here $\mathcal{L}, \mathcal{F}$ are functionals of the coordinates.

We therefore find in the time-continuous limit the expression

$$\mathcal{P} = \int \mathcal{D}p \, e^{\mathcal{S}}, \qquad \mathcal{S} = \int_0^t dt'(-i\mathbf{p} \cdot (\dot{\mathbf{q}} - \mathcal{L}[\mathbf{q}, r]) + \mathcal{F}[\mathbf{q}, r]). \tag{5.108}$$

Here $\int \mathcal{D}p$ is a functional integral over all values of all components of $\mathbf{p}$ and absorbs any possibly divergent constants into the definition that can be found by normalization. The quantity $\mathcal{S}$ we call the *stochastic action* takes the form of a time integral over a function of the state coordinates and readouts. The action integrand has a dynamical term $-i\mathbf{p} \cdot \dot{\mathbf{q}}$, as well as a contribution we call the *stochastic Hamiltonian* in analogy to the Feynman path integral, although there is no direct connection with energy – rather we will see it plays the role of a constant of motion:

$$\mathcal{H} = i\mathbf{p} \cdot \mathcal{L}[\mathbf{q}, r] + \mathcal{F}[\mathbf{q}, r]. \tag{5.109}$$

This form of the stochastic path integral is reminiscent of the quantum description of a fictitious mechanical system with Hamiltonian $\mathcal{H}$, coordinates $\mathbf{q}$, and momentum $\mathbf{p}$. However, there are a number of differences. Since this action describes diffusive motion, the action can be made real, after rotating the contour of the p integrals along the imaginary axis. The dimension of the fictitious mechanical system is the same as the Hilbert space of the system. Further, while Feynman paths are taken through configuration space, it is essential to note here the paths of the path integral are through the Hilbert space of the quantum system (more generally the quantum state space). The interpretation of the momentum-like variables $\mathbf{p}$ is analogous to momentum in the sense that they generate translations of the canonically conjugate coordinates $\mathbf{q}$, which do have a direct physical meaning. Further, the theory has a canonical structure, so changes of variables of the **q** coordinates must be accompanied by changes of **p** coordinates that are canonically invariant – that is, the Poisson brackets must be preserved. Physically, the stochastic Hamiltonian $\mathcal{H}$ encodes the detailed physics of the quantum measurement process – the nature of the probability distribution of the results, as well as the form of quantum backaction that is appropriate to the kind of measurement one is making. If the kind of measurement that is being carried out changes, the form of the stochastic action will also change, just like the right-hand side of the stochastic Schrödinger or master equation will change.

From the master probability distribution function $\mathcal{P}$, various quantities can be calculated, such as correlation functions of the form

$$\langle A(x_1, x_2, \ldots x_M) \rangle = \int \mathcal{D}q\mathcal{D}q\mathcal{D}r A(x_1, x_2, \ldots x_M) e^{\mathcal{S}}, \tag{5.110}$$

where the x variables represent any function of **q** or r at times $t_1, t_2, \ldots, t_M$. More generally, we can also account for additional constraints of a conditional nature. For example, if we also wish to impose a final boundary condition, this can done by only allowing quantum trajectories in the averages that end at a desired end point $\mathbf{q}_f$. In future chapters, we will explore a natural application of this formalism that is completely hidden in the stochastic differential equation approach: the concept of the most likely path the quantum system takes between final and initial conditions. The most likely path is analogous to the emergence of the classical path from the Feynman path integral in the limit of small $\hbar$, which gives rise to a new kind of action principle for a continuously measured quantum system.

We now give a few example applications of this formalism.

Application to the Case of Continuous Dispersive Qubit Measurement

Let us return to the continuously measured qubit example explored earlier in this chapter. The ingredients we need are the qubit coordinates – here simply the Bloch coordinates used in Eqs. (5.72, 5.73, 5.74), as well as the functionals $\mathcal{L}, \mathcal{F}$. The associated pseudo-momenta we call p_x, p_y, p_z and are canonically conjugate to x, y, z. To find $\mathcal{F}$, we recall that the distribution of result r is given by Eq. (5.55). Expanding the logarithm of $P(r|\mathbf{q})$ to first order in Δt, we find

$$\ln P(r|\mathbf{q}) \approx -\frac{\Delta t}{2\tau_m}(r^2 - 2rz + 1), \tag{5.111}$$

where we drop higher-order terms in Δt and absorb the constant term of order $\ln \Delta t$ into the normalization of the path integral. Consequently, $\mathcal{F} = (-r^2 + 2rz - 1)/(2\tau_m)$. To find $\mathcal{L}$, we make the expansion of the state disturbance equation to first order in Δt and find Eqs. (5.72, 5.73, 5.74). We can also add in dephasing from an unmonitored environment (which is equivalent to loss or inefficient measurement) by putting in additional decay terms of the form $-\gamma(x,y)$ for the coherence equations of motion. Putting all this together, we find the stochastic action to be given by

$$S = \int_0^t dt'[-ip_x(\dot{x} + \gamma x + \epsilon y + xzr/\tau_m) - ip_y(\dot{y} + \gamma y - \epsilon x + \Delta z + yzr/\tau_m)$$
$$- ip_z(\dot{z} - \Delta y - (1 - z^2)r/\tau_m) - (r^2 - 2rz + 1)/(2\tau_m)]. \quad (5.112)$$

From this form of the action, any statistical quantity of interest can now be calculated in principle.

5.6 Diffusive Measurement with Continuous Variables

In the example just given, we focused on a finite-dimensional system to illustrate the formalism and main results. However, in quantum physics we are often interested in continuous systems, such as a particle bound in a potential well, or in a scattering problem. The stochastic path integral approach has recently been applied to Gaussian state evolution in a harmonic potential undergoing joint position and momentum measurement (Karmakar et al., 2022). How continuous measurements of such systems can be described in a more general context will complete this chapter on diffusive continuous measurement.

Let us make a connection also with the Feynman path integral by considering the quantum description of a continuous measurement of the position of a particle of mass m in a potential well $V(x)$. We consider a meter that weakly measures the position operator $\hat{x}$ of the particle, repeatedly at intervals of time Δt. The measurement operator is taken to be

$$\hat{\Omega}_k = \left(\frac{\pi\sigma^2}{\Delta t}\right)^{1/4} \exp\left[-\frac{\Delta t}{4\sigma^2}(r_k - \hat{x})^2\right]. \quad (5.113)$$

Here, r_k is the measurement result at time interval k. We can find the probability density of result r_k by using the results from Chapter 3, assuming that at the previous time, the quantum wavefunction of the particle is given by $\psi_k(x)$. This yields

$$P(r_k|\psi_k) = \int_{-\infty}^{\infty} dx |\psi_k(x)|^2 \sqrt{\frac{\pi\sigma^2}{\Delta t}} \exp\left(-\frac{\Delta t}{2\sigma^2}(r_k - x)^2\right). \quad (5.114)$$

The mean of this distribution is given by $\langle\psi_k|\hat{x}|\psi_k\rangle$. The variance is given by $\text{Var}[r_k] = \sigma^2/\Delta t + \text{Var}[\hat{x}]$, where the variance of $\hat{x}$ is with respect to the state $|\psi_k\rangle$. In the limit where Δt is small, the weak measurement limit, the variance associated with the imprecise measurement, $\sigma^2/\Delta t$, is much larger than the position variance of the wavefunction, so the latter may be neglected. In this limit, we can treat the squared wavefunction like a delta function under the integral, centered at the expectation value of the position operator, with respect to the wavefunction, so the probability distribution may be well approximated as

$$P(r_k|\psi_k) \approx \sqrt{\frac{\pi\sigma^2}{\Delta t}} \exp\left(-\frac{\Delta t}{2\sigma^2}(r_k - \langle x\rangle_k)^2\right), \qquad (5.115)$$

where $\langle x\rangle_k = \langle\psi_k|\hat{x}|\psi_k\rangle$.

The new wavefunction $\psi_{k+1}(x)$ is given by

$$\langle x|\psi_{k+1}\rangle = \frac{\langle x|\hat{U}_k\hat{\Omega}_k|\psi_k\rangle}{\sqrt{P(r_k|\psi_k)}}. \qquad (5.116)$$

Here we have added in the unitary dynamics described by the Hamiltonian $H = \hat{p}^2/(2m) + V(\hat{x})$ as $\hat{U}_k = \exp(-i\Delta t\hat{H}_k/\hbar)$. The operator ordering of $\hat{U}$ and $\hat{\Omega}$ does not matter since the commutator is of order Δt^2. Inserting a complete set of position and momentum states, we find

$$\psi_{k+1}(x_{k+1}) = \int \frac{dx_k dp_k}{2\pi\hbar} \exp[ip_k(x_{k+1} - x_k)/\hbar - i\frac{p_k^2\Delta t}{2m\hbar} - i\frac{V(x_k)\Delta t}{\hbar}$$
$$- \frac{\Delta t}{4\sigma^2}(x_k - \langle x\rangle_k)^2 + \frac{\Delta t}{2\sigma^2}r_k(x_k - \langle x\rangle_k)]\psi_k(x_k). \qquad (5.117)$$

This process can be repeated to generate the final state $\psi_N(x_N)$, given by a generalized propagator,

$$\psi(x', t) = \int dx \mathcal{M}(x', x, t)\psi(x, 0), \qquad (5.118)$$

where the generalized propagator is given by

$$\mathcal{M}(x', x, t) = \int \mathcal{D}x\mathcal{D}p e^{\mathcal{S}}, \qquad (5.119)$$

$$\mathcal{S} = \int_0^t dt'\left[\frac{i}{\hbar}(p\dot{x} - \frac{p^2}{2m} - V(x)) - \frac{x^2 - \langle x\rangle^2}{4\sigma^2} + \frac{r(x - \langle x\rangle)}{2\sigma^2}\right].$$

Here the boundary conditions are $x(0) = x$ and $x(t) = x'$ and $\mathcal{D}x\mathcal{D}p = \prod_k \frac{dx_k dp_k}{2\pi\hbar}$. The function $\langle x\rangle(t)$ is the time-local expectation value of position, given the state at that time. The function $r(t)$ is the stochastic readout, which is a random time-continuous variable drawn from the distribution

$$P[r(t)] = \mathcal{N} \exp\left(-\int_0^t \frac{dt}{2\sigma^2}(r - \langle x\rangle)^2\right), \qquad (5.120)$$

where $\mathcal{N}$ is a normalization constant. That is, the variable r is a Gaussian random variable, with mean $\langle x \rangle$ at that time. It can be expressed as

$$r(t) = \langle x \rangle(t) + \sigma \xi, \tag{5.121}$$

where ξ is a delta-correlated random variable with strength 1. Indeed, we can eliminate the readout variable r from the system discussion and reexpress the stochastic action for the system as

$$\tilde{S} = \int_0^t dt' \left[\frac{i}{\hbar}(p\dot{x} - \frac{p^2}{2m} - V(x)) - \frac{(x - \langle x \rangle)^2}{4\sigma^2} + \frac{(x - \langle x \rangle)\xi}{2\sigma} \right]. \tag{5.122}$$

This result may be interpreted as the usual Feynman path integral in Hamiltonian form, but with two additions. The first (Gaussian) term provides a (real) contribution to the action that suppresses paths that wander more than a distance σ from the mean value of the position. The feature is reminiscent of Mensky's "restricted path integral," where corridors to restrict the particle's motion are put in by hand in order to rule out paths forbidden by the measurement results (Mensky, 1994). The second additional term to the action is also real but can be interpreted as a stochastic force causing the particle's position to be randomly kicked as the result of the measurement. These results are related to those of Caves and Milburn (1987). We will explore this interpretation more in the analysis of the most likely path.

While the time-continuous integrals provide an elegant representation, when going to a stochastic differential equation, confusion can arise. The corresponding stochastic differential equation to the update rule (5.116) can be derived by expanding to first order in the time-step to find

$$\frac{d|\psi\rangle}{dt} = \left[-i\hat{H}/\hbar + \frac{r}{2\sigma^2}(\hat{x} - \langle x \rangle) - \frac{1}{4\sigma^2}(\hat{x}^2 - \langle x \rangle^2) \right] |\psi\rangle. \tag{5.123}$$

Here, if we replace $r(t)$ by the signal and noise decomposition (5.121), where $\xi = dW/dt$, we obtain

$$\frac{d|\psi\rangle}{dt} = \left[-i\hat{H}/\hbar + \frac{\xi}{2\sigma}(\hat{x} - \langle x \rangle) - \frac{1}{4\sigma^2}(\hat{x} - \langle x \rangle)^2 \right] |\psi\rangle. \tag{5.124}$$

Here, $\langle x \rangle = \langle \psi | \hat{x} | \psi \rangle$ at time t. We must interpret the preceding stochastic differential equation in the Stratonovich sense. The deeper reason for this is that the equation is derived from physical considerations, therefore the stochastic noise function ξ obtained by subtracting the expected position from the measured signal can never be perfect white noise. This goes back to the previous discussion, where we insisted upon the separation of timescales between the correlation time of the noise and the dynamical timescales of the system of interest. In reality, every such physical approximation to white noise has more regular properties. In two important papers, Wong and Zakai proved that the solution to the *ordinary differential equation*, which is obtained from the stochastic differential equation by replacing the noise term with a continuous approximation to the Brownian motion, converges to that of the Stratonovich stochastic differential equation in the limit where the approximated Brownian motion becomes better and better (Wong and Zakai, 1965a,b). Thus, while the stochastic path integral is perfectly well defined as we constructed it, the corresponding differential equations need their proper interpretation. For completeness, the Itô form of the stochastic differential equation is found by expanding (5.116) to second order in $\xi = dW/dt$ and using Itô's rule as discussed earlier in this chapter to find

$$\frac{d|\psi\rangle}{dt} = \left[-i\hat{H}/\hbar + \frac{\xi}{2\sigma}(\hat{x} - \langle x\rangle) - \frac{1}{8\sigma^2}(\hat{x} - \langle x\rangle)^2 \right] |\psi\rangle. \tag{5.125}$$

where the coefficient of the drift term is changed.

Exercises

Exercise 5.6.1 Prove that the states $|\psi_\pm\rangle = (|L\rangle \pm |R\rangle)/\sqrt{2}$ describe the ground and first excited states of the DQD, respectively. Use two different methods:

(i) In the spatially symmetric case, the states should have definite parity, so under exchange of $L \leftrightarrow R$, the eigenstates are invariant up to an overall sign. Show that the $\langle x|\psi_+\rangle$ wavefunction has no nodes and must therefore be the ground state, while the $\langle x|\psi_-\rangle$ wavefunction has one node and must therefore be the first excited state.

(ii) If an electron tunnels from the left to the right well with rate $\Delta/\hbar$, we can model an effective two-state system with a Hamiltonian $\hat{H} = \Delta\hat{\sigma}_x$ in the left/right basis. Therefore $|\psi_\pm\rangle$ diagonalizes the effective Hamiltonian.

Exercise 5.6.2 Prove Eq. (5.2) is correct by solving the time-dependent Schrödinger equation.

Exercise 5.6.3 Work out the case described in Section 5.3.1, where the qubit state is mixed, and show that if we average over both outcomes, the degree of decoherence is $C_{LR} = t_L t_R^* + r_L r_R^*$, which has norm less than 1.

Exercise 5.6.4 Show that starting from Eq. (5.22), the dispersive form of the Jaynes–Cummings Hamiltonian (5.25) can be found by making use of the unitary transformation

$$\hat{U} = \exp[(g/\Delta)(\hat{a}^\dagger\hat{\sigma}_- - \hat{a}\hat{\sigma}_+)], \tag{5.126}$$

on the original Hamiltonian, and expanding to second order in g/Δ.

Exercise 5.6.5 Consider the case of monitored spontaneous emission of a two-level atom. Consider a single field mode with no photons, $|0\rangle$, and the two-level atom prepared in the state $\phi|g\rangle + \zeta|e\rangle$, where $|e\rangle, |g\rangle$ are the excited and ground states of the atom, and ϕ, ζ are complex amplitudes. After a short time δt, the atom can decay to the ground state and emit a photon with a rate γ, so the emission probability is $\epsilon = \gamma \delta t$. Show that the state at time δt is given by

$$|\psi\rangle = \sqrt{1 - \epsilon} \zeta |e, 0\rangle + \sqrt{\epsilon} \zeta |g, 1\rangle + \phi |g, 0\rangle, \qquad (5.127)$$

where $|1\rangle$ denotes a single photon emitted into the field mode. See Jordan et al. (2016) for further discussion.

Exercise 5.6.6   Suppose the field mode is detected with a homodyne detection scheme, described by projection of the field mode state onto the operator $\hat{X} = (\hat{a} + \hat{a}^\dagger)/\sqrt{2}$, where $\hat{a}$ and $\hat{a}^\dagger$ are the creation and annihilation operators of the optical field. Let $|X\rangle$ be the eigenstate of this operator. Show

$$\langle X|0\rangle = \pi^{-1/4} e^{-X^2/2}, \quad \langle X|1\rangle = \pi^{-1/4} \sqrt{2} X e^{-X^2/2}. \qquad (5.128)$$

Exercise 5.6.7   Combine the previous two exercises to show that the measurement operator for amplifying the spontaneous emission of the homodyne detection is given by

$$\hat{M}_X = e^{-X^2/2} \begin{pmatrix} \sqrt{1-\epsilon} & 0 \\ \sqrt{2\epsilon} X & 1 \end{pmatrix}, \qquad (5.129)$$

where X is the outcome of the quadrature measurement. Hint: Make an analogy to the quantum harmonic oscillator.

Exercise 5.6.8   Scaling the readout $X \to \sqrt{dt/2} r$ in the previous problem, show that the random variable r has a mean $\sqrt{\gamma} x$, and variance $1/dt$.

Exercise 5.6.9   For the scaling of X in the previous problem, show that the analogous equations of motion to Eqs. (5.72, 5.73, 5.74) are given by

$$\dot{x} = \frac{\gamma}{2} xz + \sqrt{\gamma} r(1 + z - x^2) - \epsilon y, \qquad (5.130)$$

$$\dot{y} = \frac{\gamma}{2} yz - \sqrt{\gamma} rxy + \epsilon x - \Delta z, \qquad (5.131)$$

$$\dot{z} = \frac{\gamma}{2} (z^2 - 1) - \sqrt{\gamma} r(1 + z)x + \Delta y. \qquad (5.132)$$

6

Continuous Measurement: Quantum Jump Case

In the previous chapter, we developed a powerful formalism for treating generalized measurements in quantum mechanics, including the description of a continuous weak measurement which bears many similarities to classical diffusion. We now treat the case of continuous measurement where the interaction with an environment causes a quantum system to abruptly evolve or jump from one state to another. From a historical perspective, the entry of measurement into the canon of quantum theory started with the introduction of such "quantum jumps," which were more of an ad hoc feature of the formalism rather than a well-detailed phenomenon. Building on the ideas of Max Planck and Niels Bohr in the latter's famous 1913 paper entitled "On the Constitution of Atoms and Molecules," Bohr (2016) asserted that objects at the atomic scale have a disposition that is not suited to classical physics! The laws of physics governing such objects must therefore be different at the atomic scale. Among Bohr's famous predictions were the rules associated with atomic energy levels and light emission. Not all orbits are stable because angular momentum, and thus the radius of a given orbit, has to be quantized. There is a lowest orbit, which represents the ground state. In a wave-oriented picture, one can get a general sense of this scheme by requiring that a given orbit must contain an integer number of wavelengths to satisfy periodic boundary conditions.

The transitions between these energy levels were a tremendous source of debate. On one hand, if one associated the energy difference between these levels with the frequency of an emitted photon emission, a remarkable agreement with spectroscopic experiments was obtained. However, there was a conceptual price to be paid, as these quantum jumps were viewed as instantaneous, random, and outside of the systematic machinery developed by Schrödinger and Heisenberg to evolve a system according to unitary operators. In fact, they were despised by Schrödinger, who preferred a more continuous, wavelike treatment. For many decades, the question of whether such jumps actually occurred or were a mathematical device useful only for calculations remained unanswered given the absence of advanced measurement tools in both atomic and solid-state physics. Breakthroughs in the 1980s and onward have enabled a variety of basic quantum measurement experiments to unravel the quantum ensemble to probe quantum dynamics one jump at a time. The theoretical possibility of a pure state description of quantum jumps was realized as far back as 1975 by Benjamin Mollow at the University of Massachusetts at Boston for the phenomenon of resonance fluorescence (Mollow, 1975). The formalism of partial measurements and Kraus operators developed thus far can be readily applied to the case of random, rapid transitions between quantum states, as we will show in the following sections.

6.1 Blinking Atoms and Their Emitted Photons

The first observation of quantum jumps, which happened in trapped ion systems, made a big splash in 1986, drawing the attention of the popular press, including a prominent article in the *New York Times* (Gleick, 1986). The technology to trap individual ions had reached a level of maturity that now enabled sophisticated experiments with coherent laser sources to be conducted on individual atoms rather than an ensemble. This led to many interesting philosophical questions. Is it really possible to observe a single atom, given that a measurement apparatus collapses the wave function? Is the mathematical formalism at the heart of quantum mechanics that so beautifully predicts ensemble behavior really the law of the land at the single-particle scale? Along with these open questions was the longstanding, twofold technological challenge of both directing the photons emitted during an atomic transition efficiently toward a detector and then registering their arrival with high efficiency.

Three experimental teams reported the observation of quantum jumps at nearly the same time. One group was based at the University of Washington (Nagourney et al., 1986), one at the National Bureau of Standards (Bergquist et al., 1986), and the third at the University of Hamburg (Sauter et al., 1986). All three experiments used a combination of electric fields and laser excitation to trap and radiatively cool single ions, the same basic technology that is being used today to advance ion-trap quantum computing systems. The basic operating principle in these ion-trap quantum jump experiments is illustrated in Fig. 6.1(a). A set of energy levels is chosen to have a "V"-type level structure with a single ground state $|G\rangle$ and two excited states, $|D\rangle$ and $|B\rangle$. The teams of Nagourney, Sandberg, and Dehmelt, and Sauter, Neuhauser, Blatt, and Toschek both used the $6^2S_{1/2} - 6^2P_{1/2} - 5^2D_{5/2}$ levels in the Ba^+ ion. Bergquist, Hulet, Itano, and Wineland used similar levels in Hg^{2+}. In both systems, the radiative lifetimes of the two excited states are very different. The $|G\rangle$ to $|B\rangle$ transition is bright, and the ion strongly interacts with the

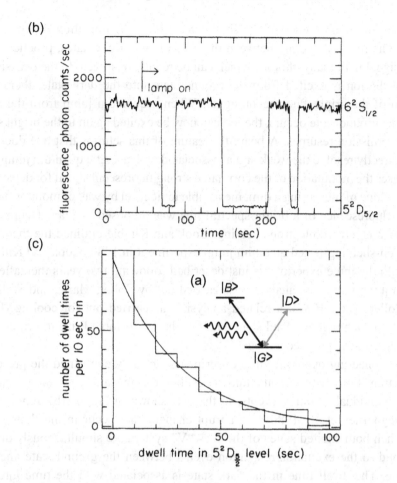

Figure 6.1 (a) The typical "V"-type level structure used to observe quantum jumps between the ground state $|G\rangle$ and a dark state $|D\rangle$. Such a transition stops the rapid cycling, and associated fluorescence, between $|G\rangle$ and the bright state $|B\rangle$. (b) In the Ba$^+$ experiments, the $5^2D_{5/2}$ is the dark state and the emitted light signal drops to a low value when it is populated. When the ion jumps back to the ground state after a varying dwell time, fluorescence resumes. (c) The dwell times in the dark state can be represented in a histogram that shows that the excited state population decays exponentially, as predicted by quantum theory (Nagourney et al., 1986). Reproduced with permission from the American Physical Society.

environment and readily absorbs and emits photons, while the other, dark state $|D\rangle$ only occasionally undergoes atomic transitions.

The scheme to observe the individual, infrequent jumps associated with the long-lived transition from $|D\rangle$ to $|G\rangle$ starts with exciting both arms of the "V" while monitoring the fluorescence of the short-lived $|B\rangle$ to $|G\rangle$ transition. Exciting the latter results in rapid cycling between the ground and bright states, producing an intense signal with many photons which can be readily observed. In the occasional case that the ion is excited from the ground state into the dark state, there is a cessation of the bright fluorescent signal. When a quantum jump from the dark state to the ground state occurs, the ion can now be excited again to the bright state and light emission resumes. A beautiful feature of this setup is that one does not have to directly resolve the weak signal associated with a single quantum jump and instead uses the modulation of the correlated strong fluorescent signal for detection. This technique of measuring a dim, metastable ionic level by way of monitoring the resonant fluorescence of a bright spectral line was proposed by Hans Dehmelt in 1975 and was termed electron shelving. Cook and Kimble outlined the theory of an electron-shelving-based quantum jump experiment in 1984 (Cook and Kimble, 1985) with the three experiments just described following two years thereafter. A quantum jump pure state analysis was carried out by Zoller, Marte, and Walls in 1987 (Zoller et al., 1987); a related analysis was carried out for cooling (Blatt et al., 1986). Dehmelt and Paul shared half of the 1989 Nobel Prize in Physics for developing ion-trapping technology.

These pioneering quantum jump experiments not only confirmed the presence of rapid transitions between quantum states, but also revealed the random nature of these individual quantum events. In the data shown in Fig. 6.1(b), reproduced from (Nagourney et al., 1986), the abrupt changes that occur in the fluorescent signal when both excited states of the Ba^+ "V" system are simultaneously driven correspond to the excitation/relaxation events between the ground state and the dark state. The dwell time in the dark state is associated with the time interval where no fluorescence is detected. As seen in the figure, this time is of variable length, and indeed quantum theory does not permit the observer to calculate the duration of a specific instance of an excitation event, but rather returns an average lifetime. The link between the observed dwell times from shot to shot of the experiment and the calculated average lifetime is made by generating a histogram of these dwell times, as shown in Fig. 6.1(c). We see an exponential decay with a characteristic time that matches the observed and predicted emission lifetime of the dark state, revealing a beautiful consistency between the properties of a single measurement and an ensemble of many. The last piece of Bohr's quantum jump model that remained unexplored was the apparently instantaneous nature of the transition. This task was well suited to the tool set afforded by superconducting circuits, and a series of quantum jump experiments in that platform are described next.

6.2 Quantum Jumps in Superconducting Qubits

The anatomy of a superconducting transmon qubit embedded in a microwave frequency cavity was detailed in the previous chapter. We can now consider the details of different measurement protocols in this platform, particularly ones that probe the random, ostensibly discrete but ultimately continuous nature of quantum jumps. While phase qubit experiments can signal a single transition from the ground state to the excited state via the onset of a steady-state voltage as described in Chapter 4, such measurements destroy the qubit and latch the circuit onto a dissipative state. This is very convenient for a high-fidelity quantum state readout at the end, say, of a quantum computation, but it does not permit the continuous monitoring of individual transitions between the levels of the qubit. A cQED architecture with the qubit and cavity detuned in frequency provides an elegant, quantum non-demolition route to continuously probe such dynamics. Moreover, the microwave photons resulting from these qubit transitions are not emitted into the many possible modes associated with free space. On the contrary, they are efficiently coupled into a one-dimensional waveguide realized by a coaxial cable that feeds into the amplification circuit. The challenge, however, in directly observing quantum jumps in this simple setup where an effective two-level atom is coupled to a measurement cavity revolves around the detection of weak microwave signals with high quantum efficiency. This task became possible around 2010 with the proliferation of robust, near-quantum-limited, superconducting parametric amplifiers (paramps). The detailed operation of these devices and other quantum measurement tools will be discussed in the next chapter.

Vijay et al. (2011) used a superconducting parametric amplifier to continuously monitor the microwave emission of a transmon qubit, readily identifying the quantum jumps associated with individual spontaneous emission events. The measurement scheme is illustrated in Fig. 6.2(a). Signals predominantly propagate in one direction through the use of microwave circulators. Circulators typically contain magnetic elements that allow for the construction of nonreciprocal circuit elements. To observe jumps, the qubit is first excited with a short resonant pulse to populate the excited state, as shown in Fig. 6.2(b). The readout cavity is then populated with on average thirty photons to realize a projective measurement of the qubit state. The microwave pulse scattered from the cavity acquires a qubit-state-dependent phase, which is recorded in a homodyne scheme that uses a mixer to translate the information encoded in the oscillating signal into a static voltage.

This voltage output is shown in Fig. 6.2(c). The curves correspond to three exemplar voltage-time traces: two after the application of a π pulse that populates the excited state and one after a 2π pulse that returns the system to the ground state.

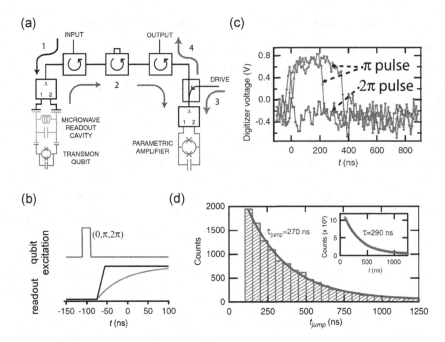

Figure 6.2 (a) Circuit diagram indicating the microwave components used in a superconducting circuit experiment to observe single quantum jumps. The signal path is indicated by numbered arrows, and circulators (square elements with a circular arrow) allow for directional signal propagation. (b) The pulse sequence used to generate single shot measurement traces, shown in (c), where a high (low) value of the homodyne voltage indicates population of the qubit excited (ground) state. (d) A histogram of excited state dwell times agrees well with the measured T_1 of the qubit, shown in the inset (Vijay et al., 2011). Adapted with permission from the American Physical Society.

The latter shows a steady signal at a low voltage value corresponding to occupancy of the ground state, while the π pulses initially result in a high voltage value, associated with the qubit excited state, followed by a rapid fall to the ground state, corresponding to a single quantum jump of the qubit. Much like the data generated in the ion trap experiments, the dwell time of the excited state is distributed over a set of values that average to yield an exponential decay in time. This can be seen in Fig. 6.2(d). The characteristic time extracted from the jump data closely match the ~ 290 ns lifetime of the qubit, measured in a standard T_1 sequence.

Another operational mode was explored in this experiment to reveal the effect of quantum backaction in a measurement. In panel (a) of Fig. 6.3, a measurement protocol is described where the measurement cavity is first populated, followed by a long excitation pulse applied to the qubit. In this case, the qubit is being simultaneously measured and excited. The measurement apparatus attempts to pin

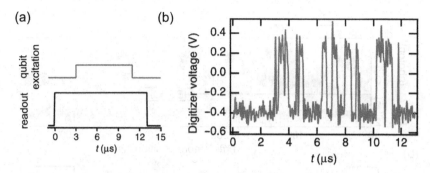

Figure 6.3 (a) Applying both cavity measurement and qubit excitation tones simultaneously suppresses the evolution of the qubit, with the system mostly populating the ground state, as shown in (b) (Vijay et al., 2011). Adapted with permission from the American Physical Society.

the qubit in the ground state – a phenomenon known as the Quantum Zeno Effect, as discussed later in this chapter. This is seen in Fig. 6.3(b) where the qubit is spending most of its time in the ground state rather than cycling back and forth between its two states under the effect of continuous excitation. The control pulse is only occasionally able to excite the qubit; the stronger the measurement is relative to the control pulse, the greater the suppression of state transitions.

6.3 Continuous Nature of Quantum Jumps

The observation of the Quantum Zeno Effect just described illustrates how measurement and control can both play a dramatic role in determining the evolution of a quantum system. We now describe an experiment that leveraged advanced quantum control protocols not only to observe quantum jumps, but also to guide their evolution. The experiment of Minev and coworkers (Minev et al., 2019) was performed in the group of Michel Devoret and revealed the continuous nature of a quantum jump, bringing a resolution of the Bohr–Schrödinger debate in a way that satisfies both parties! The setup consists of two superconducting transmon qubits strongly coupled to realize a "V" system with a ground state $|G\rangle$ and two excited states $|B\rangle$, $|D\rangle$ where the former is strongly coupled to the environment and bright and the latter is dark. This is analogous to the original configuration used by ion trappers in the first quantum jump experiments. The three-level system is dispersively coupled to a readout cavity such that a transition to the $|B\rangle$ state causes a ~ 5 MHz decrease in the bare ~ 9 GHz cavity frequency. The shift associated with the $|D\rangle$ state is more than ten times smaller, rendering the cavity effective in determining whether or not the $|B\rangle$ is occupied while leaving $|D\rangle$ isolated. As such, the coherence time of the dark state is greater than 100 μs. The coupling strengths and cavity line widths are chosen so that, on average, five photons populate the cavity

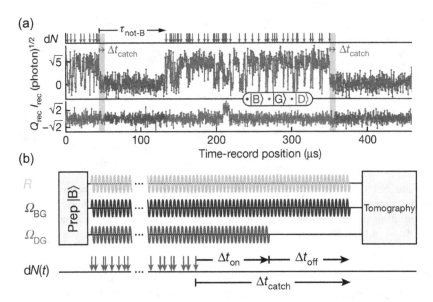

Figure 6.4 (a) Quantum jump data showing rapid transitions between the ground and bright states, with long periods of quiescence in the homodyne quadrature voltage signals, indicating population of the dark state. The little arrows on the top bar indicate where jumps between the bright and ground states are inferred. (b) Pulse sequence used to capture a quantum jump mid-flight and subsequently tomographically reconstruct the state shown in Fig. 6.5. The readout pulse is indicated by R, with Ω_{BG} and Ω_{DG} indicating the drive pulses applied to the bright and dark transitions, respectively (Minev et al., 2019). Adapted with permission from the Springer Nature.

when the coupled transmon circuit is in the $|B\rangle$ state. The standard jump protocol involves continuously driving both excited states on resonance. This results in long periods of cessation in the cavity emission, as seen in Fig. 6.4(a), when a quantum jump to the dark state $|D\rangle$ takes place. Much like the experiment of Slichter et al., the data here confirm that quantum jumps occur at random times, as predicted by Bohr and observed in all previous experiments.

Unique to the microwave platform employed in the Minev et al. experiment over the optical technology used in ion tap systems is the ability to continuously monitor the occupancy of the $|B\rangle$ state, as inferred by the cavity fluorescence, with a very small latency set primarily by the cavity line width. There is also no appreciable measurement dead time commonly associated with optical detectors. These features allow the experimentalist to faithfully unravel the time dynamics of a single quantum jump. Specifically, one can stop the driving pulses after a variable amount of time Δt_{catch} after the cavity emission goes dark. This can then be followed by pulses to tomographically reconstruct the quantum state in terms of the Bloch coordinates, X, Y, and Z. Such a protocol is described in Fig. 6.4(b) and

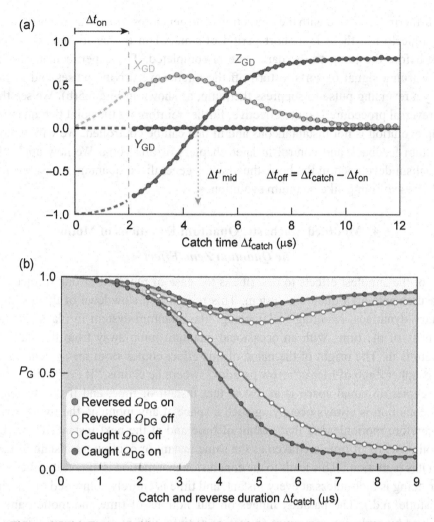

Figure 6.5 (a) Tomographic reconstruction of the state using the pulse sequence in Fig. 6.4(b). (b) After applying a reversal pulse during Δt_{on}, the jump evolution can be suppressed. Comparison is made when the system is simply caught versus reversed. Traces also indicate the role of the dark state excitation during the jump evolution (Minev et al., 2019). Adapted with permission from the Springer Nature.

essentially catches a quantum jump mid-flight, and then repeats this process many times to visualize the state. In the tomography data shown in Fig. 6.5(a) we see a very smooth path associated with the $|G\rangle$ to $|D\rangle$ transition, corresponding to an evolution of $Z = -1$ to $Z = +1$. These data clearly demonstrate that the dynamics of a single quantum jump are coherent and continuous, even if the onset time of each jump is random.

Furthermore, in line with the predictions one generates from an analysis of quantum trajectories, there is a short period of time when the cavity emission goes dark before the jump to the dark state is completed. The experimenter can use this warning signal of sorts to turn off the microwave driving pulses and in fact apply a reversing pulse to suppress the jump, as shown in Fig. 6.5(b). We see that the reversal procedure is quite effective, further reinforcing the point that quantum jump evolution is not instantaneous and in fact can be controlled. We will explore quantum feedback and control in later chapters of this book. We now apply the formalism developed in the book thus far for generalized quantum measurements to the case of jump-like quantum evolution.

6.4 Modified Stochastic Quantum Equations of Motion

The Quantum Zeno Effect

One of the simplest effects to describe is the case of repeated, strong (projective) measurements on a quantum system. This results in a slowdown of the quantum unitary dynamics, resulting in "freezing" the quantum system in place the vast majority of the time, with an occasional quantum jump away from the state the system is in. The origin of the name of this effect comes from the ancient Greek philosopher Zeno of Elea's "arrow paradox," where he claims, "If everything when it occupies an equal space is at rest at that instant of time, and if that which is in locomotion is always occupying such a space at any moment, the flying arrow is therefore motionless at that instant of time and at the next instant of time but if both instants of time are taken as the same instant or continuous instant of time then it is in motion." This leads to the conclusion that motion is impossible, because everything is motionless at every instant, and time is entirely composed of instants (Aristotle, n.d.). The paradox hinges on our notions of time and motion and is resolved by replacing motion at an instant of time with motion at nearby instants of time, as systematized in the mathematics of calculus. However, Zeno may have the last laugh in quantum measurement dynamics.

Consider a system with Hamiltonian $\hat{H}$ prepared in an initial quantum state $|\psi_0\rangle$. Let us consider the simple case of a two-outcome projective measurement, with projection operators $\hat{\Pi}_0 = |\psi_0\rangle\langle\psi_0|$ and $\hat{\Pi}_\perp = \mathbf{1} - \hat{\Pi}_0$. The quantum state in the absence of measurement will evolve according to the Schrödinger equation,

$$|\psi(t)\rangle = e^{-i\hat{H}t/\hbar}|\psi(0)\rangle, \tag{6.1}$$

where we assume no explicit time dependence in the Hamiltonian. The state will develop in time and move away from the initial state. However, if after a short amount of time, the preceding projective measurement is made, the system will be projected back into its initial state with probability

$$P_z = |\langle\psi_0|e^{-i\hat{H}t/\hbar}|\psi_0\rangle|^2 \approx |1 - it\langle\psi_0|\hat{H}|\psi_0\rangle/\hbar - (t^2/2\hbar^2)\langle\psi_0|\hat{H}^2|\psi_0\rangle + \ldots|^2.$$
$$(6.2)$$

We expand the absolute square to find, at small time,

$$P_z = 1 - t^2\Delta H^2/\hbar^2, \quad\quad (6.3)$$

where $\Delta H^2 = \langle\psi_0|\hat{H}^2|\psi_0\rangle - \langle\psi_0|\hat{H}|\psi_0\rangle^2$ is the variance of the Hamiltonian in the initial state.

Suppose now this procedure is repeated in time, for a total time T. We make N projective measurements at intervals of $t = T/N$. The probability of staying in the same initial state is then given by the probability P_z raised to the N power, because each attempt is independent of the last. We thus find that the probability of staying in the initial state is given by

$$P_{z,N} = (1 - T^2\Delta H^2/(N^2\hbar^2))^N \approx \exp(-T^2\Delta H^2/(\hbar^2 N)). \quad\quad (6.4)$$

From this expression, we see that as the number of measurements N is increased, the probability of staying in the initial state limits to 1, effectively freezing the system in place. This is the origin of the term "Quantum Zeno Effect."

Lindblad Master Equation

A widely used description of open quantum system is the master equation, derived by Lindblad (1976) and Gorini et al. (1976), followed by the work of Kraus (1981), Kraus et al. (1983), and others.

The formalism is rooted in the density matrix picture, where the Schrödinger equation for closed quantum systems is generalized to include an additional term that can account for dissipation and decoherence as a result of the environment,

$$\frac{d\hat{\rho}}{dt} = -i[\hat{H}, \hat{\rho}] + \mathcal{L}[\hat{\rho}], \quad\quad (6.5)$$

where $\mathcal{L}$ is defined as the Lindbladian. The Lindbladian is sometimes called a superoperator because it acts on operators. The entire right-hand side of the above equation is also called the Liouvillian superoperator, operating on the density operator, incorporating both the von Neumann Hamiltonian dynamics and the Lindbladian dynamics. The action of $\mathcal{L}$ on the density matrix is defined as

$$\mathcal{L}[\hat{\rho}] = \hat{L}\hat{\rho}\hat{L}^\dagger - \frac{1}{2}\{\hat{L}^\dagger\hat{L}, \hat{\rho}\}, \qu\quad (6.6)$$

and $\hat{L}$ is the Lindblad operator, proportional to the system operator that couples to the environmental degrees of freedom, and $\{\ldots,\ldots\}$ is the anti-commutator. Here we have given the simplest version of the Lindblad master equation; more complicated versions can be had by including a set of Lindblad operators $\{\hat{L}_j\}$ and summing over each Lindbladian $\mathcal{L}_j$ for each Lindblad operator. This reflects that there can be multiple ways of coupling the environment to the system.

The preceding equation is still a first-order equation in time and is an accurate description of environments that rapidly carry away any correlation with the system, so there are no memory effects. To justify this equation, a microscopic model can be considered, with coupling to the environment with potential V, followed by tracing out the environment degrees of freedom. Moving to the interaction picture with respect to the free Hamiltonians of the system and environment, the Lindblad equation capturing the evolution induced by the coupling to the environment can be derived by going to second order in the interaction term of the Hamiltonian and making a time coarse-graining of the exact joint quantum dynamics followed by tracing over the environment. The approximations of no memory of the system [Markov] and weak coupling [Born] are made to find

$$\mathcal{L}[\hat{\rho}] = -\frac{1}{\Delta t} \text{Tr}_E \int_t^{t+\Delta t} ds \int_0^\infty d\tau [[\hat{V}_I(s), [\hat{V}_I(s-\tau), \hat{\rho}(s) \otimes \hat{\rho}_{E,eq}]], \qquad (6.7)$$

where $\hat{V}_I$ refers to the coupling potential in the interaction picture, and we take the environment to be in equilibrium with state $\hat{\rho}_{E,eq}$.

The explicit form of the Lindbladian $\mathcal{L}$ can be derived using the decomposition of $\hat{V}_I$ in the basis of eigenoperators of $\hat{H}_{sys}$ and with an assumption regarding the respective magnitude of the energy transitions in the system, the correlation time of the bath, and the typical coupling strength to the bath, via the appropriate secular approximation. This approximation consists in neglecting the terms oscillating in time at frequencies larger than $1/\Delta t$. We refer the interested reader to several treatments of this problem in the literature (Cohen-Tannoudji et al., 1992; Manzano, 2020; Breuer and Petruccione, 2002).

Rather than make a detailed microscopic analysis, it is equally instructive to return to the perspective of Eq. (3.42) and the discussion around it. We recall that the density matrix can be updated as

$$\hat{\rho}(t+dt) = \frac{\hat{\Omega}_k \hat{\rho} \hat{\Omega}_k^\dagger}{P_k}, \qquad (6.8)$$

where $\hat{\Omega}_k$ are the measurement operators and $P_k = \text{Tr}[\hat{\Omega}_k^\dagger \hat{\Omega}_k \hat{\rho}]$. When passing to a time-continuous picture, the measurement operators can be expanded to leading order in time. For the Lindbladian to be first order, in general, we require terms of order $\sqrt{dt}$. For the sake of specificity, let us consider leading order expansion of $\hat{\Omega}$ for a continuous measurement process, using Itô calculus,

$$\hat{\Omega} \propto 1 - i\hat{H}dt - \hat{A}dt + \hat{L}dW, \qquad (6.9)$$

where $\hat{H}$ is the system Hamiltonian, $\hat{A}$ is a Hermitian operator to be determined; we will see that $\hat{L}$ is the Lindblad operator, and dW is the Wiener increment. In Exercise 6.4.2, you will show that $\hat{A} = -(1/2)\hat{L}^\dagger\hat{L}$. With the expansion of $\hat{\Omega}$, the normalized equation of motion is given in the Itô interpretation by

$$d\hat{\rho} = -i[\hat{H}, \hat{\rho}] + \hat{L}\hat{\rho}\hat{L}^\dagger - (1/2)(\hat{L}^\dagger\hat{L}\hat{\rho} + \hat{\rho}\hat{L}^\dagger\hat{L}) \tag{6.10}$$
$$+ \left[(\hat{L}\hat{\rho} + \hat{\rho}\hat{L}^\dagger) - \mathrm{Tr}[\hat{L}\hat{\rho} + \hat{\rho}\hat{L}^\dagger]\hat{\rho} \right] dW.$$

With this result, we can average over the Wiener increment dW to recover the (unconditional) Lindblad master equation mentioned at the beginning of this section. To make connection with the previous chapter, we can consider the two-level system, and let the Lindblad operator be $\hat{L} = \hat{\sigma}_z/(2\sqrt{\tau_m})$. The equations in (5.101) are then recovered. The readout of the measurement device is given by $r = \mathrm{Tr}[(\hat{L} + \hat{L}^\dagger)\hat{\rho} + dW/dt$, here proportional to z.

Lindblad to Quantum Jumps

We can now easily transition to the formalism for quantum jumps. Jump statistics are not Gaussian, but Poissonian. Poissonian statistics describe the probability of a given number of events occurring in a fixed interval of time at a fixed rate. This is typical for processes like spontaneous emission of light from an atom, nuclear decay, and the statistics of electrical current in a tunnel junction (Plenio and Knight, 1998).

Considering a discrete event, like the arrival and detection of a photon at a click-detector, we observe that most of the time, nothing happens, and only occasionally an event occurs. Let the typical rate for such an event be γ. The timescale of the registration of the event is taken to be much smaller than γ^{-1}. We consider a Lindblad master equation, where the first (no result) outcome is taken to be of the form $\hat{\Omega}_0 = 1 - i\hat{H}dt + \hat{A}dt$, where $\hat{A}$ is some Hermitian operator that will be determined, and to any other event we assign a measurement operator that scales in time like $\sqrt{dt}$, $\hat{\Omega}_j = \sqrt{dt}\hat{L}_j$. We can now check that the condition $\sum_j \hat{\Omega}_j^\dagger\hat{\Omega}_j = 1$ implies the condition $\hat{A} = -(1/2)\sum_{j\neq0} \hat{L}_j^\dagger\hat{L}_j$. Expanding Eq. (6.8) to first order in dt, we find two types of dynamics,

$$\hat{\rho}^{nj}(t + dt) = \hat{\rho}^{nj} - i[H, \hat{\rho}] - \frac{dt}{2}\{\hat{\rho}, \sum_j \hat{L}_j^\dagger\hat{L}_j\} + dt \sum_j \mathrm{Tr}[\hat{L}_j^\dagger\hat{L}_j\hat{\rho}]\hat{\rho}, \tag{6.11}$$

which corresponds to the (conditional and *deterministic*) "no-jump" dynamics, and

$$\hat{\rho}_j = \frac{\hat{L}_j\hat{\rho}\hat{L}_j^\dagger}{\mathrm{Tr}[\hat{L}_j^\dagger\hat{L}_j\hat{\rho}]}, \tag{6.12}$$

which corresponds to the jump that occurs when result j occurs. Notice that there is no time-scale dt here, or Hamiltonian dynamics – it cancels out in the state norm; the new state $\hat{\rho}_j$ abruptly changes, which is the prediction for the new state after the jump. The probabilities of these events are given by $P_{nj} = \text{Tr}[\hat{\Omega}_0 \hat{\rho} \hat{\Omega}_0^\dagger]$; here we specify the "event" to be a no-jump duration of time Δt. Similarly, a jump occurs with probability $P_j = \text{Tr}[\hat{\Omega}_j \hat{\rho} \hat{\Omega}_j^\dagger]$, and is proportional to dt, since this is a rate process.

In this case, the quantum trajectory is a list of events, indexed by the time of occurrence. When no event occurs, the quantum state does *not* remain unchanged, or even evolve according to the Schrödinger equation alone – it changes according to Eq. (6.11). When an event j occurs at a given time, the state abruptly changes according to Eq. (6.12), where the state $\hat{\rho}$ is the state at the time just before the jump event registers.

Example: Resonance Fluorescence

The preceding prescription for the quantum dynamics is better understood with an example. Here we discuss the case of a two-level atom that can spontaneously emit from its excited state (Kimble and Mandel, 1976). We assume that the emitted photon is deterministically channeled to an efficient photo detector for simplicity. While challenging for real atoms in an optical cavity, superconducting cavities can be coupled to microwave frequency transmission lines, which allows a high capture probability (Campagne-Ibarcq et al., 2016; Naghiloo et al., 2016). We define the energy splitting to be ϵ between the excited state $|e\rangle$ and the ground state $|g\rangle$.

In this case, the Lindblad operator corresponds to emission,

$$\hat{L} = \sqrt{\gamma/2}\,\hat{\sigma}_-, \tag{6.13}$$

where $\hat{\sigma}_- = |g\rangle\langle e| = (\hat{\sigma}_x - i\hat{\sigma}_y)/2$, and γ is the rate of emission of the atom. If the two-level atom is continuously excited with a drive Hamiltonian $H_d = \Delta\hat{\sigma}_x/2$, in the $e-g$ basis, then the detector can collect a series of photons in time. Each detection event corresponds to the state collapse,

$$\hat{\rho}_j \propto \hat{L}\hat{\rho}\hat{L}^\dagger = |g\rangle\langle g| \tag{6.14}$$

down to the ground state, which is then reexcited by the drive. The no-jump dynamics is more involved. Let us recall Eq. (6.11). The combination of the operators $\hat{L}^\dagger\hat{L}$ is given here by $|e\rangle\langle e| = \hat{\Pi}_e$, the projector onto the excited state. This gives the no-jump equation of motion as

$$\dot{\hat{\rho}} = -i[H,\hat{\rho}] - \frac{\gamma}{2}[\hat{\rho}\hat{\Pi}_e + \hat{\Pi}_e\hat{\rho}] + \gamma\rho_{ee}\hat{\rho}. \tag{6.15}$$

Here ρ_{ee} is the population of the qubit in the excited state – one of the diagonal matrix elements. Let us express this equation in the basis defined by the excited and ground states, noting that the density matrix is kept normalized ($\rho_{ee} + \rho_{gg} = 1$):

$$\dot{\rho}_{ee} = -\Delta \mathrm{Im}\rho_{eg} - \gamma\rho_{ee}(1 - \rho_{ee}), \tag{6.16}$$

$$\dot{\rho}_{eg} = -i\epsilon\rho_{eg} + i\frac{\Delta}{2}(\rho_{ee} - \rho_{gg}) - \frac{\gamma}{2}\rho_{eg} + \gamma\rho_{ee}\rho_{eg}. \tag{6.17}$$

These equations specify the dynamics completely, noting that $\rho_{ge} = \rho_{eg}^*$ and $\rho_{gg} = 1 - \rho_{ee}$. These equations should be solved to find the quantum state at some later time before a jump occurs. For the special case $\Delta = 0$, you will show in Exercise 6.4.4 that the solution is

$$\rho_{ee}(t) = \frac{\rho_{ee}(0)e^{-\gamma t}}{\rho_{gg}(0) + \rho_{ee}(0)e^{-\gamma t}}, \tag{6.18}$$

for the excited state population, and

$$\rho_{eg}(t) = \rho_{eg}(0)\frac{e^{-i\epsilon t - \gamma t/2}}{\rho_{gg}(0) + \rho_{ee}(0)e^{-\gamma t}}, \tag{6.19}$$

for the coherence. Notice that this result is similar to the case we investigated earlier in Section 4.2. However, in this case, the jump dynamics is not destructive – the jump/no-jump dynamics can go on indefinitely.

The probability of a no-tunneling event in the preceding case is given by

$$P_{nj}(t) = \rho_{gg}(0) + \rho_{ee}(0)e^{-\gamma t}. \tag{6.20}$$

This equation has the simple interpretation of the law of total probability – the system is in the ground state with probability $\rho_{gg}(0)$ and will never jump, or the system is in the excited state and its probability is weighted with the Poissonian probability of waiting a time t without a jump. In Exercise 6.4.5 you will prove this formula by time-slicing the problem and finding the probability of successive no-jump events. Figure 6.6 illustrates the experimental procedure involved in measuring the fluorescence quantum trajectories. If a microwave photon detector were present, quantum jumps of the artificial atom would be observed. Instead, in this experiment in the group of Benjamin Huard, a heterodyne detector was used, so the trajectories are in fact diffusive, despite the emission of photons via spontaneous emission (Campagne-Ibarcq et al., 2016).

Example: The Quantum Tunnel Junction Detector

Let us discuss a second example of quantum jumps, taken from condensed matter physics. We revisit our discussion of the quantum point contact, but in the limit where the transmission of the QPC is much smaller than one, $\mathcal{T} \ll 1$. In this

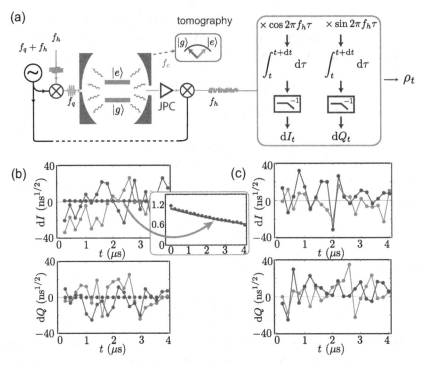

Figure 6.6 Illustration of resonance fluorescence experiments. (a) A qubit is excited into its excited state. It then spontaneously decays, emitting a microwave frequency photon. The radiation is collected with a microwave coaxial cable and then mixed with a local oscillator and amplified to produce an electric signal. Demodulating the high-frequency signal yields the two quadrature signals, giving a phase-preserving measurement of the field quadrature of the emitted radiation. (b) The measured quadrature signals I and Q are plotted versus time for two realizations of the experiment, starting in the state $|+x\rangle$. The smooth curve corresponds to the average record on all experiments, and the blue line corresponds to an exponential fit as a decay process. A zoom-in is shown as an inset. (c) The same as in panel (b), but starting in the excited state (Campagne-Ibarcq et al., 2016). Adapted with permission from the American Physical Society.

case, the potential between the source and the drain has an energy much higher than the energies of the electrons, so the system exhibits *quantum tunneling* – the wavefunction is exponentially suppressed into the classically forbidden region of the potential, but nevertheless, the electron can occasionally pass from one side of the potential barrier to the other.

In this case, the rate of tunneling is state-dependent. Let us define the rate of tunneling, if the qubit is in state 0 or 1, to be $\gamma_{0,1}$. The tunneling rate then determines the electrical current as $I_{0,1} = e\gamma_{0,1}$, with e being the electron charge. The Schottky formula gives the state-dependent shot noise power as $S_{0,1} = 2eI_{0,1}$. We take the previously discussed qubit Hamiltonian, $\hat{H} = (\epsilon/2)\hat{\sigma}_z + (\Delta/2)\hat{\sigma}_x$, diagonal in the

0, 1 basis. In this case, the Lindblad operator for the state-selective tunnel junction is given by

$$\hat{L}_{tj} = \sqrt{\gamma_0}|0\rangle\langle 0| + \sqrt{\gamma_1}|1\rangle\langle 1|. \tag{6.21}$$

We can now apply the formalism of the previous section to find the jump equations, determining the new quantum state of the qubit when an electron is found to have jumped across the barrier and registered in the drain part of the circuit. We express them as elements of the density matrix in the 0, 1 basis as

$$\rho'_{00} = \frac{\gamma_0 \rho_{00}}{\gamma_0 \rho_{00} + \gamma_1 \rho_{11}}, \tag{6.22}$$

$$\rho'_{01} = \frac{\sqrt{\gamma_0 \gamma_1} \rho_{01}}{\gamma_0 \rho_{00} + \gamma_1 \rho_{11}}, \tag{6.23}$$

where we see again that any Hamiltonian is neglected during the jump.

The no-jump dynamics can be found from the equation of motion, Eq. (6.11),

$$\dot{\rho}_{00} = -\Delta \operatorname{Im}\rho_{eg} - \gamma_0 \rho_{00} + (\gamma_0 \rho_{00} + \gamma_1 \rho_{11})\rho_{00}, \tag{6.24}$$

$$\dot{\rho}_{01} = -i\epsilon\rho_{01} + i\frac{\Delta}{2}(\rho_{00} - \rho_{11}) - \frac{\gamma_0 + \gamma_1}{2}\rho_{01} + (\gamma_0 \rho_{00} + \gamma_1 \rho_{11})\rho_{01}. \tag{6.25}$$

The reader can verify that in the case $\Delta = 0$, the following expressions solve these differential equations:

$$\rho_{00}(t) = \frac{\rho_{00}(0)e^{-\gamma_0 t}}{\rho_{00}(0)e^{-\gamma_0 t} + \rho_{11}e^{-\gamma_1 t}}, \tag{6.26}$$

$$\rho_{01}(t) = \frac{\rho_{01}(0)e^{-(\gamma_0+\gamma_1)t/2+i\epsilon t}}{\rho_{00}(0)e^{-\gamma_0 t} + \rho_{11}e^{-\gamma_1 t}}. \tag{6.27}$$

Let us go a step further, and consider the limit of many detected electrons, $N(t)$. The electrical current flowing through the tunnel junction is given by $I(t) = edN(t)/dt$, where e is the electron charge. We can derive a different kind of stochastic master equation by introducing the increment of the number of detected electrons, $dN(t)$. This classical random variable is analogous to the Wiener increment for diffusive processes. However, there are important differences. The stochastic increment, $dN(t)$, can only be 0 or 1 – either no electrons are detected in the drain (the typical case), or 1 electron is additionally detected at time t. Therefore, we have the important property $dN(t)^2 = dN(t)$, implying that all moments of this quantity take the same value, as is characteristic of the Poisson process. This identity is the point process analogue of the Wiener increment rule $dW^2 = dt$. With this random variable, we can compactly capture both the jump and no-jump processes as a stochastic master equation:

$$\hat{\rho}(t+dt) = dN(t)\frac{\hat{L}\hat{\rho}(t)\hat{L}^\dagger}{\mathrm{Tr}[\hat{L}\hat{\rho}(t)\hat{L}^\dagger]} + (1 - dN(t))\hat{\rho}^{nj}(t+dt). \qquad (6.28)$$

This equation indicates that if the increment $dN = 1$, the quantum jump density matrix is selected, whereas if $dN = 0$, then the no-jump master equation is selected. From the preceding discussion, the expectation of dN is given by $\langle dN(t) \rangle = (\gamma_0 \rho_{00} + \gamma_1 \rho_{11})dt$, the probability of a jump event in time dt in this case (Goan et al., 2001).

Let us now consider the case of many electrons that pass the tunnel junction. To proceed, let us index the density matrix with an index (n), indicating how many electrons have been collected in a time t, defining the electrical current. We can now combine the jump and no-jump dynamical equations to obtain an infinite set of equations for the set of density matrices $\hat{\rho}^{(n)}$. Before doing so, it is convenient to learn a useful trick in applying these methods to solve practical problems. We first note that a density matrix $\hat{\rho}$ can always be replaced by another unnormalized density matrix $\hat{\sigma}$, so long we are careful to renormalize at the end, $\hat{\rho} = \hat{\sigma}/\mathrm{Tr}[\hat{\sigma}]$. Using this trick, we can always drop any overall factor that multiplies each density matrix element. The statistical independence of each time-step implies that this norm will multiply at each time step. Consequently, we can find the probability of each quantum trajectory by simply finding the norm of the state at the end of the process. In this case, we will shortly see that that it is useful to map $\hat{\rho} \to \hat{\sigma}$ and change the norm of the updated state from $[1 - \gamma_0 \rho_{00}^n dt - \gamma_1 \rho_{11}^n dt]$ to $[1 - \gamma_0 \sigma_{00}^{n-1} dt - \gamma_1 \sigma_{11}^{n-1} dt]$. Here the label n refers to the state with n electrons collected (after a jump) and $n-1$ refers to when only $n-1$ electrons are collected (before the jump). One can explicitly check that $\sigma_{ij}/(\sigma_{00} + \sigma_{11})$ is independent of this altered norm. With the redefinition of the state and its norm, the no-jump equation of motion (6.24) becomes

$$\dot{\sigma}_{00}^n = -\Delta\mathrm{Im}[\sigma_{01}^n] - \gamma_0\sigma_{00}^n + (\gamma_0\sigma_{00}^{n-1} + \gamma_1\sigma_{11}^{n-1})\sigma_{00}^n, \qquad (6.29)$$

$$\dot{\sigma}_{01}^n = -i\epsilon\sigma_{01}^n + i\frac{\Delta}{2}(\sigma_{00}^n - \sigma_{11}^n) - \frac{\gamma_0 + \gamma_1}{2}\sigma_{01}^{n-1} + (\gamma_0\sigma_{00}^{n-1} + \gamma_1\sigma_{11}^{n-1})\sigma_{01}^n.$$

We can now link the state after the jump to the state before the jump with Eqs. (6.22) to find our final form of the equations of motion (Gurvitz, 1997),

$$\dot{\sigma}_{00}^n = -\Delta\mathrm{Im}[\sigma_{01}^n] - \gamma_0\sigma_{00}^n + \gamma_0\sigma_{00}^{n-1}, \qquad (6.30)$$

$$\dot{\sigma}_{01}^n = -i\epsilon\sigma_{01}^n + i\frac{\Delta}{2}(\sigma_{00}^n - \sigma_{11}^n) - \frac{\gamma_0 + \gamma_1}{2}\sigma_{01}^{n-1} + \sqrt{\gamma_1\gamma_2}\sigma_{01}^{n-1}.$$

This set gives an infinite number of coupled equations for each integer n. The redefined norm makes these equations linear, so they can be solved exactly in the special case where $\Delta = 0$. The solution can be verified directly, or solved using generating function methods (see Exercise 6.4.7) (Korotkov, 2001),

$$\sigma_{00}^n(t) = \sigma_{00}^n(0)\frac{(\gamma_0 t)^n}{n!}e^{-\gamma_0 t}, \tag{6.31}$$

$$\sigma_{00}^n(t) = \sigma_{00}^n(0)\frac{(\gamma_1 t)^n}{n!}e^{-\gamma_1 t}, \tag{6.32}$$

$$\sigma_{01}^n(t) = \sigma_{01}^n(0)\frac{(\sqrt{\gamma_0 \gamma_1}t)^n}{n!}e^{-(\gamma_0+\gamma_1)t/2-i\epsilon t}. \tag{6.33}$$

The probability of receiving n electrons is given by the norm of the solution,

$$P(n) = \sigma_{00}^n(t) + \sigma_{11}^n(t), \tag{6.34}$$

and the normalized solution is found by dividing the matrix elements by $P(n)$. Notice we can again view this result as the law of total probability, where $\rho_{ii}^0(t) = \sigma_{ii}^0(t)$ is the initial probability of occupying state $i = 0$ or 1, and is weighted by a Poissonian distribution of events, with a rate of either γ_0 or γ_1. The coherent matrix element solution is purely quantum and the quantity

$$\mathcal{M} = |\sigma_{01}|^2/\sigma_{00}\sigma_{11}, \tag{6.35}$$

is a constant of motion, called "murity" in Jordan and Korotkov (2006), in analogy to purity, which is kept constant by unitary operations. Finally, we note that the dephasing rate of the ensemble-averaged measurement can be found by averaging the off-diagonal matrix element over all events,

$$D(t) = \sum_n P(n)|\rho_{01}^n| = \sum_n |\sigma_{01}^n| = \exp(-(\sqrt{\gamma_1} - \sqrt{\gamma_2})^2 t/2). \tag{6.36}$$

Here we see a different form of the ensemble averaged dephasing rate, $\sqrt{\gamma_1} - \sqrt{\gamma_2})^2/2$, since the shot noise is proportional to the average current for both of the quantum dot states (Gurvitz, 1997; Korotkov, 1999; Goan et al., 2001).

The Transition from Quantum Diffusion to Quantum Jumps

The organization of the last two chapters has been first to discuss diffusive quantum trajectories, and then to discuss quantum jumps. This presentation may lead one to view these as two very different types of quantum dynamics. In fact, these are not totally distinct types of quantum dynamics, and it is possible to have a continuous transition between the two types of quantum physics. We will illustrate that transition in this section by returning to quantum diffusive dynamics in the presence of quantum tunneling, described by Eqs. (5.101). We note that there are two important time scales in the dynamical equations of motion, the characteristic measurement time τ_m and the Rabi oscillation period, $\tau_q = \hbar/\Delta$.

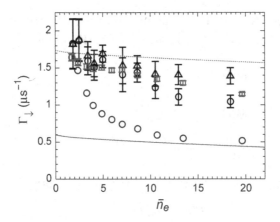

Figure 6.7 The qubit excited-to-ground jump rate $\Gamma_\downarrow$ is plotted versus average cavity photon occupation $\bar{n}_e$, when the qubit is in the excited state. The squares are measured values of the qubit decay rate, while the triangles and circles are values derived from numerical simulations; the former distinguishes between the excited state and higher energy levels in the rate extraction process, while the latter does not (as is the case for our experimental data). The results demonstrate the theoretical expectation that decreasing the characteristic measurement time (controlled by increasing $\bar{n}_e$) decreases the jump rate. Error bars represent 95 percent confidence intervals (Slichter et al., 2016). Reproduced with permission from IOP Publishing.

In the limit where $\tau_q \ll \tau_m$, the measurement is very weak, and this is a small perturbation to the Rabi oscillations. The measurement adds diffusion to the phase of the Rabi oscillation and leads to a random phase acquired in each realization of the measurement process.

In the opposite limit, $\tau_q \gg \tau_m$, the measurement rapidly collapses the quantum state to an eigenstate of the measurement. If τ_q is divergent, then the quantum state will simply stay in that collapsed state. However, from our discussion of the Zeno effect in Section 6.4, the strong continuous measurement is analogous to rapid repeated projective measurements. Consequently, we expect in this limit the quantum state will switch randomly between eigenstates of the measurement on a time-scale much longer than either τ_m or τ_q. The rate of jumps Γ_j is ideally given by

$$\Gamma_j = \tau_m/\tau_q^2. \tag{6.37}$$

See, for example, Chantasri et al. (2013) for a derivation. The dependence on the Zeno jump rate between levels on the measurement rate is illustrated in Fig. 6.7. In the experimental data given there, adding more photons into the readout cavity decreases the characteristic measurement time (5.54), resulting in a decreasing of the interlevel qubit jump rate (Slichter et al., 2016). The jump rates may be extracted from a statistical analysis of random telegraph signals resulting from the jumps back and forth between the pinned qubit levels. Greater insight into this transition from jumps to diffusion comes from the most likely path analysis, discussed in Chapter 9.

Don't Let Your Master Equation Unravel

We have seen now two different ways that different kinds of detection can give rise to the same unconditional master equation. However, one could have given this analysis in the opposite order, first writing down the unconditional master equation, and then adding on the stochastic terms to account for the different kinds of detection. Indeed, the historical treatment followed the latter route. However, in the literature this has led to a number of misunderstandings and misconceptions about the theory and practice of quantum trajectories. The most unfortunate of these is that the unconditional master equation is believed to be a kind of objective quantum property of the system, and it must be "unraveled" to give a stochastic master equation. The number of possible ways of unraveling this master equation is infinite, so this sometimes leads to the accusation that the stochastic Schrödinger equation is not unique, effectively rendering it useless, except as a calculational tool to first simulate the quantum system as an ensemble of pure quantum trajectories, after which one can average over the quantum trajectories, in order to make a mixed state at some time. Since one unraveling is as good as any other for that purpose, there is no physical significance to any given unraveling.

There are several comments that can be made about this line of thinking. First, is that this approach has merit if the environment is unmonitored. Indeed, since the theory of master equations was developed to describe open quantum systems where the environment consists of many degrees of freedom of a macroscopic environment, it is natural to expect this is the best description of such an open system. However, if we allow the possibility that the environment is itself monitored, then new possibilities emerge. Indeed, the design of quantum detectors involves carefully designing the type of interaction and limiting the number of uncontrolled degrees of freedom. Second, in the experiments described in the previous sections, it is possible to make additional predictions about the dynamics of the quantum state beyond the unconditional master equation. That is, after all, the physical content of this theory for continuous quantum measurements. The predictions of the theory can be experimentally verified as the outcomes expected by a given quantum state, conditioned on a measurement record. This gives the predictive advantage of the theory. Third, it is critical in the theory of quantum trajectories to specify the role of the detector. That is, what kind of measurement is being done on the system or its meter degree of freedom. In the case of spontaneous emission of an atom, one can measure the particle-like nature of the electromagnetic field with a photodetector, determining whether a photon arrived or did not arrive at a given time. That gives rise to a quantum jump of the state of the atom, as described earlier in this chapter. If, however, a wave-like aspect of the field is measured with a homodyne measurement or a heterodyne measurement, then one measures a continuous, noisy detector signal, and the quantum dynamics of the atom is a diffusive quantum trajectory. It should be stressed that the physical process of what the atom is doing is exactly the same in both cases; however, the type of detection one makes gives rise to qualitatively different types of quantum trajectories. Thus, the problem of the uniqueness of which unraveling one uses is solved by specifying what the detector is measuring.

We can make a helpful analogy to this phenomenon by appealing to the quantum state collapse of a bipartite system that is spatially separated, similar to the EPR/Bell setup described in Chapter 1. Supposing we have two parties, Alice and Bob, measuring and controlling the local systems, we let the combined state be described by a pure state, $|\Psi\rangle$, that can be entangled. If Bob only has access to measurement data on his part of the state, then Bob's state is equivalent to

$$\hat{\rho}_B = \text{Tr}_A |\Psi\rangle\langle\Psi|, \tag{6.38}$$

as described in Appendix B on mixed states. Thus, every possible statistical prediction that Bob can make on his measurement outcomes is described only by the preceding mixed state. However, if Alice also makes a measurement on her part of the system and communicates it to Bob, then additional information has entered the discussion and process of physical prediction. If, for example, both systems are qubits and the state $|\Psi\rangle$ is a singlet state, then a measurement of Alice in the Z-basis with result ± 1 then indicates Bob's state should in fact be replaced by

$$|\psi\rangle_B = |\mp\rangle_Z, \tag{6.39}$$

the anti-correlated Z-eigenstate to Alice's state. This is a pure state, so it allows Bob to make a confident physical prediction about any subsequent measurements on his part of the system. Without this knowledge, his state is simply

$$\hat{\rho}_B = \frac{1}{2}\text{Tr}[|+-\rangle - |-+\rangle][\langle +-| - \langle +-|] = \frac{1}{2}[|+\rangle\langle +| + |-\rangle\langle -|], \tag{6.40}$$

the maximally mixed state, which has minimal predictive information in it.

Suppose next that Alice instead decided to measure her system in the X-basis, with result ± 1. In that case, Bob's state is now given by

$$|\psi\rangle_B = |\mp\rangle_X, \tag{6.41}$$

the anti-correlated state in the X-basis. In either case, if Bob takes those states and averages over Alice's results (with probabilities $p_\pm = 1/2$), he will find a density matrix,

$$\hat{\rho}_B = \sum_{j=+,-} p_j |j\rangle_{ZorX}\,_{ZorX}\langle j| = \frac{1}{2}[|+\rangle\langle+| + |-\rangle\langle-|], \qquad (6.42)$$

which is the same density matrix found earlier by tracing over Alice's part of the state.

While both possible choices of Alice give rise to the same density matrix for Bob when knowledge of her actions is hidden from him, it would be silly to say that the fully mixed density matrix is Bob's proper state. Rather, the state assignment is subject to Bob's available information, and by making only a fully mixed state assignment when he also knows Alice's basis and result, he is missing out on making the best physical prediction. From this analogy we can apply the lesson to the quantum master equation description of the dynamics of the state. The unconditional master equation is the best statistical description one can make of the system for an unmonitored environment. However, if additional information is available in the form of a measurement record, then the stochastic master equation is both unique (once the type of measurement is specified) and makes the optimal physical prediction about measurement statistics of the quantum system. We note that the same kind of spooky measurement backaction present in the EPR thought experiment is also present here (this is also known as "quantum steering"). Just as the state assignment of Bob should take place no matter far away Alice is, and is quite different depending on what Alice chooses to measure, so too in the measurement case, we can delay the choice of the measurement type until well after the system dynamics has occurred. For example, in the spontaneous emission case, we can put the emitted radiation into a waveguide in an optical fiber and send that light across the ocean. At the other end of the optical fiber, one could choose to make either photo-counting or -dyne measurement. The result will be totally different kinds of quantum trajectories of the original atom, depending on your choice of measurement!

Exercises

Exercise 6.4.1 See the quantum Zeno effect in action for yourself. Prepare a qubit in state $|0\rangle$. Given a Hamiltonian $\hat{H} = \Delta\hat{\sigma}_x/2$, let the qubit coherently evolve for a time t. Now, measure projectively in the $|0\rangle, |1\rangle$ basis. Repeat this process N times, and find the probability the qubit has collapsed to its original state $|0\rangle$ after every measurement, given a total time $T = Nt$, where N is taken to be large.

Exercise 6.4.2   For the measurement operator to be properly normalized (i.e. $\sum_k P_k = 1$ for all states), show that $\hat{A} = -(1/2)\hat{L}^\dagger\hat{L}$, using Itô calculus.

Exercise 6.4.3   Show that Eq. (6.10) follows from the expansion (6.9) and (6.8) using Itô calculus.

Exercise 6.4.4   Solve the no-jump dynamics in the case where there is no jump for some time t, and for simplicity, take the case of $\Delta = 0$. Given the initial conditions to be $\hat{\rho}(0)$, you should recover the results (6.18, 6.19). Hint: Use $1/[a(1-a)] = 1/a + 1/(1-a)$, as well as $\ln(f(x))' = f'(x)/f(x)$.

Exercise 6.4.5   Find the probability of a no-tunneling event in the case described in exercise 6.4.4, noting in this case $P_{nj}(dt) = 1 - i[H, \hat{\rho}]dt - \gamma dt \rho_{ee}(t)$. Time slice the problem, and note that the probability of a sequence of no-jump events is given by the probability of no-jump at the current time, times the probability of no-jump at the previous time. This logic leads to

$$P_{nj}(t) = P(t)P(t - dt)P(t - 2dt)\ldots P(dt) = \prod_{j=1}^{N}(1 - \gamma dt \rho_{ee}(jdt))$$

$$\approx \exp\left(-\gamma \int_0^t dt \rho_{ee}(t)\right). \tag{6.43}$$

You should find the result (6.20).

Exercise 6.4.6   Check that the ensemble average of the update rule (6.28) recovered the Lindblad master equation for this process with the identified Lindblad operator.

Exercise 6.4.7   Prove that Eq. (6.31, 6.32, 6.33) is the correct solution to the equation set (6.30) using the method of generating functions. Define the generating function

$$Z_{ij}(\lambda) = \sum_{n=0}^{\infty} e^{i\lambda n}\sigma_{ij}^n. \tag{6.44}$$

Z is called a generating function because

$$\frac{\partial^k Z}{i^k \partial\lambda^k}\bigg|_{\lambda=0} = \sum_{n=0}^{\infty} n^k \sigma_{ij}^n, \tag{6.45}$$

that is, the nth moment of σ_{ij}. If $i \neq j$, this is a generalized moment because σ_{ij} does not have an interpretation as a probability distribution.

Multiply both sides of equations (6.30) by $\exp(i\lambda n)$ and sum over n. Derive an equation of motion for $Z_{ij}(\lambda)$ and solve it. The function σ_{ij}^n can be found from an inverse Fourier transform,

$$\sigma_{ij}^m = \frac{1}{2\pi} \int_0^{2\pi} e^{-i\lambda m} Z_{ij}(\lambda)d\lambda. \tag{6.46}$$

7

Linear Detectors

In this chapter, we introduce some of the experimental tools used in quantum measurements, in particular those used to probe semiconducting and superconducting electrical circuits using microwave photons. For quantum systems based on naturally occurring atoms, the characteristic energies associated with optical transitions are much larger than their GHz frequency cousins, and a robust set of single photon sources and detectors have been in use for decades. For microwave measurements, on the other hand, it is necessary to approach the quantum limit of sensitivity to observe coherent effects and to not be dominated by the background noise of the amplifiers and detectors needed to record weak signals. The design and operation of these devices is a nontrivial task, and a key part of quantum measurement science at microwave frequencies. In a nutshell, the basic challenge is to take the information contained in a coherent, quantum mechanical system and faithfully register it, preferably with minimal perturbation to the quantum system, onto a classical meter which has many degrees of freedom. In this chain of information transfer, most protocols involve first mapping the quantum state of the system of interest onto a "pointer," which can then be coupled to a classical measurement device. Later on in this chapter, we will give examples of how a resonant cavity and a quantum point contact can act as dispersive and dissipative detectors, respectively, that bridge information between a quantum bit and classical electronics.

We start our discussion of measurement devices by noting that a detector or amplifier obeys the same quantum mechanical principles that govern the operation of a qubit or other coherent circuit (see Fig. 7.1). Crucially, this includes the Heisenberg uncertainty principle that limits the precision with which two noncommuting observables can be interrogated. For the well-studied case of a particle in motion, a measurement of the position at a given time is associated with a random perturbation of the momentum that adds noise to subsequent position measurements. This effect is known as the quantum backaction associated with the measurement. We note that this principle does not limit the accuracy that can be

(a)

(b)

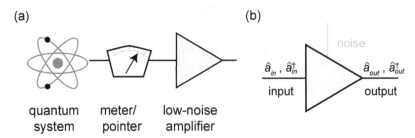

quantum   meter/   low-noise
system   pointer   amplifier

Figure 7.1 The generic detection methodology for quantum-limited measurements. (a) The quantum system interacts with a meter degree of freedom. An influence of the meter (or pointer) is then amplified by a low-noise amplifier. (b) The amplifier converts inputs into outputs and will usually add power and noise to the output to achieve gain. Bosonic mode a describes the input signal and mode $\hat{a}_{out} = \sqrt{G}\hat{a}_{in}$ is the amplified output, where G is the power gain. Additional modes must be included to retain the commutation relations, giving rise to added noise.

obtained when probing a single observable, such as the position or the momentum of a particle. For electrical signals, the uncertainty principle also imposes analogous quantum limits to sensitivity. Specifically, if a linear amplifier that preserves the phase of an input signal operates with large power gain, then it must add at least one half of a quantum of noise to the output signal.

While manifestations of this type of standard quantum limit to sensitivity ultimately trace their origin to the commutator algebra of incompatible observables, looking at specific physical systems and the detailed quantum equations governing them gives us valuable insight into general principles by which to construct low-noise detectors and amplifiers. Additional insight is gained when analyzing a general amplifier using a linear response formalism. The basic relations quantifying the measurement gain and backaction obtained using this approach are developed. We apply the formalism to a common voltage amplifier.

In this chapter, we also analyze the case of a resonant cavity coupled to a quantum spin. In this setup, the state of the spin results in a dispersive shift, as described in previous chapters, of the cavity. The interaction between the cavity and spin can be described in terms of two rates: the measurement rate Γ_m, and the dephasing rate Γ_ϕ, which parameterizes the quantum backaction. We then apply a similar analysis to a quantum point contact. It turns out that the quantum limit is reached when the ratio of these quantities is unity, describing an ideal setup where information is being extracted at the same speed as the system is being perturbed. This fundamental principle can be applied in general to other measurement systems. It also illustrates how more complex devices can appear to have additional noise if there are modes in the detector that have quantum backaction but whose information content is either inaccessible or unmonitored.

7.1 Quantum Noise and Measurement Limits

Adding Half a Photon

A very compact and general derivation of the standard quantum limit for linear amplifiers (Caves, 1982) starts by considering a narrow-band detector operating at frequency ω and consisting of bosonic modes at the input and output. We can describe these modes by annihilation and creation operators, $\hat{a}_{in}, \hat{a}_{out}$ and $\hat{a}_{in}^{\dagger}, \hat{a}_{out}^{\dagger}$, respectively. By construction, these operators must observe the canonical commutation relations $[\hat{a}_{in}, \hat{a}_{in}^{\dagger}] = [\hat{a}_{out}, \hat{a}_{out}^{\dagger}] = 1$. Now, we envision that the amplifier operates with some photon number gain G, which drains power supplied to the amplifier through an auxiliary port so as to magnify the input signal such that $\hat{a}_{out} = \sqrt{G}\hat{a}_{in}$ and $\hat{a}_{out}^{\dagger} = \sqrt{G}\hat{a}_{in}^{\dagger}$. We immediately see that our simple two-mode description of the amplifier is inadequate since the canonical commutation relations for $\hat{a}_{in}$ and $\hat{a}_{out}$ cannot be satisfied for $G > 1$.

To proceed, we posit that there must be another operator, which we label by $\hat{\mathcal{F}}$, that is needed to power and operate the amplifier. This additional degree of freedom is also an extra source of fluctuations. We now add the contribution of $\hat{\mathcal{F}}$ to the amplifier output $\hat{a}_{out}$. Thus, $\hat{a}_{out} = \sqrt{G}\hat{a}_{in} + \hat{\mathcal{F}}$ and $\hat{a}_{out}^{\dagger} = \sqrt{G}\hat{a}_{in}^{\dagger} + \hat{\mathcal{F}}^{\dagger}$. We now calculate the commutation relation for the output mode $\hat{a}_{out}$ in terms of $\hat{a}_{in}$ and $\hat{\mathcal{F}}$ and obtain

$$[\hat{a}_{out}, \hat{a}_{out}^{\dagger}] = [\sqrt{G}\hat{a}_{in} + \hat{\mathcal{F}}, \sqrt{G}\hat{a}_{in}^{\dagger} + \hat{\mathcal{F}}^{\dagger}] \tag{7.1}$$
$$= G[\hat{a}_{in}, \hat{a}_{in}^{\dagger}] + [\hat{\mathcal{F}}, \hat{\mathcal{F}}^{\dagger}] + \sqrt{G}[\hat{a}_{in}, \hat{\mathcal{F}}^{\dagger}] + \sqrt{G}[\hat{\mathcal{F}}, \hat{a}_{in}^{\dagger}].$$

We make the assumption here that there are no correlations between the input $\hat{a}_{in}$ and the operator $\hat{\mathcal{F}}$. As such, commutators involving these two operators vanish. Requiring $[\hat{a}_{out}, \hat{a}_{out}^{\dagger}]$ to be unity yields

$$[\hat{\mathcal{F}}, \hat{\mathcal{F}}^{\dagger}] = 1 - G. \tag{7.2}$$

To quantify the amount of noise added by the act of amplification, we can calculate the variances of the input and output fields. We define the variance of input operator $\hat{a}_{in}$ using the following symmetrized expression, which can be compactly expressed in terms of the usual anti-commutator,

$$(\Delta\hat{a}_{in})^2 = \left\langle \frac{\hat{a}_{in}\hat{a}_{in}^{\dagger} + \hat{a}_{in}^{\dagger}\hat{a}_{in}}{2} \right\rangle - |\langle \hat{a}_{in} \rangle|^2 = \frac{1}{2}\langle \{\hat{a}_{in}, \hat{a}_{in}^{\dagger}\} \rangle - |\langle \hat{a}_{in} \rangle|^2. \tag{7.3}$$

An analogous expression can be written for the output mode $\hat{a}_{out}$. Expressing the variance of $\hat{a}_{out}$ in terms of $\hat{a}_{in}$ and $\hat{\mathcal{F}}$, we obtain,

$$(\Delta\hat{a}_{out})^2 = G(\Delta\hat{a}_{in})^2 + (\Delta\hat{\mathcal{F}})^2. \tag{7.4}$$

The second term can be simplified using the definition of the variance in terms of the anti-commutator given and noting that the mean of $\hat{\mathcal{F}}$, which is associated with added noise at the output, can be taken to be zero. This results in the following expression for the variance of $\hat{a}_{out}$,

$$(\Delta \hat{a}_{out})^2 = G(\Delta \hat{a}_{in})^2 + \frac{1}{2}\langle\{\hat{\mathcal{F}}, \hat{\mathcal{F}}^\dagger\}\rangle. \tag{7.5}$$

We can further simplify the second term on the right-hand side by explicitly expanding the anti-commutator and using the triangle inequality, which states that for two complex numbers u and v, $|u| + |v| \geq |u + v|$. Expressing in terms of the commutator of $\hat{\mathcal{F}}$, we obtain

$$(\Delta \hat{a}_{out})^2 \geq G(\Delta \hat{a}_{in})^2 + \frac{1}{2}|\langle[\hat{\mathcal{F}}, \hat{\mathcal{F}}^\dagger]\rangle| \geq G(\Delta \hat{a}_{in})^2 + |G - 1|/2. \tag{7.6}$$

If we divide the output by the gain and take the limit $G \gg 1$, we obtain the well-known result that a linear, phase-preserving amplifier adds 1/2 photon of noise referenced to the input signal (Caves, 1982).

This result is typically called the standard quantum limit on measurement sensitivity. It is important to note again that this mandatory added noise does not apply when measuring a single quadrature of an input signal, in the same way that measuring either the position or the momentum by itself does not trigger restrictions on account of the uncertainty principle. While a fundamental and powerful result, the above derivation does not *a priori* give one guidance on how to construct a measurement device that saturates the lower bound with respect to the noise added by $\hat{\mathcal{F}}$. We can get some insight, however, by considering a minimal model for $\hat{\mathcal{F}}$ that consists of a single bosonic mode $\hat{b}$ at zero temperature, such that $\hat{\mathcal{F}} = \sqrt{G-1}\hat{b}^\dagger$ and $\hat{\mathcal{F}}^\dagger = \sqrt{G-1}\hat{b}$. This formulation results in the minimum added noise of half a photon – adding more modes introduces additional noise. In the next chapter, we explicitly derive the equations governing the operation of a parametric amplifier and show that it exhibits this type of mode structure and is in principle able to reach the quantum limit of sensitivity. It is important to note that this type of scattering approach is well suited to optical devices, particularly when reflections from the input of the detector can be ignored. In the next section, we introduce linear response theory, common in solid-state physics, which is convenient for describing both noise added to the output of a detector and the quantum backaction on the source being measured.

7.2 Linear Response Theory

To model an amplifier using linear response theory (Clerk et al., 2010), we start by first considering the general approach to tackling a Hamiltonian with a static component and a small time-varying perturbation. We can write such an operator as

$$\hat{H} = \hat{H}_0 + \theta(t - t_0)\delta\hat{H}(t), \tag{7.7}$$

where $\hat{H}_0$ is time-independent and θ is the Heaviside step function whose value is unity for a positive argument, zero otherwise. We now consider an operator $\hat{A}$ associated with a physical observable and calculate its expectation value. For time $t < t_0$, we can describe the state of the system as an equilibrium distribution of the energy eigenstates associated with $\hat{H}_0$, $|\alpha\rangle$ with energies E_α^0,

$$\langle\hat{A}\rangle_0 = \frac{1}{Z_0}\sum_\alpha e^{-\beta E_\alpha^0}\langle\alpha|\hat{A}|\alpha\rangle, \tag{7.8}$$

where β is the inverse temperature in energy units and $Z_0 = \sum_\alpha e^{-\beta E_\alpha^0}$ is the partition function.

For time $t > t_0$, we must now take into account the perturbation resulting from $\delta\hat{H}(t)$. For a small perturbation, we can assume that the distribution of thermal weights has not changed, and can, in the Schrödinger representation, simply write our expectation value in terms of time-evolved states,

$$\langle\hat{A}(t)\rangle = \frac{1}{Z_0}\sum_\alpha e^{-\beta E_\alpha^0}\langle\alpha(t)|\hat{A}|\alpha(t)\rangle. \tag{7.9}$$

It becomes convenient to work in the interaction representation where one removes the time evolution due to $\hat{H}_0$. In this case, the states $|\alpha(t)\rangle_I = \hat{U}(t, t_0)|\alpha(t_0)\rangle_I$ where $\hat{U}$ propagates the state under the action of $\delta\hat{H}(t)$ from the time the perturbation is turned on at t_0. Operators also evolve in this representation and $\hat{A}_I(t) = e^{i\hat{H}_0 t}\hat{A}e^{-i\hat{H}_0 t}$. Rewriting the expression for the time-dependent expectation value of $\hat{A}$ in the interaction representation, we obtain

$$\langle\hat{A}(t)\rangle = \frac{1}{Z_0}\sum_\alpha e^{-\beta E_\alpha^0}{}_I\langle\alpha(t_0)|\hat{U}^\dagger(t, t_0)\hat{A}_I\hat{U}(t, t_0)|\alpha(t_0)\rangle_I. \tag{7.10}$$

To simplify this equation, we first note that at $t = t_0$, the eigenstates of the system are simply those due to the static Hamiltonian $\hat{H}_0$. This is readily seen by writing out the explicit time dependence of $|\alpha(t_0)\rangle_I = e^{i\hat{H}_0 t_0}|\alpha(t_0)\rangle = e^{i\hat{H}_0 t_0}e^{-i\hat{H}_0 t_0}|\alpha\rangle$. Next, we can express $\hat{U}(t, t_0)$ in integral form to first order in $\delta\hat{H}$ as $\mathbb{1} - \frac{i}{\hbar}\int_{t_0}^t \delta\hat{H}(t')dt'$. We can substitute this relation into the equation above and expand to linear order. This gives us

$$\langle\hat{A}(t)\rangle = \frac{1}{Z_0}\sum_\alpha e^{-\beta E_\alpha^0}\langle\alpha|\hat{A}|\alpha\rangle + i/\hbar\frac{1}{Z_0}\sum_\alpha e^{-\beta E_\alpha^0}\int_{t_0}^t dt'\langle\alpha|\delta\hat{H}(t')\hat{A}_I - \hat{A}_I\delta\hat{H}(t')|\alpha\rangle. \tag{7.11}$$

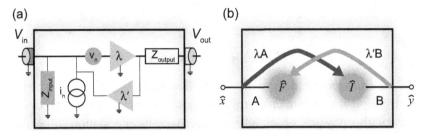

Figure 7.2 Illustration of linear response theory. (a) Electrical circuit representation of a classical amplifier. With an input impedance Z_{input}, an input voltage V_{in} gets amplified by λ and generates an output voltage V_{out} across an output impedance Z_{output}. Fluctuations at the input of the amplifier that contribute to the output noise are modeled by a voltage source V_{in}. The amplifier also has some gain in the reverse direction parameterized by λ'. Fluctuations at the output that couple noise to the amplifier input are modeled as a current source i_{in}. (b) The basic quantum mechanical model of a linear amplifier. Operators $\hat{I}$ and $\hat{F}$ represent the generalized output measurement current and the quantum backaction force asserted on the input, respectively. These operators couple to input and output properties $\hat{x}$ and $\hat{y}$ respectively via coupling constants A and B. In this model, the input variable is amplified by a gain λ and coupled to the output operator $\hat{I}$ and recorded by way of measurement of $\hat{y}$. The output circuit can couple to the signal source by way of a reverse gain λ' that affects the input operator $\hat{F}$ and exerts a force on the physical property described by $\hat{x}$.

We recognize the first term on the right-hand side of the equation as the expectation value with respect to the static Hamiltonian. Writing the two terms in the integrand as a commutator, we obtain the Kubo formula for the first-order change in the expectation value of an operator under a perturbation $\delta\hat{H}$ (Mahan, 2013),

$$\langle\hat{A}(t)\rangle = \langle\hat{A}\rangle_0 - \frac{i}{\hbar}\int_{t_0}^{t} dt'\,\langle[\hat{A}_I(t), \delta\hat{H}(t')]\rangle_0. \tag{7.12}$$

Note that the states that are used to calculate the expectation values are the unperturbed eigenstates, as is the case in other formulations of first-order perturbation theory in quantum mechanics. We also remark here that the product structure in the expectation values can be readily related to correlation functions and spectral densities. The quantum spectral density of an operator $\hat{F}(t)$ is defined, in close analogy with classical functions, as $S_{FF}(\omega) = \int_{-\infty}^{+\infty} dt\, e^{i\omega t}\langle\hat{F}(t)\hat{F}(0)\rangle$. We will make use of these relations to write expressions for the gain of a linear amplifier and quantum constraints on noise.

Now, to start analyzing a quantum amplifier, we consider the basic model illustrated in Fig. 7.2. In panel (a), a basic equivalent circuit for a classical measurement device that adds noise of two flavors is shown. A voltage source models noise added to the output signal measured from the amplifier while a current source acts on the source of the signal itself. The quantum translation of this model is shown in panel (b) where the operators $\hat{I}$ and $\hat{F}$ represent the generalized measurement current and the quantum backaction force asserted on the object, respectively. These operators couple to input and output circuits by way of physical properties described by $\hat{x}$ and $\hat{y}$ with coupling strengths A and B, respectively. In this model, the input variable $\hat{x}$ produces a signal that is amplified by a gain λ and is coupled to the output operator $\hat{I}$ and recorded by way of measurement of $\hat{y}$. The output circuit can couple to the signal source by way of a reverse gain λ' that affects the input operator $\hat{F}$ and exerts a force on the physical property described by $\hat{x}$. As we shall see shortly, it is desirable to eliminate the reverse gain so that additional backaction from the output measurement circuit does not add noise to the amplification process beyond that mandated by quantum mechanics. Note that in the presence of reverse gain, even vacuum fluctuations at the output are a nonnegligible source of noise for a quantum-limited measurement device.

Let us start by proposing a linear interaction Hamiltonian that captures the coupling of the input variable $\hat{x}$ and the amplifier input port, described by $\hat{F}$, thus giving $\hat{H}_{int} = \epsilon \hat{x} \hat{F}$, where ϵ is taken to be small enough to justify linearization. Following the general pattern outlined in the derivation of the Kubo formula, we first calculate the time-dependent expectation value of the amplifier output operator $\hat{I}$,

$$\langle \hat{I}(t) \rangle = \langle \hat{I}(t) \rangle_0 + \epsilon \lambda \langle \hat{x} \rangle_0. \tag{7.13}$$

Here, the first term describes the signal at the amplifier output in the absence of an input signal, and the second term describes the perturbation generated by the input signal. Both expectation values are taken with respect to the equilibrium distributions of the amplifier and the source. Applying the Kubo formula, we can express the gain in terms of $\hat{I}$ and $\hat{F}$,

$$\lambda = \frac{-i}{\hbar} \int_0^\infty dt' \langle [\hat{I}(t'), \hat{F}(0)] \rangle_0. \tag{7.14}$$

Similarly, we can define the reverse gain λ',

$$\lambda' = \frac{-i}{\hbar} \int_0^\infty dt' \langle [\hat{F}(t'), \hat{I}(0)] \rangle_0. \tag{7.15}$$

To link these linear response coefficients to noise spectral densities, we can expand the commutators and express them in terms of the imaginary part of a correlation function, which for the forward gain becomes

$$\lambda = \operatorname{Im} \frac{2}{\hbar} \int_0^\infty dt' \langle \hat{I}(t') \hat{F}(0) \rangle_0. \tag{7.16}$$

With the gains defined, we can now turn our attention to the noise processes present in amplification. We start by defining a series of spectral densities that characterize fluctuations in the operators $\hat{I}$ and $\hat{F}$, as well as cross-correlations. For simplicity and for catering to situations most commonly encountered in measurement, we can look at the zero frequency value of the noise correlators, assuming that the detector is able to instantly respond to variations in the input signal. The reason this zero-frequency assumption is important is that we want the detector to be able to respond much faster than any timescale of the measured system of interest. More precisely, we assume there is a large separation of timescales between the correlation time of the detector and the dynamical timescales of the system. This point will become clearer in the examples given in the following sections. We thus have

$$S_{II} = 2 \int_{-\infty}^{+\infty} dt' \, \langle \hat{I}(t')\hat{I}(0)\rangle_0, \tag{7.17}$$

$$S_{FF} = 2 \int_{-\infty}^{+\infty} dt' \, \langle \hat{F}(t')\hat{F}(0)\rangle_0, \tag{7.18}$$

$$S_{IF} = 2 \int_{-\infty}^{+\infty} dt' \, \langle \hat{I}(t')\hat{F}(0)\rangle_0, \tag{7.19}$$

$$S_{FI} = 2 \int_{-\infty}^{+\infty} dt' \, \langle \hat{F}(t')\hat{I}(0)\rangle_0. \tag{7.20}$$

Note that here we have implicitly subtracted the mean from the expectation values of $\hat{I}$ and $\hat{F}$.

These spectral densities can be used to describe the noise at the output and input of an amplifier. We pen a measurement rate $\Gamma_M = \epsilon^2\lambda^2/S_{II}$, which essentially factors in the coupled signal gain and a diminution in information extraction due to the presence of noise. This expression can be derived in a number of ways, including a direct calculation of the measurement rate of a qubit or by a more general information-theoretic approach (Clerk et al., 2003; Averin and Sukhorukov, 2005). Correspondingly, a quantum measurement must induce fluctuations in the variable conjugate to the input. We can capture this as a generalized dephasing process $\Gamma_\phi = \epsilon^2 S_{FF}/\hbar^2$. We see that the quantum backaction due to fluctuations in $\hat{F}$ dephases the system, with ϵ again quantifying the coupling between the input and the amplifier. For an ideal measurement, we would measure as fast as we dephase, and thus $\Gamma_M/\Gamma_\phi = 1$. Any entanglement to degrees of freedom that are not measured would cause an increased level of dephasing while providing the experimentalist no additional information, and hence increase the noise associated with the amplifier.

Using the preceding expressions, the ratio of the measurement and dephasing rates is compactly given by

$$\frac{\Gamma_M}{\Gamma_\phi} = \frac{\hbar^2 \lambda^2}{S_{II} S_{FF}}. \tag{7.21}$$

In general, additional processes result in this ratio being less than 1. An illustrative way to see this involves comparing the product of S_{II} and S_{FF} to their cross-correlator using the Cauchy–Schwartz inequality, which expresses the idea that the cross-correlations can at most be as strong as the product of the self-correlations,

$$S_{II} S_{FF} \geq |S_{IF}|^2 = \hbar^2 (\lambda - \lambda')^2 + (\mathrm{Re}\, S_{IF})^2. \tag{7.22}$$

Substituting into the previous equation, we obtain an important relation that elegantly expresses the standard quantum limit for an amplifier:

$$\frac{\Gamma_M}{\Gamma_\phi} = \frac{\hbar^2 \lambda^2}{\hbar^2 (\lambda - \lambda')^2 + (\mathrm{Re}\, S_{IF})^2} \leq 1. \tag{7.23}$$

Examining this relation, we see that to reach the maximum ratio of 1, we need (i) the reverse gain λ' and (ii) the real part of the cross-correlator S_{IF} to vanish. These conditions describe a situation when noise from the output circuit does not induce backaction and there is a great deal of similarity between the states of $\hat{I}$ and $\hat{F}$, indicating that there are no extraneous internal degrees of freedom that could extract information from the measurement circuit. More details are given in the literature (Averin, 2003; Korotkov and Averin, 2001; Jordan and Büttiker, 2005).

7.3 Quantum Limited Pointer States

We now go ahead and apply the concepts we have developed so far to analyze a few common measurement devices and protocols used in solid-state quantum information systems. The basic archetype involves first coupling a quantum system to some type of pointer state that typically converts the quantum degree of freedom to a directly measurable quantity. The pointer variable is then amplified and recorded by classical electronics. In the electrical domain, one can perform a dispersive or dissipative measurement; we give here an example for each. As an example of a dispersive measurement, we detail the operation of a resonant cavity coupled to a qubit. The frequency of the detector changes in a digital fashion dependent on the qubit state and is probed by way of microwave or optical frequency reflectometry. For the dissipative case, we discuss the operation of a quantum point contact detector where capacitively coupling a semiconductor-based qubit to a mesoscopic conductor changes the conductance of the latter. Quantum state readout of the qubit can be performed by applying a static voltage to the point contact detector and monitoring the time-dependent electrical current.

Resonant Cavity Detector

A very common measurement architecture involves parametrically coupling a quantized system to a high-quality-factor (Q) cavity. In a typical dispersive measurement scheme, the resonant frequency of the cavity is shifted depending on the state of the quantum system. Depending on the specific circuit topology to be used, one then illuminates the cavity with a measurement signal near resonance and records changes in the scattered amplitude and/or phase. The interaction Hamiltonian can be simply expressed as

$$H_{int} = \hbar\omega \, A\hat{z}\hat{a}^\dagger\hat{a}, \tag{7.24}$$

where A parameterizes the coupling strength; $\hat{z}$ is the input quantum variable, such as the state of a qubit, which can be expressed as a Pauli operator, say $\hat{\sigma}_z$; and the creation and annihilation operators $\hat{a}, \hat{a}^\dagger$ describe a bosonic mode of the resonant cavity.

Much like with a quantum point contact, we can treat the cavity as a scattering element. We assume that the dynamics of the quantum system are slow relative to the relaxation rate of the cavity. Let us consider a single-sided cavity where one mirror is partially transmissive and the other is fully reflective. For an incoming wavepacket incident from the left, there is complete reflection of the input power, albeit with a phase shift, which results from the combination of waves that are reflected from each surface. The reflection coefficient r can be readily calculated using simple scattering theory and is given by

$$r = \frac{-(1 + 2iAQ\hat{z})}{(1 - 2iAQ\hat{z})}. \tag{7.25}$$

As expected, this quantity has unity modulus, and the decomposition into polar coordinates yields a quantum state–dependent phase shift that can be approximated as $\theta \approx 4QA\hat{z}$. As a note, this quantity can also be related to the Wigner–Smith delay matrix described later for a point contact, and it simply reduces to the Wigner delay time for the cavity, $4/\kappa$, where $\kappa = \omega/Q$ is the cavity decay rate. The phase shift is then simply the cavity frequency shift multiplied by this delay time.

To probe the cavity response, we illuminate it with a coherent state beam with photon number N outside the cavity. The photon number inside the cavity is n. We denote the flux of the incident beam as $\dot{N}$, and averages are denoted with a bar. A coherent state obeys Poisson statistics and thus the variance of the number of photons in the beam is related to the average, $(\Delta N)^2 = \bar{N}$. For large $\bar{N}$, we can readily translate the usual position-momentum uncertainty relation into one for the number and phase of a coherent state. This is expressed as

$$\Delta N \Delta \theta = 1/2. \tag{7.26}$$

Let us now calculate an uncertainty relation for the product of the spectral density of fluctuations in phase and photon number. The variance of the phase is simply expressed as $(\Delta \theta)^2 = 1/(4\bar{N})$. When detecting photons in such a beam, the variance in the photon number grows linearly in time, $(\Delta N)^2 = S_{\dot{N}\dot{N}}t$, where the spectral density of photon number flux fluctuations is directly related to the average flux,

$$S_{\dot{N}\dot{N}} = \bar{\dot{N}}. \tag{7.27}$$

For the phase, we can consider a homodyne measurement where the oscillation electromagnetic field is multiplied by a local oscillator of the same frequency. In this scheme, one can measure either the amplitude or phase of the coherent state. When measuring the phase, which has a value of θ_0, say, we consider an output signal I, integrated over a time t, to be given by $I = \theta_0 t + \int_0^t d\tau \delta\theta(\tau)$. Here $\delta\theta(\tau)$ represents the fluctuations in θ due to shot noise in the output field. The variance of the phase $(\Delta \theta)^2$ decreases with time and can be expressed through the spectral density of phase fluctuations to yield

$$(\Delta \theta)^2 = S_{\theta\theta}/t = 1/(4\bar{\dot{N}}t). \tag{7.28}$$

Solving for the spectral density of phase fluctuations and multiplying by the spectral density of number fluctuations just obtained, we arrive at the following relation,

$$S_{\theta\theta} S_{\dot{N}\dot{N}} = \frac{1}{4}. \tag{7.29}$$

Armed with these relations, let us now apply them to the response of a single-sided resonant cavity. Fluctuations in the phase result in an imprecision noise at the output when trying to measuring the quantum observable $\hat{z}$ given by

$$S_{zz} = S_{\theta\theta} \left(\frac{\kappa}{4A\omega}\right)^2. \tag{7.30}$$

In addition to noise in the output signal, quantum mechanics dictates that there will be a backaction force, with its own fluctuations, in the variable conjugate to $\hat{z}$. We can express this force by negating the derivative of the interaction Hamiltonian with respect to $\hat{z}$. The random component of this force is thus given by $-A\hbar\omega(\hat{n} - \bar{n})$. The corresponding spectral density of backaction fluctuations, S_{FF}, is therefore

$$S_{FF} = (A\hbar\omega)^2 S_{nn}, \tag{7.31}$$

where S_{nn} is the spectral density of photon number fluctuations in a cavity and can be expressed in terms of the cavity decay rate,

$$S_{nn}(\omega) = 2 \int_{-\infty}^{\infty} dt \langle n(t)n(0) \rangle,$$

$$= 2\bar{n} \int_{-\infty}^{\infty} dt \exp(-\kappa|t|/2 - i(\omega + \Delta)t),$$

$$= \frac{\bar{n}\kappa}{(\omega + \Delta)^2 + (\kappa/2)^2}. \tag{7.32}$$

Here Δ is the detuning from resonance. The cavity essentially filters the photon shot noise, and at frequencies much smaller than the inverse decay rate,

$$S_{nn} \approx 4\bar{n}/\kappa. \tag{7.33}$$

We can calculate the mean photon number in the cavity by multiplying the incident photon flux and the delay time in the cavity, yielding $\bar{n} = 4\bar{N}/\kappa$. We can substitute this result into the preceding relation for the spectral density of backaction fluctuations to obtain

$$S_{FF} = (4A\hbar\omega/\kappa)^2 S_{\dot{N}\dot{N}}. \tag{7.34}$$

The product of $S_{FF}S_{zz}$ equals $\hbar^2/4$. This result is a very important one and affirms that a parametrically coupled resonant cavity can function as quantum-noise limited detector – the product of these noise functions is minimal.

We can gain some additional practical insight by catering to the case where a cavity is used to measure the state of a qubit. In that setting, the observable $\hat{z}$ takes on the values ± 1, and the scattered signal phase $\theta_0 = \pm 4A\omega/\kappa$. The variance of the homodyne output signal I can again be expressed in terms of phase fluctuations, $(\Delta I)^2 = S_{\theta\theta}t$. Dividing the square of the mean value of the output signal to its variance defines a power signal-to-noise ratio equal to $\theta_0^2 t/S_{\theta\theta}$. Here we have defined the mean of the output signal as $\theta_0 t$. Note that as time advances, the signal grows linearly in t and the fluctuations as $\sqrt{t}$, resulting in an overall improvement in measurement accuracy with time. The measurement rate Γ_M can be related to this signal-to-noise ratio by calculating the rate at which information is obtained from the cavity, yielding

$$\Gamma_M = \frac{\theta_0^2}{2S_{\theta\theta}} = \frac{1}{2S_{zz}}. \tag{7.35}$$

Now, let us focus on the measurement-induced dephasing. We can make use of the result developed in the beginning of the chapter where we connected the dephasing rate Γ_ϕ to the spectral density of backaction fluctuations, assuming that the accumulated phase after a long time is Gaussian distributed. For the present case,

$$\Gamma_\phi = \frac{2}{\hbar^2} S_{FF} = 2\theta_0^2 S_{\dot{N}\dot{N}}. \qquad (7.36)$$

The ratio of the dephasing rate to the measurement rate is thus

$$\frac{\Gamma_\phi}{\Gamma_M} = \frac{4}{\hbar^2} S_{zz} S_{FF} = 1, \qquad (7.37)$$

again representing the minimum uncertainty mandated by quantum mechanics where one dephases as fast as one measures.

The setup we have been describing, in an ideal implementation, wastes no information. There is no dissipation to carry away valuable signal into inaccessible modes or coupling to other degrees of freedom that are not monitored. About this last point, this fact becomes apparent if we consider the case of a two-sided cavity where the incident photon beam is not only fully reflected but also partially transmitted. Here the qubit-induced phase shift is a factor of two smaller than the case of a single-sided cavity, and the measurement rate is one half of the dephasing rate, causing the detector to miss the quantum limit of sensitivity by a factor of two. In this case, the reflected signal carries half of the total information. An interesting trick is that one can still approach the quantum limit in a transmission measurement by making one of the cavity mirrors highly opaque, but not fully reflective. In this case, very few photons enter the cavity but nearly all of the ones that do exit carry the full amount of information available. One has to illuminate the cavity with a large enough photon flux to obtain a sufficiently intense transmitted signal.

Quantum Point Contact

Let us consider a typical lateral quantum dot geometry, as illustrated previously in Fig. 5.2. A two-dimensional layer of mobile electrons (electron gas) is produced by modulation doping of a planar semiconductor heterostructure, and gate electrodes are used to isolate a pool or "dot" of electrons of varying number. To reduce the energy scale and achieve finer control of the electronic level structure, a double-dot structure can be employed where two individual dots are capacitively coupled and can be tuned into and out of resonance to control the avoided level crossing. The double-dot itself is then coupled to a one-dimensional conductor, also defined by way of lateral metallic gate electrodes.

To evaluate the quantum efficiency of this one-dimensional conductor or point contact as a detector, we calculate its measurement and dephasing rates, as defined in the previous section. We can employ a scattering approach to determine the effect of the electronic state of the double-dot on the point contact. We first concentrate on calculating the dephasing induced on the double-dot qubit by the detector, resulting from its charge fluctuations. The point contact is coupled to two electronic reservoirs labeled by α, β, previously labeled by S and D in Chapter 5. The scattering matrix $S_{\alpha\beta}$ couples the incoming and outgoing states of the point contact. Restricting oneself to considering slow charge fluctuations, the time delays associated with a wavepacket traversing the point contact can be associated with the eigenvalues of the so-called Wigner–Smith time delay or density-of-states matrix,

$$\mathcal{N}_{\delta\gamma} = \frac{1}{2\pi i} \sum_\alpha S_{\alpha\delta}^\dagger \frac{dS_{\alpha\gamma}}{dE}, \tag{7.38}$$

where E is the energy of the wavepacket and the indices refer to the reservoirs in the point contact. The trace of the matrix describes the density of states N of the point contact, and the off-diagonal elements describe charge fluctuations (Wigner, 1955; Smith, 1960; Büttiker et al., 1993; Blanter and Büttiker, 2000).

Specifically, we can express the spectral density of charge fluctuations at frequency ω and zero applied voltage as $S_{QQ} = 2\hbar|\omega|R_q C_\mu^2$. The quantity C_μ is the so-called electrochemical capacitance, which has contributions from both the density of electronic states and the geometric capacitance, $C_\mu^{-1} = C^{-1} + (e^2 N)^{-1}$ where e is the electron charge. The fluctuations are set by the effective resistance R_q, which is related to $\mathcal{N}$ via the following relation,

$$R_q = \frac{h}{2e^2 N^2} \sum_{\gamma\delta} \mathrm{Tr}\, \mathcal{N}_{\gamma\delta} \mathcal{N}_{\gamma\delta}^\dagger. \tag{7.39}$$

Note that the quantity $R_q C_\mu$ determines the charge relaxation time of the conductor. Similar to the preceding expression, we can express the charge fluctuations due to the presence of an applied voltage V in terms of an effective resistance R_v, $S_{QQ} = 2e|V|R_v C_\mu^2$. Here, R_v is given by

$$R_v = \frac{h}{e^2 N^2} \mathrm{Tr}\, \mathcal{N}_{21} \mathcal{N}^\dagger{}_{21}. \tag{7.40}$$

Note that only the off-diagonal elements are involved when considering transport effects. Given explicit expressions for the spectral density of charge fluctuations in the point contact, we can now combine the two preceding into an expression for the dephasing rate,

$$\Gamma_\phi = \left(\frac{e}{\hbar}\right)^2 \frac{\epsilon^2}{\Omega^2} \frac{C_\mu^2}{C_i^2} (R_q kT + R_v e|V|). \tag{7.41}$$

Here ϵ is the energy difference between the two dots in the qubit and $\Omega = \sqrt{\epsilon^2 + \Delta^2}$ is the full qubit splitting with Δ being the tunnel coupling between the two dots. We see that there is a contribution from both the temperature and the voltage bias across the point contact. For starters, a good detector must overcome the thermal noise, so $eV \gg kT$. Further conditions will be discussed shortly.

Moving on now to the measurement rate, we can express Γ_M in terms of the current in the point contact I as

$$\Gamma_M = \frac{(\Delta I)^2}{4 S_{II}}. \tag{7.42}$$

We remind the reader that $\Delta I = I_L - I_R$ is the average current difference depending on the two qubit states. The spectral density S_{II} is given by the shot noise of a mesoscopic conductor with n conduction channels of transmission probability T_n and reflection probability $R_n = 1 - T_n$. We thus generalize the results in Chapter 5 to many channels,

$$S_{II} = e|V| \frac{e^2}{\pi \hbar} \sum_n R_n T_n. \tag{7.43}$$

The modulation of the transport current ΔI due to the qubit state is expressed in terms of the change in the potential ΔU of the point contact,

$$\Delta I = \frac{e^2}{h} |V| \sum_n \frac{dT_n}{dE} (e \Delta U). \tag{7.44}$$

The change in potential ΔU is given by

$$\Delta U = \frac{C_\mu}{eN(C_i - C_\mu)}. \tag{7.45}$$

Thus, in the limit that $C_i \gg C_{1,2}$, the measurement rate equals

$$\Gamma_M = \frac{C_\mu^2}{C_i^2} R_m \frac{e|V|}{\hbar}, \tag{7.46}$$

where R_m is given by

$$R_m = \frac{1}{16\pi N^2} \frac{\left(\sum_n \frac{dT_n}{dE} \right)^2}{\sum_n R_n T_n}. \tag{7.47}$$

With expressions for the dephasing and measurement rates in hand, we can ask the question under what conditions does one realize an ideal detector with the former approaching the latter. For starters, we require the decoherence due to thermal energy to be negligible, thus mandating that $kT \ll e|V|$. We also want Δ to vanish. Additional constraints arise from a deeper examination of the scattering matrix S. In a nutshell, information should be channeled into scattering parameters that are probed. As such, the presence of time-reversal symmetry, inversion symmetry in the scattering potential of the quantum point contact, and separability of the potential into spatially independent pieces aid in this respect. The last condition is particularly important in suppressing chaotic behavior in systems with multiple transmission channels. For a saddle-point potential for the point contact, these quantities are evaluated in (Pedersen et al., 1998). Further discussion and interpretation of these scattering quantities is given in references (Pilgram and Büttiker, 2002; Clerk et al., 2003).

Exercises

Exercise 7.3.1 Consider two detectors measuring the same quantum observable. Average the detector outputs: $I = (I_1 + I_2)/2$. Using linear response theory, show that on one hand, the measurement is faster, but on the other hand, for statistically independent detectors, the measurement-induced dephasing rate is simply the sum of the individual dephasing rates. Show the detection scheme is efficient for twin detectors that are themselves quantum limited.

Exercise 7.3.2 Show that, for a continuously monitored qubit undergoing Rabi oscillations, the time autocorrelation function $K = \langle I(\tau + t)I(t)\rangle$ of the detector output is given by

$$K(\tau) = S_{II}\delta(\tau) + \frac{\lambda^2}{4}\text{Tr}[\rho\sigma_z(0)\sigma_z(\tau)]. \tag{7.48}$$

Exercise 7.3.3 Use the unconditional master equation to calculate the preceding correlation function. Show that in a Fourier-transformed frequency picture, the maximum value of the noise spectral density is bounded as $S_{max} \leq 4S_I$, the so-called Korotkov–Averin bound (Korotkov and Averin, 2001; Averin, 2001). Hint: Use linear response theory.

Exercise 7.3.4 Show that by using the twin detectors described just described, their cross-correlation can violate the Korotkov–Averin bound.

Exercise 7.3.5 Calculate the ensemble-averaged dephasing rate Γ_ϕ for a saddle-point model of a single-channel quantum point contact, where the transmission function has the form (5.5).

Exercise 7.3.6 Calculate the measurement rate Γ_M for a saddle-point model of a single-channel quantum point contact, where the transmission function has the form (5.5).

Exercise 7.3.7 Show that the single-channel quantum point contact, where the transmission function has the form (5.5), is an efficient detector, satisfying the equality (7.23).

8

Quantum Amplification

8.1 Quantum-Noise-Limited Operation

The realization of efficient measurement devices that allow the experimentalist who lives in the classical world to faithfully observe and control quantum phenomena is an active area of research. In the previous chapter, we described means by which information can be mapped onto a pointer that interfaces with the external environment. We also discussed some quantum mechanical limitations that are inherent to the measurement process. Reaching the sensitivity limit imposed by quantum uncertainty on an amplification device is a challenging task that requires – among other criteria – dissipation be engineered carefully to minimize added fluctuations, sufficient gain be present to overcome ambient noise, and information extracted from the quantum system be channeled only into monitored degrees of freedom and thus not be wasted.

To this end, a parametric amplifier, or paramp, is a simple, elegant, pedagogical, and practical device for the quantum engineer. On the theoretical level, the Hamiltonian for the paramp can be readily penned and has the minimum degrees of freedom to reach the quantum limit of sensitivity. It can be operated in a mode that adds a half photon of noise when the phase of the input signal is faithfully mapped (phase-preserving) onto the output signal, or with no added noise if the gain is phase dependent (phase-sensitive), thereby amplifying one quadrature of an input signal at the expense of suppressing the other. The amplifier, as its name implies, basically uses a nonlinear element to make a parameter of the device dependent on the input signal. For example, the refractive index of a material can vary with power, the frequency of a physical pendulum varies with excitation amplitude, and so on. These basic physical processes can be readily designed into optical and electrical systems to produce high-quality amplifiers.

To expand on the two modes of operation just discussed, recall the minimalistic model of a quantum limited amplifier introduced in the previous chapter

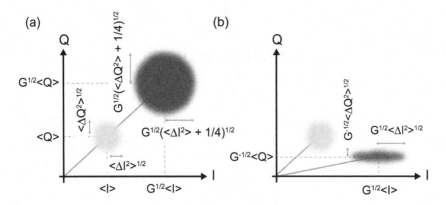

Figure 8.1 Phase-space distributions for (a) phase-preserving and (b) phase-sensitive amplification. A coherent state input signal has quantum fluctuations equally distributed in both quadratures, depicted as a light gray disk at the tip of the phasor. A phase-preserving amplifier increases both the magnitude of the mean signal amplitude in both quadratures while adding an additional half photon of noise. The orientation of the signal in the I-Q plane is preserved. A phase-sensitive amplifier magnifies one quadrature and squeezes the other, preserving the total area of the noise disk. No noise is thus added in the amplification process, at the expense, however, of losing phase information.

where signals at the input and output ports are described by bosonic annihilation (creation) operators $\hat{a}_{in}, \hat{a}_{out}$ $(\hat{a}^{\dagger}_{in}, \hat{a}^{\dagger}_{out})$. An additional mode, $\hat{b}, \hat{b}^{\dagger}$, is needed to satisfy canonical commutation relations and naturally imposes the standard quantum limit. If we take this additional mode $\hat{b}$ to be a distinct mode, then $\hat{a}_{out} = \sqrt{G}\,\hat{a}_{in} + \sqrt{G-1}\,\hat{b}^{\dagger}$ where G is the power gain, and the amplifier has the effect of transforming a coherent state at the input, described as a phasor in Fig. 8.1(a), into a state with a greater amplitude and one where the fluctuations in both the coordinate and momentum quadratures are equally amplified. If the input state only has quantum fluctuations, then the area of the circular region at the terminus of the coherent state phasor is simply given by the Heisenberg uncertainty relation. After amplification, this area grows both due to an enhancement of quantum fluctuations of the input signal, and due to the addition of noise associated with the mode $\hat{b}$. Thus, the amplifier preserves the phase information of the input signal and necessarily adds a half photon of additional noise.

In contrast, we can consider a device where the additional mode is degenerate in frequency with the input signal, resulting in an output field that depends on the signal phase: $\hat{a}_{out} = \sqrt{G}\,\hat{a}_{in} + e^{i\phi}\sqrt{G-1}\,\hat{a}^{\dagger}_{in}$ where ϕ is measured with respect to the pump signal that powers the amplifier. In this situation, a coherent state at the input is squeezed so that the length of the phasor is increased, but the fluctuations at the output are enhanced in one quadrature and suppressed in the other. The amplifier thus does not preserve the phase information of the input signal, but has the advantage of operating with a noise below the standard quantum limit in one quadrature, as shown in Fig. 8.1(b).

In such a device, one typically measures the quadrature components of the output signal using a homodyne setup where the output signal is multiplied by a local oscillator at the same frequency with an adjustable phase. The basic pieces of a homodyne detector are sketched here. We consider two electric fields $\hat{\mathbf{E}}_1$ and $\hat{\mathbf{E}}_2$ incident on a beam slitter with transparency η. We can write the electric fields in terms of bosonic operators $\hat{a}$ and $\hat{b}$: $\hat{\mathbf{E}}_1 = i\mathbf{E}_0(\hat{a}e^{i(\mathbf{k}\cdot\mathbf{r}-\omega t)} - \hat{a}^\dagger e^{-i(\mathbf{k}\cdot\mathbf{r}-\omega t)})$ and $\hat{\mathbf{E}}_2 = i\mathbf{E}_0(\hat{b}e^{i(\mathbf{k}\cdot\mathbf{r}-\omega t)} - \hat{b}^\dagger e^{-i(\mathbf{k}\cdot\mathbf{r}-\omega t)})$. Assuming the same polarization for both waves, the combined electric field can be expressed as $\hat{\mathbf{E}}_{\text{total}} = i\mathbf{E}_0(\hat{c}e^{i(\mathbf{k}\cdot\mathbf{r}-\omega t)} - \hat{c}^\dagger e^{-i(\mathbf{k}\cdot\mathbf{r}-\omega t)})$ where $\hat{c} = \sqrt{\eta}\hat{a} + i\sqrt{1-\eta}\hat{b}$. Note that we have taken the phase associated with reflection into account here. Using a photodetector, we would measure the mean photon number $\langle\hat{c}^\dagger\hat{c}\rangle$. Multiplying through, we get

$$\langle\hat{c}^\dagger\hat{c}\rangle = \eta\langle\hat{a}^\dagger\hat{a}\rangle + (1-\eta)\langle\hat{b}^\dagger\hat{b}\rangle - i\sqrt{\eta(1-\eta)}(\langle\hat{a}\rangle\langle\hat{b}^\dagger\rangle - \langle\hat{a}^\dagger\rangle\langle\hat{b}\rangle). \tag{8.1}$$

If we consider the case that $\hat{\mathbf{E}}_2$ is an intense field that can serve as a reference or local oscillator (LO), we can model that as a coherent state of size β. The mean intensity is then simply

$$\langle\hat{c}^\dagger\hat{c}\rangle = (1-\eta)|\beta|^2 + |\beta|\sqrt{\eta(1-\eta)}\langle\hat{X}_{\theta+\pi/2}\rangle, \tag{8.2}$$

where $\hat{X}_\theta = \hat{a}e^{-i\theta} + \hat{a}^\dagger e^{i\theta}$. If we subtract the intensity associated with the LO, we see that we can measure the intensity of the field quadratures by changing the phase of the pump θ through $\pi/2$. In a typical microwave experiment, one can directly measure the quadratures of total electric field. As a note, measuring the phase of the electric field canonically conjugate to $\hat{c}^\dagger\hat{c}$ is nontrivial and discussed in Chapter 10. Heterodyne detection uses a small frequency mismatch between the LO and the signal to be measured to periodically rotate through all values of θ.

8.2 Superconducting Josephson Tunnel Junction Circuits

We focus in this chapter on superconducting paramps, which can be operated in either phase-preserving or phase-sensitive amplification mode and have recently been used in many quantum information–related experiments. Our goal is to present the basic foundations of the operation and optimization of these devices, leaving detailed calculations to works we will reference. We note, of course, that there is also a rich body of literature on optical parametric amplification in the context of nonlinear quantum optics, and the basic formulae derived here are identical to those found in such works; see, for example, Mandel and Wolf

(1995). The basic starting point is the identification of a nonlinear element. In superconducting devices, one can start with an oscillator in which the inductance is provided by the tunneling of Cooper pairs in a Josephson tunnel junction or an array of such junctions. It is also possible to form nonlinear elements using a constriction or a wire with a large kinetic inductance. We cater here to the tunnel junction case. The choice of nonlinear element allows one to tune the input power needed for amplification and the dynamic range over which linear operation is achieved. Therefore, a particular amplifier architecture can be tailored in an application-specific manner. Crucially, these structures having vanishingly small dissipation at temperatures well below the superconducting transition temperature, and thus have minimal intrinsic dissipation.

Next, the amplifier must satisfy energy and momentum conservation. Typically, devices operate in either three- or four-wave mixing mode. Energy from an intense pump signal at frequency ω_P is transferred by way of the nonlinear element to a weak input signal at frequency ω_s. As shown in Section 7.1 and recapped earlier in this chapter, a basic amplifier must have at least one additional mode, and in this case, it is called the "idler" mode ω_i. In a three-wave mixing process, a single pump photon is converted in signal and idler excitations such that $\omega_P = \omega_s + \omega_i$. In a four-wave mixing process, two pump photons are involved, such that $2\omega_P = \omega_s + \omega_i$. Additionally, either process can have signal and idler modes at the same (degenerate) or different (nondegenerate) frequencies. This gives rise to four basic configurations that are described in Fig. 8.2.

In terms of momentum matching, we consider two basic structures that are used to enhance interaction of the nonlinear element with the input signal: a resonant cavity and a nonlinear transmission line. In the case of the former, momentum matching is readily achieved as the device operates over a narrow frequency band dictated by the properties of the cavity. For the latter, dedicated schemes to efficiently transfer power from the pump wave to the signal along the length of the transmission line must be engineered. Cavity-based amplifiers are well suited for measuring the signal of, say, a single qubit. Traveling-wave devices based on transmission lines have much larger bandwidth and power handling capacity, enabling the detection of many qubits in parallel. Moreover, as one tries to optimize the gain, sensitivity, and dynamic range of a measurement solution for a quantum system with many degrees of freedom, a suitable combination of both types of amplifiers is often needed.

When designing an amplifier, we are typically concerned with a few standard figures of merit that are direct analogues of quantities used in classical signal processing: signal gain, signal bandwidth, added noise, and saturation threshold. The first two quantities describe the frequency range over which amplification occurs and its frequency-dependent magnitude. The latter two describe degradation of the signal due to added noise and a saturation of the power gain when the input signal

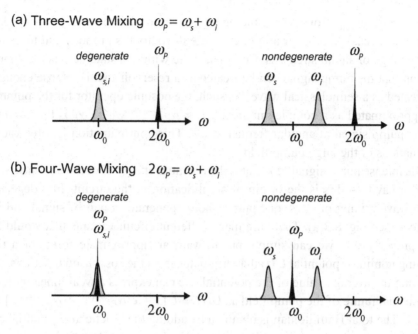

Figure 8.2 The basic operating modes of a parametric amplifier. In (a) three-wave mixing, one pump photon is converted into a signal and an idler photon. In (b) four-wave mixing, two pump photons are converted into a signal and an idler photon. In either scheme, if the signal and idler tones are at different frequencies, the amplifier is nondegenerate. Conversely, a degenerate parametric amplifier has both the signal and idler tones evolving in the same mode.

is too intense, respectively. The basic procedure used here to arrive at these quantities follows from a quantum mechanical approach, followed by approximations. We start by writing a Hamiltonian for the isolated amplifier circuit, simply tallying the energies of the modes involved. We then couple the circuit to the outside world with its fluctuations by way of input-output theory and solve for the dynamics of the circuit using a quantum Langevin or master equation approach. Detailed calculations for Josephson paramps exist in many places (Wahlsten et al., 1978; Mygind et al., 1979; Yurke et al., 1989), and our goal here is to outline the key steps for an archetypal circuit – a degenerate three-wave mixing device – to illustrate the basic operation, design trade-offs, and nonidealities in these amplifiers. Other design variants are enumerated later in the chapter along with examples of recent realizations with Josephson circuits.

8.3 Degenerate Parametric Amplifier

We start by deriving the Hamiltonian of the amplifier (Boutin et al., 2017). The input to the device will consist of a weak signal to be amplified in tandem with an energetic pump from which the energy for amplification will be derived. We consider the case of a linear amplifier where the output is proportional to its input over a range of signal strengths. Commensurate with this statement is the approximation that the pump signal can be treated as a reservoir that is intense enough to be treated as a semiclassical wave. As such, the bosonic operator for the pump can be approximated as a complex number, $\hat{a}_P \approx \alpha_P e^{-i\omega_P t}$ where α_P is the amplitude of the pump signal at angular frequency ω_P. This approximation ignores vacuum fluctuations of the large pump field.

The intense pump signal modulates the potential energy U of the nonlinear oscillator or cavity – this is the origin of amplification in the circuit. In a degenerate three-wave mixing process, one pump photon generates a pair of signal and idler photons; see Fig. 8.3(a). There are many different circuit variants that would have this property, and we can simply put forward an approximate form for a time-varying nonlinear potential $U(t)$ that originates from the Josephson junctions. With $\bar{U}$ being the average value of the potential, we can express a weak linear coupling to one quadrature of the pump field as $U(t) = \bar{U}[1 + \epsilon(\alpha_P e^{-i\omega_P t} + \alpha_P^* e^{+i\omega_P t})]$ with $\epsilon \ll 1$. The total Hamiltonian is obtained by adding the kinetic and potential energies of the pump and circuit oscillators, and thus has a contribution from the pump, the nonlinear oscillator, and an interaction term:

$$\hat{H}_{amp} = \hbar\omega_P \alpha_P^* \alpha_P + \hbar\omega_0 \hat{a}^\dagger \hat{a} + \frac{\hbar\epsilon Z\bar{U}}{2}(\alpha_P e^{-i\omega_P t} + \alpha_P^* e^{+i\omega_P t})(\hat{a} + \hat{a}^\dagger)^2. \quad (8.3)$$

Here $\hat{a}$ describes the intracavity field and ω_0 is the cavity frequency. Notice that the coefficient of the nonlinear oscillator potential energy depends on ϵ, the intensity of the pump oscillator, and the magnitude of the zero-point fluctuations in the position coordinate, which scale as $\hbar Z/2$ where Z is the impedance of the cavity. Here $(\hat{a} + \hat{a}^\dagger)$ corresponds to the coordinate of the oscillator. To further simplify the Hamiltonian, we can apply the rotating-wave approximation in a frame of reference moving at *half* the pump frequency, $\omega_P/2$ (Exercise 8.5.1). This gives us the standard form of the Hamiltonian of the degenerate parametric amplifier,

$$\hat{H}_{amp} = \hbar\Delta\hat{a}^\dagger\hat{a} + \frac{\hbar\lambda}{2}\hat{a}^{\dagger 2} + \frac{\hbar\lambda^*}{2}\hat{a}^2. \quad (8.4)$$

In this expression, $\Delta = (\omega_0 - \omega_P/2)$ is the detuning of the pump from the nonlinear oscillator frequency. The scale of the nonlinear terms is set by $\lambda = (\epsilon Z\bar{U})\alpha_P$.

With the Hamiltonian in hand, we can solve for the dynamics of the intracavity field using the usual quantum Langevin or master equation approach, which has a term arising from the Heisenberg equation of motion and dissipative terms to describe coupling to the environment. The input and output signals to the amplifier are described by bosonic operators $\hat{a}_{in}$ and $\hat{a}_{out}$, respectively. The idler mode is described by the operator $\hat{b}$. The coupling between the nonlinear oscillator and the input port is parameterized by κ, and that of the idler port by γ. The time derivative of the intracavity field $\hat{a}$ is given by Collett and Gardiner (1984)

$$\frac{d\hat{a}}{dt} = \frac{i}{\hbar}[\hat{H}_{amp}, \hat{a}] - \frac{\kappa + \gamma}{2}\hat{a} + \sqrt{\kappa}\,\hat{a}_{in}(t) + \sqrt{\gamma}\,\hat{b}_{in}(t). \tag{8.5}$$

To solve this equation, we make use of the intuitive boundary condition for the output field: $\hat{a}_{out} = \sqrt{\kappa}\,\hat{a} - \hat{a}_{in}$. The equation can then be readily solved in Fourier space by defining $\tilde{a}(\omega) = \int_{-\infty}^{\infty} dt\, e^{-i\omega t}\hat{a}(t)$.

An efficient route to arriving at a closed-form solution starts by expressing the pair of equations for $\hat{a}$ and $\hat{a}^\dagger$ in terms of a matrix relation. As such, we define a vector quantity

$$\mathbf{a} = \begin{pmatrix} \hat{a} \\ \hat{a}^\dagger \end{pmatrix}. \tag{8.6}$$

We define similar relations for the input, output, and idler modes. The dynamics of the cavity can then be compactly written as

$$\dot{\mathbf{a}} = \overline{\overline{\mathbf{L}}}\mathbf{a} + \sqrt{\kappa}\,\mathbf{a_{in}} + \sqrt{\gamma}\,\mathbf{b_{in}}. \tag{8.7}$$

Here, the matrix $\overline{\overline{\mathbf{L}}}$ is given by

$$\overline{\overline{\mathbf{L}}} = \begin{pmatrix} -(i\Delta + \frac{\kappa+\gamma}{2}) & -i\lambda \\ i\lambda^* & (i\Delta - \frac{\kappa+\gamma}{2}) \end{pmatrix}. \tag{8.8}$$

We can express this equation in Fourier space, converting differentiation to multiplication, to obtain

$$\tilde{\mathbf{a}}(\omega) = (\overline{\overline{\mathbf{L}}} + i\omega\overline{\overline{\mathbf{I}}})^{-1}\,(\sqrt{\kappa}\,\tilde{\mathbf{a}}_{in}(\omega) + \sqrt{\gamma}\,\tilde{\mathbf{b}}_{in}(\omega)), \tag{8.9}$$

where $\overline{\overline{\mathbf{I}}}$ is the identity. We can now apply the boundary condition given earlier to obtain a relation for the output field:

$$\tilde{a}_{out}(\omega) = g_s\tilde{a}_{in}(\omega) + g_i\tilde{a}_{in}^\dagger(-\omega) + \sqrt{\frac{\gamma}{\kappa}}[(g_s + 1)\tilde{b}_{in}(\omega) + g_i\tilde{b}_{in}^\dagger(-\omega)]. \tag{8.10}$$

In this expression, g_s and g_i are the frequency-dependent amplitude gains for the signal and idler modes and are given by the following expressions:

$$g_s = \frac{\frac{\kappa(\kappa+\gamma)}{2} - i\kappa(\Delta + \omega)}{\Delta^2 + (\frac{\kappa+\gamma}{2} - i\omega)^2 - |\lambda|^2} - 1 \tag{8.11}$$

and

$$g_i = \frac{-i\kappa\lambda}{\Delta^2 + (\frac{\kappa+\gamma}{2} - i\omega)^2 - |\lambda|^2}. \tag{8.12}$$

We have made use here of the relation $\tilde{a^\dagger}(\omega) = \tilde{a}^\dagger(-\omega)$.

The preceding relations for the amplifier gain are very instructive and give insight into the general behavior of paramps based on a nonlinear oscillator. Here, we make some remarks about the gain, power handling, and noise performance of the paramp. The gain is large when the signal is near the oscillator frequency and drops off with detuning. The bandwidth available for amplification depends on the gain, with the product of these two quantities being bound. The gain diverges when the denominator in the preceding gain expression vanishes, and paramps are operated close to, but below, this threshold, which corresponds to $\lambda < \lambda_C = \sqrt{\Delta^2 + (\kappa + \gamma)^2/4}$. When operating close to this threshold, the effective signal bandwidth of the amplifier is simply related scales as $\kappa/|g_s|$. Pumping the oscillator above λ_C results in spontaneous photon generation due to quantum and thermal fluctuations.

In terms of saturation, the signal is much weaker than the pump – a factor of 100 in power is a typical ratio. This gives one a sense of the maximum input power the amplifier can handle. As such, a weaker nonlinearity will require larger pump power and is thus compatible with larger input signals, such as in an astronomical array detector. Conversely, the signal from a single superconducting qubit is quite weak, so a small pump power is sufficient for amplification and the deleterious effects of pump photons leaking out of the amplifier circuit are minimized. Finally, in terms of added noise, the paramp can operate in either phase-sensitive or phase-preserving mode depending on whether or not the idler signal is measured. In the case of the latter, a half photon of noise is added, as discussed earlier.

8.4 Standing-Wave Amplifier Circuits

Superconducting tunnel junction circuits and passive microwave circuit elements provide a powerful toolbox for realizing amplifiers operating near the quantum limit (Aumentado, 2020). We describe here four amplifier types, spanning the space of degenerate and nondegenerate devices operating in three- and four-wave mixing mode. The basic circuits here make use of resonant cavities to strengthen the interaction between microwave signals and the nonlinear medium. This makes the bandwidth of these amplifiers dependent on the linewidth of the cavity. Simplified electrical circuits for four types of Josephson paramps that have been used to realize the different mixing modes described earlier are presented in Fig. 8.3. Specific details of each circuit variant can be found in the references cited throughout the chapter. Finally, effects that involve high-order terms beyond those present in the ideal Hamiltonian for these Josephson devices are briefly mentioned at the end of this section.

The archetypal paramp, namely a degenerate three-wave mixing circuit, has been detailed earlier in the chapter, and we now consider the circuit implementation

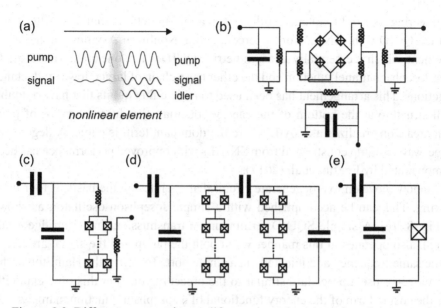

Figure 8.3 Different parametric amplifier variants based on Josephson tunnel junctions. Each circuit has a nonlinear element supporting either three-wave or four-wave mixing, as shown in (a). In the ring structure shown in (b), three modes are independently addressed via inductively coupled resonators. This circuit realizes a nondegenerate, three-wave mixing device. In (c), a nonlinear oscillator where an array of SQUIDs acts as a magnetic-flux-tunable inductor can be used to realize a degenerate, three-wave mixer where the pump is applied at twice the signal frequency via the inductively coupled flux port. Coupling two oscillator circuits, as shown in (d), can be used to realize nondegenerate, four-wave mixing. In panel (e), a simple, single-junction oscillator is used to achieve degenerate four-wave mixing.

of this amplifier. An early implementation of this design was presented in Zhou et al. (2014) and consists of a transmission line terminated with a SQUID (Superconducting QUantum Interference Device) or an array of SQUIDs, as shown in Fig. 8.3(c). The transmission line inputs the small signal of interest to the Josephson circuit and also serves as a conduit for the reflected amplified output. The pump wave is provided through another inductively coupled transmission line, which carries both a static current to magnetically flux bias the SQUID near a point of high sensitivity, and a microwave frequency current at twice the resonant frequency of the nonlinear oscillator. This implements the resonant case of the Hamiltonian presented in Eq. (8.4) where $\Delta = 0$.

A challenge with SQUID devices is that the Josephson potential is symmetric and a fourth-order Kerr type nonlinear term is present. This contribution results in a drive-induced frequency shift of the oscillator, which can result in different modes of saturation that limit the dynamic range of an amplifier, and complex engineering to avoid frequency collisions in a multijunction circuit. An alternative to the SQUID is the SNAIL, or Superconducting Nonlinear Asymmetric Inductive eLement, which is essentially an asymmetric SQUID where one branch has a small-area Josephson tunnel junction and the other has a chain of larger, lower-inductance junctions. This arrangement has been used to realize flux qubits that have a double well structure at the bottom of the energy potential. For different ratios of junction areas, one realizes a SNAIL where the dominant term is cubic. A degenerate three-wave paramp constructed from SNAILs with improved performance has been demonstrated by Frattini et al. (2017).

We now move on to a degenerate circuit that operates on the basis of four-wave mixing. This can be accomplished with a simple Josephson oscillator, as shown in Fig. 8.3(e). A simple SQUID terminating a transmission line (Hatridge et al., 2011), also operates in this manner when both the pump and the signal are nearly at the same frequency and injected via a single port. To derive the Hamiltonian for this case, we use a procedure similar to the preceding one but this time explicitly use the cosine form of the energy functional of a Josephson junction, namely,

$$U(t) = -E_J \cos(\sqrt{\hbar Z/2}(\hat{a} + \hat{a}^\dagger)). \tag{8.13}$$

This can then be expanded to quartic order followed by the application of the rotating-wave approximation (Exercise 8.5.2). The leading-order Hamiltonian is then

$$\hat{H}_{amp} = \hbar \Delta \hat{a}^\dagger \hat{a} + \frac{\lambda}{2}\hat{a}^{\dagger 2} + \frac{\lambda^*}{2}\hat{a}^2, \tag{8.14}$$

which has the same form as the three-wave mixing case, but notably here $\Delta = \omega_0 - \omega_p - 2E_C|\alpha|^2/\hbar$ and $\lambda = -E_C\alpha^2$. We have expressed Z in terms of E_J and E_C to simplify the equation for λ. Here α represents the amplitude of the coherent field in the resonator. The amplification process is resonant when the signal and pump are close to each other and commensurate with the displaced oscillator frequency. This circuit is robust but has limitations when high dynamic range is needed to minimize saturation and the generation of additional mixing products. A detailed study of this simple circuit shows how the effect of higher-order terms can deform the output field from the characteristic ellipse of an ideal squeezer to patterns where information is mixed between quadratures. Different pumping schemes can help suppress the effect of these higher-order terms – specifically applying symmetrically spaced pump tones about the resonator frequency or flux pumping can eliminate cubic corrections (Boutin et al., 2017).

Now let us move on to nondegenerate amplifiers. A three-wave circuit can be realized by the JRM or Josephson ring modulator (Bergeal et al., 2010a,b). The basic topology, as shown in Fig. 8.3(b), consists of four Josephson junctions arranged in a bridge-like geometry. The signal and idler modes are connected to the two opposing terminals of the bridge, and the pump tone is applied symmetrically across these two modes. A static magnetic flux of 1/2 flux quantum threads the loop. The Hamiltonian for this device closely resembles the degenerate case but now has two explicitly defined modes for the signal and the idler. It reads as follows:

$$\hat{H}_{amp} = \hbar\Delta\hat{a}^\dagger\hat{a} + \hbar\Delta\hat{b}^\dagger\hat{b} + \frac{\lambda}{2}\hat{a}^\dagger\hat{b}^\dagger + \frac{\lambda^*}{2}\hat{a}\hat{b}. \tag{8.15}$$

Here, Δ for both modes is the same as the degenerate case and is given by $\omega_0^{a,b} - \omega_p/2$. The superscripts label the two modes a and b. The coefficient λ is simply $2g\alpha_P$ where g is the light matter coupling between each pair of oscillators and α_P the pump amplitude.

Lastly, for the four-wave, nondegenerate mixing case, one approach is to couple a pumped nonlinear oscillator to another, as shown in Fig. 8.3(d). One physical implementation involves capacitively coupling two arrays of Josephson junctions (Eichler and Wallraff, 2014). The basic Hamiltonian in this case reads,

$$\hat{H}_{amp} = \hbar\Delta\hat{a}^\dagger\hat{a} + \hbar\Delta\hat{b}^\dagger\hat{b} + g(\hat{a}^\dagger\hat{b}+\hat{b}^\dagger\hat{a}) + \frac{\lambda_a}{2}\hat{a}^{\dagger 2} + \frac{\lambda_b}{2}\hat{b}^{\dagger 2} + \frac{\lambda_a^*}{2}\hat{a}^2 + \frac{\lambda_b^*}{2}\hat{b}^2. \tag{8.16}$$

Here, like the degenerate case, the detuning expression Δ involves the shifted resonator frequencies and is given by $\omega_0^{a,b} - \omega_p - 2E_{C_{a,b}}|\alpha_{a,b}|^2/\hbar$. The coefficient λ for each mode is $-E_{C_{a,b}}\alpha_{a,b}^2$.

8.5 Traveling-Wave Amplifier Circuits

We now explore another approach to increasing the interaction strength of a nonlinear medium and microwave signals: a nonlinear transmission line comprised of superconducting elements, in particular tunnel junctions (see Fig. 8.4(a)). Such amplifiers are thus traveling-wave devices where an input signal increases in magnitude as it traverses the length of a Josephson transmission line, drawing energy from a much more intense copropagating pump signal. For such an amplification scheme to be effective, the input signal and pump waves must be phase matched along the length of the line. This is not a priori achieved by placing an array of identical junctions along the length of the line, as we shall see, and well-thought-out schemes for momentum matching have been developed.

For a simple classical analysis of traveling wave devices, we can start by analyzing a simple transmission line where the inductive element has been replaced by a Josephson tunneling junction, which has a nonlinear inductance L_J and geometric capacitance C_J. A simple single-mode cavity can be described by an LC circuit, and an extant array of such elements form an LC ladder described by a linear transmission line. The substitution of a Josephson junction in each unit cell added the

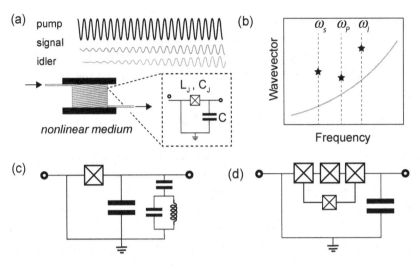

Figure 8.4 A traveling-wave parametric amplifier (a) enhances the interaction between the pump and signal tones by copropagating them along a nonlinear transmission line (made from a long array of unit cells, each defined by the elements in the dashed box), wrapped, for example, on a spool for illustration. The transmission line has dispersion arising from chromatic aberration and frequency shifts due to the Kerr effect. This results in a mismatch of the pump, signal, and idler wavevectors, as shown in (b). The pump and signal wave can destructively interfere, inhibiting the efficient amplification of the signal down the line and reducing the total amplifier gain. In (c), we have an example of dispersion engineering where an additional resonator is added in parallel in each unit cell (dashed box in (a)) to displace the pump wavevector and match it with the signal and idler. An alternate approach, shown in (d), involves using an asymmetric SQUID in each unit cell (dashed box in (a)) where either a positive or a negative dispersion from the Kerr effect can be realized and adjusted to cancel the chromatic aberration. This results in effective momentum matching and the ability to electronically tune the amplification band.

requisite nonlinearity for three- and four-wave mixing processes. We first derive a nonlinear wave equation that describes the wave solutions that propagate in this effective nonlinear medium. With these in hand, we can predict the expected gain of a parametric amplifier and factors needed to maximize it. An analysis of the noise, in particular, due to quantum fluctuations, is more complex, and different treatments can be found in the literature (Zhao and Withington, 2021).

To arrive at a wave equation for a Josephson transmission line, one can write the Lagrangian for each simple unit cell shown in Fig. 8.4(a) in terms of the energy stored in the capacitance and the inductance. Note that we have to consider the capacitance C to ground, which is fabricated as part of the amplifier circuit in addition to the junction capacitance C_J. We can define Φ_n as the magnetic flux through the nth loop and relate the potential difference to the flux at a node:

$$\frac{d\Phi_n}{dt} = V_n, \tag{8.17}$$

where V is the voltage. The Lagrangian for the line is then

$$L = \sum_n \left[E_J \cos\left(\frac{\Phi_{n+1} - \Phi_n}{\phi_0}\right) + \frac{C_J}{2}\left(\frac{d\Phi_{n+1}}{dt} - \frac{d\Phi_n}{dt}\right)^2 + \frac{C}{2}\left(\frac{d\Phi_n}{dt}\right)^2 \right]. \tag{8.18}$$

As per usual, $\phi_0 = \frac{\hbar}{2e}$ is the reduced flux quantum. We can write this expression in terms of a Lagrangian density $\mathcal{L}$ by explicitly inserting the size of the unit cell a. We obtain

$$L = \sum_n a\mathcal{L} = \sum_n a\left[\frac{E_J}{a} \cos\left(\frac{\Phi_{n+1} - \Phi_n}{\phi_0}\right) + \frac{C_J}{2a}\left(\frac{\partial\Phi_{n+1}}{\partial t} - \frac{\partial\Phi_n}{\partial t}\right)^2 \right.$$
$$\left. + \frac{C}{2a}\left(\frac{\partial\Phi_n}{\partial t}\right)^2 \right]. \tag{8.19}$$

We can now take the limit $a \to 0$ and convert the discrete sum to an integral over the length of the transmission line. This yields a Lagrangian density

$$\mathcal{L} = \frac{E_J}{a} \cos\left(\frac{a}{\phi_0}\frac{\partial\Phi}{\partial x}\right) + \frac{aC_J}{2}\left(\frac{\partial^2\Phi}{\partial x\partial t}\right)^2 + \frac{C}{2a}\left(\frac{\partial\Phi}{\partial t}\right)^2. \tag{8.20}$$

With the Lagrangian in hand, we can proceed to derive the equation of motion for the system using the Euler–Lagrange equation. This results in the following nonlinear equation for the system (Exercise 8.5.3),

$$\frac{aE_J}{\phi_0^2} \cos\left(\frac{a}{\phi_0}\frac{\partial\Phi}{\partial x}\right)\frac{\partial^2\Phi}{\partial x^2} - \frac{C}{a}\frac{\partial^2\Phi}{\partial t^2} + aC_J\frac{\partial^4\Phi}{\partial x^2\partial t^2} = 0. \tag{8.21}$$

Finally, to obtain the leading-order behavior, we can expand the cosine term to quadratic order. This results in the following wave equation:

$$\frac{\partial^2\Phi}{\partial t^2} - \frac{a^2 E_J}{C\phi_0^2}\frac{\partial^2\Phi}{\partial x^2} - \frac{a^2 C_J}{C}\frac{\partial^4\Phi}{\partial x^2\partial t^2} + \frac{a^4 E_J}{2C\phi_0^4}\left(\frac{\partial\Phi}{\partial x}\right)^2\frac{\partial^2\Phi}{\partial x^2} = 0. \tag{8.22}$$

In this equation, the nonlinear effects are contained in the last term on the right. The preceding terms describe the propagation of weakly dispersive linear waves with a position-dependent phase velocity.

We now look for traveling wave solutions of the general form $Ae^{i(kx-\omega t)}$ for the signal, pump, and idler waves,

$$\Phi = (A_p(x)e^{i(k_p x-\omega_p t)} + A_s(x)e^{i(k_s x-\omega_s t)} + A_i(x)e^{i(k_i x-\omega_i t)})/2 + c.c. \tag{8.23}$$

This ansatz can then be substituted into the wave equation, and a number of approximations can be used to obtain a pair of coupled equations that govern the spatial behavior of the signal and idler wave amplitude. These simplifications include the notion that the pump is intense enough that any diminution of its amplitude due to amplification of the signal is negligible. Additionally, higher-order derivatives are also ignored. Specific details can be found in O'Brien et al. (2014). To leading order, this process results in the following simple expression for the pump wave amplitude and a pair of coupled equations for the signal and idler wave amplitudes:

$$A_p(x) = A_p^0 e^{i\alpha_p x}, \tag{8.24}$$

$$\frac{\partial A_s(x)}{\partial x} = +i\alpha_s A_s + i\kappa_s A_i^* \, e^{i(\Delta k_L + 2\alpha_p)x}, \tag{8.25}$$

$$\frac{\partial A_i(x)}{\partial x} = +i\alpha_i A_i + i\kappa_i A_s^* \, e^{i(\Delta k_L + 2\alpha_p)x}. \tag{8.26}$$

In these equations, the propagation coefficients are defined as

$$\alpha_p = \frac{\kappa k_p^3 a^2}{LC\omega_p^2}, \quad \kappa = \frac{a^2 k_p^2 |A_p^0|^2}{16 I_0^2 L^2}, \tag{8.27}$$

$$\alpha_s = \frac{2\kappa k_s^3 a^2}{LC\omega_s^2}, \quad \kappa_s = \frac{\kappa(2k_p - k_i)k_s k_i a^2}{LC\omega_s^2}, \tag{8.28}$$

$$\alpha_i = \frac{2\kappa k_i^3 a^2}{LC\omega_i^2}, \quad \kappa_i = \frac{\kappa(2k_p - k_s)k_s k_i a^2}{LC\omega_i^2}. \tag{8.29}$$

The mismatch in the low-power limit between the pump, signal, and idler wavevectors $\Delta k_L = 2k_p - k_s - k_i$.

We can further simplify the coupled equations for the signal and idler waves, assuming solutions of the form $A_s(x) = a_s(x)e^{i\alpha_s x}$ and $A_i(x) = a_i(x)e^{i\alpha_i x}$, respectively. Upon substitution into Eqs. (8.24, 8.25, 8.26), we obtain the following pair of equations for wave amplitudes $a_{s,i}$:

$$\frac{\partial a_s(x)}{\partial x} - i\kappa_s a_i \, e^{i(\Delta k_L + 2\alpha_p - \alpha_s - \alpha_i)x} = 0 \tag{8.30}$$

and

$$\frac{\partial a_i(x)}{\partial x} - i\kappa_i a_s \, e^{i(\Delta k_L + 2\alpha_p - \alpha_s - \alpha_i)x} = 0. \tag{8.31}$$

The solutions to these equations can be readily taken from the optical parametric amplifier literature, and consist of hyperbolic trigonometric functions. The spatial dependence of the signal and idler wave amplitudes is contained in the following expressions:

$$a_s(x) = \left[a_s(0) \left(\cosh(gx) - \frac{i\Delta k}{2g} \sinh(gx) \right) + \frac{i\kappa_s}{g} a_i^*(0) \sinh(gx) \right] e^{i\Delta kx/2} \quad (8.32)$$

and

$$a_i(x) = \left[a_i(0) \left(\cosh(gx) - \frac{i\Delta k}{2g} \sinh(gx) \right) + \frac{i\kappa_i}{g} a_s^*(0) \sinh(gx) \right] e^{i\Delta kx/2}. \quad (8.33)$$

Here, we have grouped the propagation vectors such that $\Delta k = \Delta k_L + 2\alpha_p - \alpha_s - \alpha_i$. The gain coefficient g can be expressed as $g = \sqrt{\left(\kappa_s \kappa_i^* - (\Delta k/2)^2\right)}$.

Phase Matching

In Equation (8.33), we have derived an expression for the amplitude of the signal and idler waves as a function of distance traversed across a nonlinear Josephson transmission line. The gain coefficient g determines the rate of signal grown along the line and changes from a real number to an imaginary one depending on the phase mismatch Δk. Ideally, the phase mismatch should be zero, ensuring that pump energy is efficiently being transferred into the signal wave. In this case, the signal will exhibit exponential growth along the line. However, deviations from a purely linear dispersion relation, as shown in Fig. 8.4(b), prevent this from happening – chromatic aberration, self-Kerr modulation, and cross-Kerr modulation are typically present in Josephson devices. The latter two phenomena are related to a distortion of the pump and signal waves due to a power-dependent dielectric response typically seen when the line is illuminated with an intense pump tone. In this case, g can be imaginary with only quadratic growth along the line. This typically limits the gain that can be practically achieved in a traveling-wave amplifier as materials losses, say in capacitor dielectrics or on surfaces, also increase with line length.

Several approaches have been adopted to improve phase matching in superconducting traveling-wave parametric amplifiers. One general approach is to alter the dispersion relation by introducing periodic loading or resonant elements. Alternatively, one can use Josephson-junction-based circuit elements in which the sign of the Kerr interaction can be tuned in situ – this effect has been demonstrated using asymmetric SQUID elements described in the previous section on standing-wave amplifiers. A brief history of superconducting traveling-wave parametric amplifiers and a discussion of different momentum matching strategies and their efficacy are presented in Esposito et al. (2021). We highlight here two examples that have yielded high-gain amplification: matching with resonant circuit elements and reverse-Kerr phase matching with SNAILs.

The scheme developed in Macklin et al. (2015) and shown in Fig. 8.4(c) involves adding an LC resonator in parallel along the nonlinear transmission line. This opens up a resonant feature in the dispersion relation such that a stop band is opened up about the resonance. The wavevector is purely imaginary inside the stop band and can be varied over the full range of real values outside of it as one moves away from resonance. This technique permits one to change the wavevector of the pump by changing the pump frequency and pump power. This type of TWPA demonstrated near quantum-noise-limited operation with $\sim$ 20 dB gain over a GHz frequency band.

An alternate approach described in Ranadive et al. (2022) and shown in Fig. 8.4(d) involves replacing the Josephson junctions in the basic nonlinear transmission line with SQUID elements consisting of one small junction and three large ones. The current phase relation of this SNAIL device has both even and odd terms that are involved in three- and four-wave mixing processes, respectively. Threading a magnetic flux through the loop allows one to change the ratio of these terms, and in operating a four-wave mixer, the bias is adjusted to null the second-order terms and maximize the negative third-order term. This allows for effective momentum matching at certain points where the Kerr contribution is equal and opposite to the chromatic aberration. A high gain is observed over a wide band. One of the advantages of this design is that one can tune the amplification band in situ without any gaps in the amplifier transmission spectrum.

Like their standing-wave cousins, superconducting traveling-wave amplifiers also exhibit deviations from ideal behavior. We briefly enumerate a few of them here. Firstly, since the amplification band is so wide, it is possible to couple to higher-order idler modes, which contribute additional noise but not information content if left unmonitored. Engineering cutoff frequencies in the basic unit cell can help suppress these additional degrees of freedom. Additionally, the treatment just presented caters to simply sinusoidal traveling wave solutions. When considering the strong field dynamics of a Josephson nonlinear transmission line, the exact solutions are actually snoidal waves. Signals with a sinusoidal wave shape are distorted as they travel along the line and eventually develop a discontinuity, thereby limiting the pump power that can be effectively supplied to the amplifier and resulting in gain instabilities. While these differences are small, they do figure into a precision noise budget on the order of the quantum limit of noise.

Exercises

Exercise 8.5.1   Make the rotating wave transformation $\hat{a} \rightarrow \hat{a}\exp(i\omega_p t/2)$ and drop the quickly rotating terms to show result Eq. (8.4).

Exercise 8.5.2   Make the rotating wave transformation in the previous problem to show result Eq. (8.14).

Exercise 8.5.3   Use the field theory version of Lagrange's equations of motion (Goldstein, 2011)

$$\frac{\partial}{\partial t}\frac{\partial \mathcal{L}}{\partial \dot{\Phi}} + \frac{\partial}{\partial x}\frac{\partial \mathcal{L}}{\partial \Phi'} = \frac{\partial \mathcal{L}}{\partial \Phi}, \tag{8.34}$$

where the dot and prime represent time and space partial derivatives, respectively, to show result (8.22).

Exercise 8.5.4   Verify Eqs. (8.11, 8.12) are the correct expressions for the gain of the signal and idler modes.

Exercise 8.5.5   Plot the signal and idler gain versus frequency for different values of detuning. How does the gain behave as λ approaches λ_C?

Exercise 8.5.6   Using the traveling wave ansatz applied to the nonlinear wave equation (8.22), show that the equations for the amplitude of the pump, signal, and idler obey the equations (8.26, 8.25, 8.26) after suitable approximations mentioned in the text are made: Assume the pump amplitude is constant in space (no depletion of the strong pump). Neglect the second derivatives of the slowly varying amplitudes using the slowly varying envelope approximation. Neglect the first derivatives of the slowly varying amplitudes and the nonlinear polarizability term, with four spatial derivatives.

Exercise 8.5.7   Plot the gain of a phase-matched traveling wave amplifier versus signal frequency, for fixed pump frequency and (large) pump power, for device length much larger than the unit cell a.

9

Measurement-Related Phenomena and Applications

In previous chapters, you, dear reader, have learned about the formalism and many of the basic effects surrounding the physics of quantum measurement as well as supporting technologies. In this chapter, we will delve deeper into some of the phenomena associated with quantum measurement, and apply some of the methods.

9.1 Measurement Reversal

Recall that one of the cardinal properties of quantum measurement discussed in the first chapter is the irreversibility of quantum measurement. In this section, we will see that generalized quantum measurements can in fact be reversed. We can see that the measurement reversal is possible in two different ways. From the quantum Bayesian point of view, the partial collapse of the quantum state, conditioned on a measurement outcome, can be interpreted as an updating of the quantum state, given some information acquired from the measurement outcome. It stands to reason that it is possible that the measurement can give no information about the possible quantum state, in which case, the quantum state remains as it was before that no-information measurement. With this in mind, the strategy to "uncollapse" the quantum state is to perform a second measurement, such that the net information content between the first measurement and the second measurement is zero. The second way to view the reversibility of measurement is to observe that the conditioned state update is usually a one-to-one map from quantum states to quantum states. Therefore, the mathematical prescription to undo the measurement is to find a way to apply the inverse map. It turns out this cannot usually be done deterministically, and therefore, the successful reversal of a measurement can only happen with some probability. Let us see how this works with a couple of examples.

Let us return to the example we gave in Sec. 4.2, the case of selective tunneling. We recall that the realization of a no-tunneling measurement for some time implemented a generalized measurement of the phase qubit, resulting in a partial collapse toward the ground state of the system. We interpreted this effect as the ground state was more likely than the excited state. How can this measurement be reversed? If we make another measurement, it would just collapse further toward the ground state, or else tunnel. Notice, however, if a unitary operation is applied, exchanging the amplitudes of the excited and ground states, then a subsequent no-tunneling measurement can undo the effect of the first measurement. The "π-pulse" applies the operation $\hat{\sigma}_x \hat{\rho}' \hat{\sigma}_x$ to the quantum state (4.14), exchanging the amplitudes of the excited and ground states. Now, apply the same procedure, and perform the tunneling measurement for the same time t as the first measurement. Assume that again, a null measurement occurs.

This remarkable effect does not depend on any knowledge of what the initial state $\hat{\rho}$ is. We may give the following interpretation: the first measurement gives us a hint that the state is more likely ground than excited, leading toward partial collapse toward the ground state. However, the exchange of amplitudes, putting the excited state amplitude in the place of the ground state amplitude, and vice versa, followed by the second null measurement of the same duration, gives us exactly the opposite hint as the first measurement. Consequently, the two hints together give no information about the quantum system, leaving the state exactly as if it had been untouched the entire time!

However, there is a catch. There is no guarantee that the second measurement will give the no-tunneling result we want. We can apply the formalism we have learned to calculate how probable it is to reverse the measurement. Recall that the probability of the null result measurement is $P_d = \text{Tr}(\hat{\rho}E_d)$ where E_d is defined in Eq. (4.11). In Exercise 9.4.2 you will show, using the state $\hat{\sigma}_x \hat{\rho}' \hat{\sigma}_x$, that the uncollapse probability is given by

$$P_{uc} = \frac{e^{-\Gamma t}}{\rho_{gg} + \rho_{ee} e^{-\Gamma t}}, \qquad (9.1)$$

where $\hat{\rho}$ is the original state.

The uncollapse probability just presented is instructive. The probability may be viewed as a continuous function of the elapsed time t. In the limit where no time has elapsed, $t = 0$, no measurement takes place, so there is nothing to reverse, and the probability is 1. However, as the time t becomes larger than Γ^{-1}, the probability to reverse the measurement limits to zero, recovering the traditional statement of irreversibility. It is easy to see why: when the time is very long, the qubit is collapsed to nearly the ground state during the first measurement. After the π-pulse, the state is now nearly the excited state, so during the long second tunneling measurement, the qubit will almost certainly tunnel out of the metastable well. Therefore, we almost never register a no-tunneling event after the second measurement. However, for any finite strength measurement, there is some possibility to reverse it. Looking at Eq. (9.1), it does depend on the initial state, and therefore, naively, one might think that one could learn what the initial state of the qubit is by repeatedly measuring and unmeasuring the qubit. However, this idea is not correct. The answer why can be seen because the net probability of a measurement and unmeasurement sequence is given by $P_{rev} = P_d P_{uc} = e^{-\Gamma t}$, which is independent of the initial state. This result is true quite generally (Jordan and Korotkov, 2010). Phase qubit measurement reversal was experimentally verified in the group of John Martinis (Katz et al., 2008).

As a second example, let us consider the case of diffusive quantum trajectories. For the system of a quantum double dot, measured by a quantum point contact as described in Chapter 5, we recall the solution (5.58) in the simplest case of no system Hamiltonian, given in normalized form as

$$|\psi\rangle = \frac{1}{\sqrt{|\alpha|^2 e^{2\bar{r}} + |\beta|^2 e^{-2\bar{r}}}} \begin{pmatrix} \alpha e^{\bar{r}} \\ \beta e^{-\bar{r}} \end{pmatrix}, \tag{9.2}$$

where $\bar{r} = (1/2\tau_m) \int_0^t dt' r(t')$, and α, β are the $t = 0$ state coefficients. Like in the previous example, we can view the postmeasurement state as a function of $\bar{r}$. We notice that if $\bar{r} = 0$, the postmeasurement state is the same as the initial state. This suggests the following uncollapse strategy. Suppose the first measurement occurs over some time t_1, and gives rise to a measurement record $r_1(t)$, giving a time-integrated result $\bar{r}_1 = (1/2\tau_m) \int_0^{t_1} dt' r_1(t')$. We can then try to undo it by making further measurements from t_1 to some unknown time t_2, getting a measurement record $r_2(t)$, hoping that the time-integrated result of the second measurement, $\bar{r}_2 = (1/2\tau_m) \int_{t_1}^{t_2} dt' r_2(t')$, is exactly the opposite of the first measurement, $-\bar{r}_1$. If the zero-crossing of the total measurement record occurs, the measurement is stopped. This situation would result in a total measurement time-integrated measurement of $\bar{r}_1 + \bar{r}_2 = 0$, resulting in the measurement uncollapse. Just like before, this zero crossing of the time-integrated measurement result does not always happen, resulting in a probabilistic reversal of the measurement. In a similar way as the previous example, the statistics of this process can be worked out, showing similar features as the measurement strength increases (Korotkov and Jordan, 2006). This reversal of the measurement does indeed occur in experiments, as seen in Fig. 9.1(a). There, whenever the z trajectory crosses 0, the initial $x = 1$ is restored, corresponding to when $r(t)$ also crosses 0. The quantum state restoration is imperfect in the experiment because of residual decoherence from lost information.

A more general approach can be made by making a singular value decomposition of the measurement operator to be reversed and applying unitary operations

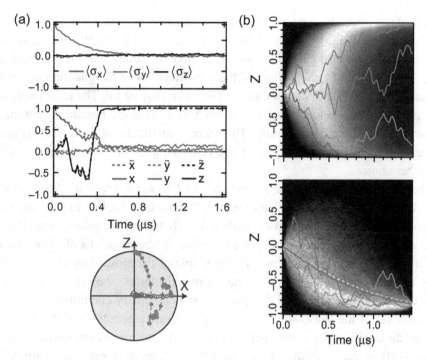

Figure 9.1 Quantum trajectories of a superconducting qubit are shown in time, represented as Bloch coordinates, panel (a). The initial condition is $x = 1$. In the upper plot, ensemble averaged trajectories are shown, showing characteristic decoherence. Quantum measurement reversal occurs in the bottom plot whenever the $z = 0$ plane is crossed. Intrinsic residual decoherence leads to imperfect state restoration. In panel (b), a histogram of trajectories is shown in shading, with representative trajectories shown as jagged lines. Continuous collapse to $z = +1, -1$ is clear. In the lower plot, a subset of trajectories is shown where the final condition is also fixed (postselection). The dashed line is the most likely path, discussed in the next section (Weber et al., 2014). Adapted with permission from Springer Nature.

together with the inverse of the singular values. More details on this generalization and more advanced applications of this approach can be read in Jordan and Korotkov (2010).

9.2 Most Likely Path

We have learned about the unconditional master equation, which assigns the time-continuous and smooth quantum state evolution. The quantum dynamics is deterministic but represents the ensemble-averaged quantum dynamics of a system coupled to an environment. We have also learned about quantum trajectories, which assigns a time-continuous, random quantum state. The trajectory is conditioned on the noisy measurement record and is diffusive in character. An intermediate description of the quantum dynamics is the most likely path. It is a smooth, deterministic path through the Hilbert space, like the solution to the unconditional master equation, but retains some of the conditionality of the quantum trajectory, because both initial and final boundary conditions may be set. The most likely path extremizes the action functional of Section 5.5 and thus is a possible quantum trajectory with optimized probability. Therefore, a small tube of fixed width centered at the most likely path will capture more fairly sampled quantum trajectories than any other choice.

The stochastic path integral formulation of Chapter 5 gives an elegant way to find the most likely path. We can apply a variational principle to find the equations of motion of the most likely path, similarly to how Hamilton's equations of motion can be found from Hamilton's principle, or the equations of ray optics can be found from Fermat's principle. The principle of the most likely path is to optimize the probability density of the quantum trajectories, subject to the quantum state disturbance relations and the quantum state boundary conditions: $\delta S = 0$, so the variation of the stochastic action is zero. Here we give the scalar q treatment for one dimension – the generalization to many dimensions is straightforward. Let us take $q(t) = q_{ml}(t) + \delta q(t)$, where we define q_{ml} as the most likely path, which satisfies the boundary condition, and δq is the variation around this path. We make similar definitions for the canonical momentum p and the readout r. Let us break the action into a dynamical path and a static part as

$$S = \int_0^t dt' [-ip\dot{q} + \mathcal{H}[q, p, r]]. \tag{9.3}$$

Here, we refer to $\mathcal{H}$ as the stochastic Hamiltonian, in analogy to Hamiltonian mechanics, which is a functional of the dynamical coordinates. Taking the variation of the action, we find

$$\delta S = \int_0^t dt' [-i(\delta p)\dot{q} - ip(\delta\dot{q}) + \delta\mathcal{H}]. \tag{9.4}$$

$$= \int_0^t dt' \left[-i(\delta p)\dot{q} + i\dot{p}(\delta q) + \left.\frac{\partial\mathcal{H}}{\partial q}\right|_0 \delta q + \left.\frac{\partial\mathcal{H}}{\partial p}\right|_0 \delta p + \left.\frac{\partial\mathcal{H}}{\partial r}\right|_0 \delta r \right]. \tag{9.5}$$

Here we have integrated the dynamical term by parts in the second equality. The notation $\ldots |_0$ indicates evaluating the derivative along the most likely path where $\delta q(t) = \delta p(t) = \delta r(t) = 0$. The variation of the action should be zero for every variation $\delta q, \delta p, \delta r$, yielding the equations of motion for the most likely path,

$$\dot{q} = \frac{\partial\mathcal{H}}{\partial(ip)}, \quad i\dot{p} = -\frac{\partial\mathcal{H}}{\partial q}, \quad \frac{\partial\mathcal{H}}{\partial r} = 0. \tag{9.6}$$

Here we have dropped the subscript ml for simplicity. These equations may be viewed as analogous to Hamilton's equations of motion of classical mechanics; they can be made entirely real by rotating $ip \to p$. Applying this principle to the action functional (5.108), and generalizing to multiple dimensions, gives the equations of motion

$$\dot{q}_j = \mathcal{L}_j[\mathbf{q}, r], \quad -i\dot{p}_j = i\mathbf{p} \cdot \frac{\partial \mathcal{L}}{\partial q_j} + \frac{\partial \mathcal{F}}{\partial q_j}. \tag{9.7}$$

Together with the constraint $\partial_r[i\mathbf{p} \cdot \mathcal{L} + \mathcal{F}] = 0$, gives a set of ordinary differential equations, that may be solved with a set of initial and final boundary conditions on the quantum state variables $\mathbf{q}$.

The action of the stochastic path integral can be expanded around the most likely path using higher variations as

$$S = S_0 + \int_0^t dt' \left[\frac{\delta S}{\delta X_i} \bigg|_0 \delta X_i + \frac{\delta^2 S}{\delta X_i \delta X_j} \bigg|_0 \delta X_i \delta X_j + \ldots \right]. \tag{9.8}$$

Here, we have introduced the combined vector of dynamical variables, $\mathbf{X} = \{\mathbf{q}, \mathbf{p}, r\}$. By construction, the first variation vanishes. The saddle point approximation is taking only the second variation and integrating the Gaussian approximation to the path integral. We note that unlike the Feynman path integral, there is not a natural expansion parameter, like $1/\hbar$ to make this approximation, analogous to a semiclassical expansion. Here, the expansions may be viewed as expanding in the strength of the measurement – in the limit where the characteristic measurement time τ_m (or inverse measurement strength) is much larger than the parameters of the inverse Hamiltonian, the measurement dynamics is a small perturbation to the Schrödinger equation dynamics, so we can view the most likely path approximation to the path integral as improving as $(\Delta \tau_m/\hbar)^{-1}$ becomes a small parameter, where Δ characterizes the energy of the system.

It is much easier to solve a set of deterministic equations of motion and fix boundary conditions than to solve the stochastic path integral or master equation for every possible realization of the stochastic measurement readout. These most likely paths form a "skeleton" upon which the diffusive process can be organized. The Hamiltonian form of the equations permits the use of many of the familiar techniques of classical mechanics, such as canonical transforms, phase space methods, constants of motion, and so on. However, it is important to remember that the coordinate, momentum, Hamiltonian, and action in this method have a very different meaning – the coordinate is really a representation of the quantum state, and the momentum only has the role of generating translations of the coordinate. The stochastic Hamiltonian is not a true energy, but is a constant of motion for time-independent systems. Nevertheless, the mapping of concepts gives us powerful tools to study continuously measured quantum systems.

Example: Dispersive Measurement of a Qubit

We can apply the preceding formalism to the case of the qubit of either a DQD/QPC system or a circuit QED system, discussed in Chapter 5, finishing at Eq. (5.112). That action takes the form of $S = \int_0^t dt'[-i\mathbf{p} \cdot \dot{\mathbf{q}} + \mathcal{H}]$, where $\mathbf{q} = (x, y, z)$, and the stochastic Hamiltonian is given by

$$\mathcal{H} = -p_x(\gamma x + \epsilon y + xzr/\tau_m) - p_y(\gamma y - \epsilon x - \Delta z + yzr/\tau_m)$$
$$- p_z(\Delta y - (1 - z^2)r/\tau_m) - (r^2 - 2rz + 1)/(2\tau_m)], \tag{9.9}$$

where we have rotated the momentum variables in the complex plane to have an entirely real stochastic Hamiltonian and equation of motion. Applying the variational principle discussed earlier, we find the equations of motion:

$$\dot{x} = -\gamma x - \epsilon y - xzr/\tau_m, \tag{9.10}$$
$$\dot{y} = -\gamma y + \epsilon x - \Delta z - yzr/\tau_m, \tag{9.11}$$
$$\dot{z} = \Delta y + (1 - z^2)r/\tau_m. \tag{9.12}$$

These have exactly the same form as the Stratonovich form differential equations (5.72, 5.73, 5.74) we found earlier, but with the important difference that the measurement result r is now the most likely measurement record, rather than the readout (5.75). This most likely measurement record can be found from $\partial_r \mathcal{H} = 0$, giving

$$r = z - p_x xz - p_y yz + p_z(1 - z^2), \tag{9.13}$$

which is indeed the "signal" z, with additional terms involving the canonical momenta. These dynamical variables are governed by the other half of Hamilton's equations,

$$\dot{p}_x = \gamma p_x + -\epsilon p_y + p_x zr/\tau_m, \tag{9.14}$$
$$\dot{p}_y = \gamma p_y + \epsilon p_x - \Delta p_z + p_y zr/\tau_m, \tag{9.15}$$
$$\dot{p}_z = \Delta p_y + (p_x x + p_y y)r/\tau_m + r(2p_z z - 1)/\tau_m. \tag{9.16}$$

These equations also depend on x, y, z, r, and the system of equations must be solved together. The six constants of integration may be determined with both the initial and final conditions of the quantum state variables x, y, z. In contrast to Hamiltonian mechanics, the canonical momentum variables have no direct physical interpretation, and their initial conditions are determined by the final conditions of the physical state variables. It is often convenient to view the dynamics of the most likely path using a phase space representation. This is because there is a constant of motion, the stochastic energy. The global dynamics follows lines of constant stochastic energy, and this representation enables one to see at a glance all possible types of dynamical behavior. We will use this picture in the next section.

We can compare these predictions of the most likely quantum path with experimental data. There are different approaches to experimentally quantifying the most likely path. The conceptually clearest notion is that of a tube of fixed width, passing through the quantum state space. The tube position that captures the largest number of trajectories and obeys the quantum state equation of motion is defined as the most likely experimental path. In practice, a simpler measure is to simply find where the quantum trajectories are the most bunched. This can be found by ranking the distance of every trajectory, according to some distance measure. We consider here the simple Euclidean distance measure

$$D_E = \sum_j \sqrt{(x_{1,j} - x_{2,j})^2 + (y_{1,j} - y_{2,j})^2 + (z_{1,j} - z_{2,j})^2}. \qquad (9.17)$$

Trajectories can be ranked by their distance, and a clustering algorithm can be used to find the highest density of trajectories in a region. Comparison to experiment is shown in Fig. 9.1(b), in the lower panel. In that experiment, there is no applied drive, the state is prepared in $| + x\rangle$, and the final boundary condition of $z_F = -0.85 \pm .03$ is applied. The characteristic measurement time is $\tau_m = 1.25$ μs. This most likely path captures the continuous collapse process on the way to one of the measurement operator eigenstates. The solid line is the theory and the dashed line is the experimental estimate. Representative trajectories that meet with boundary conditions are plotted, and the shaded background is a histogram of all trajectories that meet the boundary conditions. In the driven case, most likely paths can be seen in Weber et al. (2014).

There are many interesting phenomena associated with the most likely path. In nonlinear systems, there can be multiple most likely paths, either due to winding number multipaths (paths that reach the same final point by making different winding numbers around the quantum state space), or by the formation of caustics–points where an instability arises in the dynamics (Lewalle et al., 2017), an effect observed in resonance fluorescence experiments (Naghiloo et al., 2017). Caustic solutions are similar to those found in optics, where a Lagrange manifold of all possible solutions folds over itself, showing a transition from one to multiple solutions. See Fig. 9.2(a). In the case where the system is extended to multiple variables, or is externally driven, the system can exhibit quantum chaos, where the most likely paths exhibit exponential sensitivity to their initial conditions (Lewalle et al., 2018). Conventional studies of quantum chaos only study the quantum physics of systems that are classically chaotic – but are not chaotic themselves. However, by adding the measurement process dynamics, along with boundary conditions, bona fide chaos

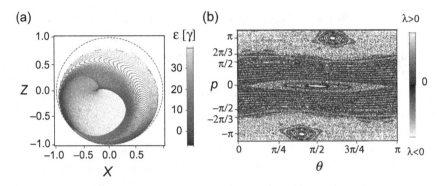

Figure 9.2 Nontrivial dynamical structures can appear in the most likely path dynamics of continuously monitored quantum systems. Panel (a): Resonance fluorescence dynamics exhibits a quantum caustic. A snapshot of the dynamics of the Lagrangian manifold is plotted inside a slice of the Bloch ball, depicting all possible trajectories starting from the qubit's excited state. The folding of the manifold over itself exhibits the caustic, and allows the possibility of multiple most likely paths for a given time. Panel (b): True quantum chaotic dynamics is shown in the phase space portrait of the most likely quantum paths. In a driven, continuously measured system, the phase space portrait of the most likely paths exhibits interleaving of regular and chaotic regions (Lewalle et al., 2017, 2018).

appears in the quantum dynamics itself. Signatures of the onset of quantum chaos can be seen in the phase portrait of the most likely paths, as shown in Fig. 9.2(b). Very similar images can be seen in the classical phase space portraits of chaotic systems such as the kicked rotor or the Chirikov standard map (Ott, 2002).

Example: Transitions from Quantum Jumps to Phase Diffusion

In the previous chapter, we ended by discussing how the difference between quantum jumps and quantum diffusion is not hard and fast; rather the transition between them can be continuous. The most likely path analysis seen via phase space portraits gives great insight into how this transition occurs. Let us consider the simplest case of a qubit with a weak drive that is pure throughout the dynamics (no intrinsic dephasing). Restricting our attention to the dynamics on the $y - z$ great circle of the Bloch sphere, we can reparameterize the state as $z = \cos\theta, y = \sin\theta, x = 0$, so the Stratonovich form differential equations (5.72, 5.73, 5.74) become simply

$$\dot{\theta} = \Delta - \frac{r\sin\theta}{\tau_m}. \tag{9.18}$$

Thus, the stochastic action of the path integral becomes

$$S = \int_0^t dt'[-p_\theta\dot{\theta} + \mathcal{H}_\theta], \tag{9.19}$$

$$\mathcal{H}_\theta = p_\theta(\Delta - r\sin\theta/\tau_m) - (r^2 - 2r\cos\theta + 1)/(2\tau_m). \tag{9.20}$$

The equations of motion for the most likely path of this system are given by (9.18), where all variables are replaced by their most likely value, and the other relations:

$$r = \cos\theta - p_\theta\sin\theta, \tag{9.21}$$

$$\dot{p}_\theta = p_\theta r\cos\theta/\tau_m + r\sin\theta/\tau_m. \tag{9.22}$$

If the most likely readout is substituted into the equations of motion (Exercise 9.4.6) or the readout is directly integrated out (Exercise 9.4.7), the resulting equations of motion are given by a new stochastic Hamiltonian

$$\mathcal{H} = a(\theta)p_\theta^2 + b(\theta)p_\theta + c(\theta), \tag{9.23}$$

where $a(\theta) = -c(\theta) = \sin^2\theta/(2\tau_m)$ and $b(\theta) = \Delta - \sin\theta\cos\theta/\tau_m$.

The resulting one-dimensional dynamics can now be tuned between the quantum jump case $\Delta\tau_m \ll 1$ and the quantum phase diffusion case $\Delta\tau_m \gg 1$ by either changing the strength of the measurement or the strength of the Rabi drive. The action has a dynamical term and a static term that in one dimension can be written as

$$S = -\int p_\theta(\theta)d\theta + Et, \tag{9.24}$$

where $\mathcal{H}(p_i, \theta_i) = E$ is a conserved quantity, fixed by the initial conditions. The presence of a constant of motion in one dimension allows the solution of the present problem in closed form for the most likely stochastic momentum as the solution of a quadratic equation,

$$p_\theta(\theta, E) = -\frac{b}{2a} \pm \sqrt{\frac{b^2}{4a^2} + 1 + \frac{E}{a}}. \tag{9.25}$$

Along this curve, the dynamical equation of motion for the angle is given by

$$\dot{\theta} = \frac{\partial\mathcal{H}}{\partial p_\theta} = \pm\sqrt{4a^2 + b^2 + 4aE}. \tag{9.26}$$

The plus and minus labels thus represent either positive or negative velocity solutions. The time T taken by a most likely path passing between boundary conditions θ_i and θ_f can also be represented as

$$T = \int d\theta/\dot{\theta} = \pm\int_{\theta_i}^{\theta_f} d\theta \frac{1}{\sqrt{4a^2 + b^2 + 4aE}}, \tag{9.27}$$

so the time taken passing from θ_i to θ_f is the same as the time taken from θ_f to θ_i.

Viewed as a dynamical set of equations, it is of great interest to notice there is a fixed point of the dynamics, defined by $\dot{\theta} = \dot{p}_\theta = 0$, corresponding to stochastic energy $E_0 = -\Delta^2 \tau_m/2$. Solving the equations for this fixed point $\theta_f, p_{\theta,f}, r_f$, we find the solution

$$\theta_f = \tan^{-1}(\Delta \tau_m), \quad p_{\theta,f} = -\Delta \tau_m, \quad r_f = \sqrt{1 + (\Delta \tau_m)^2}. \tag{9.28}$$

Linearizing around the fixed point indicates it is hyperbolic in nature, as can be seen at the intersection of the black curves in Fig. 9.3. The interpretation of this solution can be assisted by finding the associated stochastic action for the stationary solution, S_f. In the case of a system prepared at the fixed point, it will not change, so its dynamical term is 0, leaving only the stochastic energy,

$$S_f = \int_0^t dt' \mathcal{H}_{\theta,f} = -t\Delta^2 \tau_m/2. \tag{9.29}$$

The action controls the probability density of the particular most likely path. Thus, the fixed point solution is much more probable in the limit $\Delta \tau_m \ll 1$, corresponding to the quantum jump limit. Physically, this is because the case $\Delta = 0$ corresponds to the pure measurement case, where the fixed points are the eigenvalues of the measurement operator, that is, the collapse points. Turning on a small drive Δ enters the continuous version of the quantum Zeno effect, where most of the time the system is in the eigenstate, but occasionally there are jumps to the other eigenstate. Thus, to keep the system at the fixed point in the Zeno regime, the fixed point angle is $\theta_f \approx \Delta \tau_m \ll 1$, which is nearly the $z = 1$ eigenstate angle $\theta = 0$. The corresponding most likely measurement outcome $r_f \approx 1 + (\Delta \tau_m)^2/2$ is slightly above the $z = 1$ measurement outcome, reminiscent of the weak value exceeding the eigenvalue range. In this case, the anomalous measurement readout is counteracting the measurement drive to keep the system stationary. In the other limit, $\Delta \tau_m \gg 1$, staying at the fixed point becomes highly improbable, and the associated measurement readout becomes much larger than 1, also indicating this is an unlikely possibility.

Being a Hamiltonian system, all dynamical solutions conserve the stochastic energy $\mathcal{H}_\theta = E$. The most probable most likely dynamics are the ones that have zero stochastic energy, $E = 0$, because the action takes the form in Eq. (9.24). Nonzero energy solutions become very unlikely, so the action is only the dynamical part, $S = -\int_0^t p_\theta \dot{\theta} dt' = -\int d\theta p_\theta(\theta)$, where we reparameterized the time integral along the zero-energy contour.

In the Zeno limit, the zero-energy line is found by taking the positive velocity solution of the quadratic equation found from solving $\mathcal{H} = 0$ in Eq. (9.23) that connects $\theta = 0$ to $\theta = \pi$. The solution is approximately given by

$$p_{\theta,0}(\theta) \approx \begin{cases} 0, & 0 < \theta < \Delta \tau_m, \\ \frac{2(\theta - \Delta \tau_m)}{\theta^2}, & \Delta \tau_m < \theta < \pi, \end{cases} \tag{9.30}$$

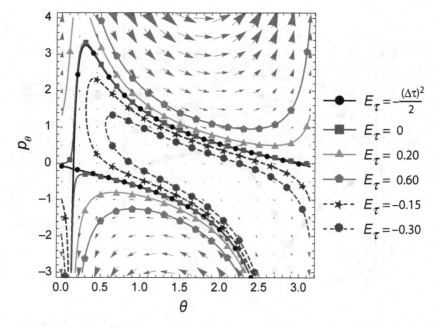

Figure 9.3 Phase space dynamics for the most likely paths. The fixed point and resulting separatrix are defined by the stochastic energy $E_\tau = \mathcal{H}\tau_m = -(\Delta\tau_m)^2/2$, plotted in circles. In this plot we take $\Delta\tau_m = 0.15$, focusing on the quantum jump dynamics. The most likely jump path between eigenstates defines by $\theta = 0$ and $\theta = \pi$ is the zero energy line, plotted in squares. Other constant energy lines are shown with positive energy trajectories given in solid lines and negative energy solutions given in dashed lines. The dynamical flow, $(\dot\theta, \dot p_\theta)$, is represented by the arrows as a vector plot.

and is plotted in squares in Fig. 9.3. We see that the state can transition from $\theta = 0$ to $\theta = \Delta\tau_m$ with nearly unit probability, but then encounters a statistical barrier that requires a large fluctuation to overcome. This result can be interpreted as the state rotating to $\theta = \Delta t$, but then collapsing back onto the $\theta = 0$ point after one measurement time $t = \tau_m$. Occasionally, the state can flip over to the opposite pole with a small rate. The area under the zero-energy curve gives the action of this path, $S_{in} \approx 2\ln(\Delta\tau_m)$, which is logarithmically divergent for small $\Delta\tau_m$. The action controls the switching rate. In this limit, the time of this path is given by $T \approx 4\tau_m \ln(1/\Delta\tau_m)$, which is also logarithmically divergent. This is a characteristic time taken by a quantum jump (Chantasri et al., 2013).

We can learn about the transition to the noisy Rabi oscillation case by taking $\Delta\tau_m \gg 1$, so the quantum effects of the measurement are only a small perturbation to the unitary evolution $\dot\theta = \Delta$. The dynamical flow diagram is plotted in Fig. 9.4 for the case where $\Delta\tau_m = 2$, and even for this case, we see the fixed point moved well away from the $p = 0$ axis, and obtain mostly horizontal flow lines, with some wiggles. The relevant solution (closer to the zero momentum, and hence the more

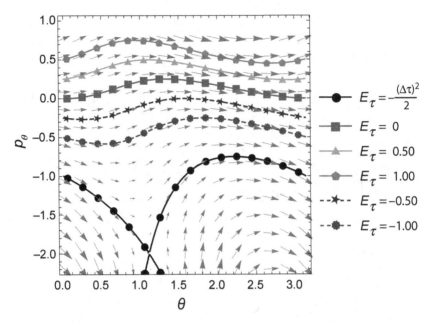

Figure 9.4 Phase space dynamics for the most likely paths. In this plot we take $\Delta\tau_m = 2$, focusing on the transition to the noise Rabi oscillation case. Here, the fixed point and resulting separatrix, defined by the stochastic energy $E_\tau = \mathcal{H}\tau_m = -(\Delta\tau_m)^2/2$, plotted in circles, are pushed into the large negative p_θ values, becoming unlikely. The most likely path between eigenstates is the zero energy line, plotted in squares. Other constant energy lines are shown with positive energy trajectories given in solid lines and negative energy solutions given in dashed lines. The dynamical flow, $(\dot\theta, \dot p_\theta)$, is represented by the arrows as a vector plot.

likely solution) to the quadratic equation $\mathcal{H} = 0$ from Eq. (9.23) is the positive root, leading to the approximate solution:

$$p_{\theta,+}(\theta) \approx \frac{\sin^2 \theta}{2\Delta\tau_m}. \tag{9.31}$$

The resulting action to start at angle θ_i and move to final angle θ_f is given by

$$S_+ = \int_{\theta_i}^{\theta_f} d\theta\, p_\theta(\theta) = \frac{1}{4\Delta\tau_m}(\theta_f - \theta_i - \frac{1}{2}(\sin(2\theta_f) - \sin(2\theta_i)). \tag{9.32}$$

The equation of motion for this solution is

$$\dot\theta \approx \Delta - \frac{\sin\theta \cos\theta}{\tau}, \tag{9.33}$$

so the solution is mostly the unitary free rotation Δt, plus a small correction from the measurement disturbance. The time T taken to move between angle θ_i and θ_f on the most likely path is given by

$$T = \int_{\theta_i}^{\theta_f} d\theta/|\dot{\theta}| = \frac{\theta_f - \theta_i}{\Delta} - \frac{1}{4\Delta^2 \tau_m}(\cos 2\theta_f - \cos 2\theta_i). \qquad (9.34)$$

As $\Delta\tau_m$ grows larger, the probability of the unitary path $\theta = \Delta t$ limits to 1.

It is of interest to consider the negative solution of the quadratic equation, corresponding to paths moving against the drive. The solution is approximately given by

$$p_{\theta,-}(\theta) \approx -\frac{2\Delta\tau_m}{\sin^2 \theta}. \qquad (9.35)$$

This solution corresponds to the flow lines below the separatrix in Fig. 9.4, which are not shown. The resulting action,

$$S_- = -\int p_{\theta,-}(\theta)d\theta = 2\Delta\tau_m(\cot \theta_i - \cot \theta_f), \qquad (9.36)$$

has divergences when θ approaches integer multiples of π, indicating that the diffusion cannot make the state evolve against the drive beyond the poles of the Bloch sphere. This is sensible because diffusion vanishes at the poles of the Bloch sphere, while the drive will continue to take the state in the positive θ direction. More about this system and how multiple extremal paths of different winding numbers are possible can be read in Lewalle et al. (2017).

9.3 Joint Measurement of Noncommuting Observables

A naive reading of the Heisenberg uncertainty principle states that one cannot jointly measure two noncommuting observables to an accuracy greater than that given by the uncertainty relation. In fact, some additional effort must be made to connect the uncertainty relation $\Delta x \Delta p \geq \hbar/2$ to an operational procedure. What this really means is that given an ensemble of measurements on identically prepared systems, if the choice is made to either measure position or momentum, the statistical spread in both measurement sets must obey the preceding relation. Standard quantum mechanics makes predictions about the outcomes of a sequence of ideal position measurement and/or momentum measurements, described as projections onto position and/or momentum projection operators, but states that a simultaneous measurement of both properties is impossible.

However, one can envision an experiment where such a simultaneous probing of both noncommuting observables can be carried out. For example, one could use two different meters, one for position $\hat{X}$ and one for momentum $\hat{P}$. They would be coupled via the coupling Hamiltonians, $\hat{H} = (g_1\hat{X}\hat{P}_1 + g_2\hat{P}\hat{P}_2)\delta(t)$, where we choose to couple to the two meters' momenta $\hat{P}_1, \hat{P}_2$ at the same time. Those meters

Figure 9.5　Illustration of the competition between the results of joint position and momentum measurements.

have their own free Hamiltonians, $\hat{H}_m = \hat{P}_1^2/2m_1 + \hat{P}_2^2/2m_2$, and their positions can later be registered as an indirect measurement of the position and momentum of the first system particle. This basic schema was described for a harmonic oscillator in Arthurs and Kelly (1965) and for qubits in Jordan and Büttiker (2005). The later paper envisioned a situation with a superconducting qubit, where both charge and flux detectors could be coupled to the same qubit – two degrees of freedom that are canonically conjugate to each other.

The essential point is that by weakening both measurements, partial information is obtained about both noncommuting observables, and the quantum backactions of both measurements are in competition with each other. If the position measurement is made stronger, the state tends to collapse to a position eigenstate, whereas if the momentum measurement is stronger, the system tends to collapse to a momentum eigenstate. However, fluctuations in either outcome can break those trends. This idea is illustrated in Fig. 9.5.

We now describe an experiment that can implement such an idea. We consider a superconducting qubit, where we can engineer two "handles" on the noncommuting pseudo-spin observables, $\hat{\sigma}_x$ and $\hat{\sigma}_z$, by allowing the qubit's electric field to couple differently to two different modes of the readout cavity it is embedded in. Rabi oscillations are driven at 40 MHz so that its Hamiltonian becomes that of an effective low-frequency qubit. As depicted in Fig. 9.6, a pair of sideband tones were applied detuned above and below the cavity frequency, which the qubit then Raman scatters to the cavity resonance. Separately, the red(blue)-detuned sidebands would induce cavity-mediated cooling (heating) of the effective qubit. When they are applied concurrently, their relative phase δ determines the interference between the up- and down-converted photons, which in turn dictates which qubit observable is encoded in the resulting signal. The net result is that it is possible to tune the Hamiltonian to the form

$$\hat{H}_{\mathit{eff}} = \chi \bar{a}_0 (\hat{a} + \hat{a}^\dagger) \sigma_\delta, \tag{9.37}$$

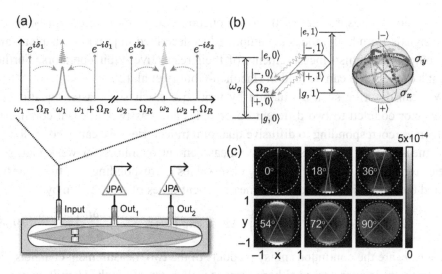

Figure 9.6 Illustration of the experiment demonstrating the joint simultaneous measurement of two noncommuting observables. (a) Two cavity modes are driven in a microwave cavity, probing one superconducting qubit, which is Rabi driven at the frequency Ω_R. Sideband tones of the same frequency Ω_R are created above and below the cavity resonances, drawn with vertical arrows on either side of the cavity resonances Ω_1, Ω_2. This situation permits the inelastic scattering of photons off the qubit, Raman scattering them either up or down in frequency to match the cavity resonance. The escaping light exits two mode-selective ports, which are then amplified with Josephson parametric amplifiers. (b) Level structure of the dressed states is indicated for one cavity mode, where Ω_q is the qubit frequency. Sideband tones drive transitions indicated by the solid lines, and undulating lines represent cavity decay, which is detected. The relative phase of the sidebands controls the effective qubit observable that is measured. The phase-tunability of the qubit observable, with its associated measurement backaction, is illustrated on the Bloch sphere. (c) Steady-state probability distribution of the density matrix, shown with the Bloch coordinates $(x; y)$, are plotted for different values of the angle between the measurement axes. The data proves that wavefunction collapse does not occur for the joint measurement of noncommuting observables (Hacohen-Gourgy et al., 2016). Adapted with permission from Springer Nature.

where χ is the qubit–cavity coupling, $\bar{a}_0$ is the sideband amplitude, $\hat{a}, \hat{a}^\dagger$ are the cavity ladder operators, and we have defined a phase-tunable qubit operator,

$$\hat{\sigma}_\delta = \cos(\delta)\,\hat{\sigma}_z + \sin(\delta)\,\hat{\sigma}_x. \tag{9.38}$$

Thus, by changing the phase difference δ between the sidebands, we can target any operator in the $\hat{\sigma}_x, \hat{\sigma}_z$ operator subspace. We note that these operators are defined in an interaction picture with respect to the original cavity and qubit Hamiltonians. Thus, using two different cavity modes and applying sideband tones to both of them, a joint measurement of the two noncommuting observables can be made. The signals from both modes are amplified as described in previous chapters, and the outputs are noisy measurements of their respective qubit operators. Further details of the setup can be read in Hacohen-Gourgy et al. (2016).

We can describe such a situation by writing a stochastic master equation for a detector coupled to two different observables and has two different continuous outcomes, corresponding to diffusive quantum trajectories. We can account for the two measurement channels with two measurement operators for weak Gaussian measurements on both the $\hat{\sigma}_x$ and $\hat{\sigma}_z$ observables corresponding to a time interval δt and the respective characteristic measurement times of τ_x, τ_z, given by

$$\hat{\Omega}_x \propto \exp\left(-(r_x - \hat{\sigma}_x)^2 \frac{\delta t}{4\tau_x}\right), \quad \hat{\Omega}_z \propto \exp\left(-(r_z - \hat{\sigma}_z)^2 \frac{\delta t}{4\tau_z}\right). \tag{9.39}$$

Here r_x, r_z are the continuous noisy readouts of the two measurement channels. We can apply these operators in either order (as well as including the Hamiltonian evolution), because the commutators are of order δt^2 and can be neglected. Following the steps leading to Eq. (5.57), the equation of motion for the unnormalized state is given by

$$\frac{d}{dt}\begin{pmatrix}\tilde{\alpha}\\\tilde{\beta}\end{pmatrix} = \left(\frac{r_z}{2\tau_z}\hat{\sigma}_z + \frac{r_x}{2\tau_x}\hat{\sigma}_x\right)\begin{pmatrix}\tilde{\alpha}\\\tilde{\beta}\end{pmatrix}. \tag{9.40}$$

Following the analysis given in Chapter 5, and adding in the different Hamiltonian for the qubit, $\hat{H}_{s,2} = (\tilde{\Delta}/2)\hat{\sigma}_y$, we find the following Itô form stochastic master equation for the qubit dynamics:

$$\frac{dx}{dt} = \tilde{\Delta}z - \frac{x}{2\tau_z} + \frac{1-x^2}{\sqrt{\tau_x}}\xi_x - \frac{xz}{\sqrt{\tau_z}}\xi_z, \tag{9.41}$$

$$\frac{dz}{dt} = -\tilde{\Delta}x - \frac{z}{2\tau_x} + \frac{1-z^2}{\sqrt{\tau_z}}\xi_z - \frac{xz}{\sqrt{\tau_x}}\xi_x. \tag{9.42}$$

Here, we have focused on just the $x - z$ dynamics and will restrict our attention to the $x - z$ plane of the Bloch sphere for simplicity. Notice that the form of the equations could have been guessed by comparing with equations (5.74), by making a suitable redefinition of the x, y, z labels of the Bloch sphere. We have expressed the measurement records r_x, r_z in terms of their respective signal and noise terms,

$$r_x = x + \sqrt{\tau_x}\xi_x, \quad r_z = z + \sqrt{\tau_z}\xi_z, \tag{9.43}$$

where ξ_x, ξ_z are uncorrelated white-noise random variables defined as

$$\langle \xi_i(t)\xi_j(t')\rangle = \delta_{ij}\delta(t - t'). \tag{9.44}$$

Notice that the equations (9.41) and (9.42) now have a symmetry between x and z, broken only by the presence of the unitary drive, and the unequal measurement characteristic times, τ_x, τ_z. A fully symmetric situation can be obtained by adding a third measurement channel on $\hat{\sigma}_y$ to generate dynamics over the entire state space (Ruskov et al., 2010).

It is instructive to restrict our attention to pure states and parameterize the state entirely by the angle θ on the $x - z$ great circle of the Bloch sphere, $z = \cos\theta, x = \sin\theta$. An equation of motion for θ cannot be directly derived from the preceding Itô form equations, because the regular rules of calculus are not valid in general with Itô calculus. Returning to the Stratonovich form of the stochastic differential equation, we can find the generalization of Eqs. (5.72, 5.73, 5.74) to be

$$\frac{dx}{dt} = \tilde{\Delta}z + \frac{r_x(1 - x^2)}{\tau_x} - \frac{r_z xz}{\tau_z}, \tag{9.45}$$

$$\frac{dz}{dt} = -\tilde{\Delta}x + \frac{r_z(1 - z^2)}{\tau_z} - \frac{r_x xz}{\tau_x}. \tag{9.46}$$

Now, the substitution in terms of the Bloch angle θ can be made to find

$$\dot{\theta} = \tilde{\Delta} - \frac{r_z}{\tau_z}\sin\theta + \frac{r_x}{\tau_x}\cos\theta. \tag{9.47}$$

To transform this equation to the Itô form, we can follow the path laid out in Chapter 5. However, it is sometimes faster to use a general result from the theory of stochastic differential equations. Consider a general system with coordinates x_j, $j = 1,\ldots N$ and stochastic noise terms ξ_k, with $k = 1,\ldots,M$. A Stratonovich differential equation of the form

$$\dot{x}_i = f_i + \sum_j L_{ij}\xi_j \tag{9.48}$$

with drift term f_i and where the diffusion term L_{ij} is equivalent to the analogous Itô stochastic differential equation of the same type (Gardiner et al., 1985), but with an additional drift term,

$$d_i = (1/2)\sum_{j=1}^{M}\sum_{k=1}^{N}\frac{\partial L_{ij}}{\partial x_k}L_{kj}. \tag{9.49}$$

Applied to our system, we have $N = 1, M = 2$. Substituting the readouts (9.43) into the equations of motion, and applying this conversion formula, we arrive at the Itô form of the stochastic differential equation

$$\dot{\theta} = \tilde{\Delta} + \sin\theta\cos\theta\left(\frac{1}{2\tau_x} - \frac{1}{2\tau_z}\right) - \frac{\sin\theta}{\sqrt{\tau_z}}\xi_z + \frac{\cos\theta}{\sqrt{\tau_x}}\xi_x. \tag{9.50}$$

The preceding equation reveals a very nice result: the equation of motion for the jointly measured qubit is simply described by diffusion on a circle. Furthermore, if the measurement strengths are equal, $\tau_x = \tau_z = \tau$, and we turn off the drive,

$\tilde{\Delta} = 0$, then there is no drift, and the process becomes entirely diffusive. In this equal strength case, it is possible to define a new stochastic variable,

$$\xi_\theta = -\sin\theta\,\xi_z + \cos\theta\,\xi_x. \tag{9.51}$$

This definition is quite nice because it has zero mean, and autocorrelation

$$\begin{aligned}\langle\xi_\theta(t)\xi_\theta(t+\tau)\rangle &= \langle(-\sin\theta\,\xi_z(t)+\cos\theta\,\xi_x(t))(-\sin\theta\,\xi_z(t+\tau)+\cos\theta\,\xi_x(t+\tau))\rangle\\ &= (\sin^2\theta+\cos^2\theta)\delta(\tau)\\ &= \delta(\tau).\end{aligned} \tag{9.52}$$

Here, we used the statistical independence of the two noise sources from each independent detector (Jordan and Büttiker, 2005). Consequently, we may view the two noise sources combining to make a single effective noise source ξ_θ, so the equation of motion is simple diffusion on a circle,

$$\dot{\theta} = \frac{\xi_\theta}{\sqrt{\tau}}. \tag{9.53}$$

Physically, this simply means that when each measurement is of equal strength, neither one is effective at collapsing the quantum state – it just diffuses aimlessly, and will never collapse. This is an important lesson: measurement does not necessarily lead to collapse!

A stochastic path integral analysis can be applied to this problem. In that case, the stochastic action can be integrated over the auxilliary variables and stochastic variables, leaving the stochastic action:

$$\mathcal{S} = -\frac{\tau}{2}\int_0^t dt'\dot{\theta}^2. \tag{9.54}$$

The Gaussian form of the action permits the exact calculation of correlation functions of the measurement readouts or quantum state variables (Chantasri et al., 2018)

For this case, it is worthwhile to mention the Fokker–Planck formalism, which is complementary. Consider an Itô form stochastic differential equation

$$\dot{x} = \mu(x) + \sigma(x)\xi(t), \tag{9.55}$$

where μ is the drift and σ is the diffusion term, both of which can depend on the variable x, and we keep the discussion to one dimension for simplicity. Let us introduce the probability distribution (density) $p(x,t)$ for the stochastic variable x. This distribution then satisfies the Fokker–Planck equation (Gardiner et al., 1985)

$$\frac{\partial p}{\partial t} = -\frac{\partial}{\partial x}(\mu p) + \frac{1}{2}\frac{\partial^2}{\partial x^2}(\sigma^2 p). \tag{9.56}$$

In the preceding case, we identify the variable x with the polar angle of the Bloch sphere θ. The simplified Fokker–Planck equation for our problem is given by

$$\frac{\partial p}{\partial t} = \frac{1}{2\tau} \frac{\partial^2 p}{\partial \theta^2}. \tag{9.57}$$

Given initial condition θ_0 at time $t = 0$, the solution is not that of diffusion on a line, because the trajectories can wind around the circle multiple times. Accounting for this possibility, the solution is given by

$$p(\theta, t) = \frac{\tau}{2\pi t} \sum_{k=-\infty}^{\infty} \exp\left[-\frac{\tau}{2t}(\theta - \theta' + 2\pi k)^2\right]. \tag{9.58}$$

At short times, only the $k = 0$ winding number is relevant, and the usual diffusive distribution function is obtained. At long time t, the distribution will spread evenly over the circle. This can be seen more easily by using the Poisson summation formula. That is, $\sum_n s(n) = \sum_k S(k)$ where the sums are over all integers and $S(k)$ is the Fourier transform of $s(n)$,

$$S(k) = \int_{-\infty}^{\infty} s(x) e^{-2\pi i k x} dx. \tag{9.59}$$

Applying this result we find

$$p(\theta, t) = \frac{1}{2\pi} \sum_{n=-\infty}^{\infty} \exp\left[in(\theta - \theta') - n^2 t/(2\tau)\right]. \tag{9.60}$$

This form of the solution is valid, as can be checked directly. For long time t, only the $n = 0$ term will contribute, corresponding to a constant solution with normalization of $1/2\pi$, a uniform distribution on the circle. In Fig. 9.6(c), the equilibrium distributions of $p(\theta)$ are shown for different choices of τ_x, τ_z. By changing the ratio of the measurement strengths, the quantum state dynamics can be continuously tuned between collapse to the eigenstates of the strongest measurement to unbiased diffusion around the Bloch sphere.

9.4 Entanglement by Measurement: Direct and Indirect

One of the most dramatic applications of the physics of quantum measurement is its role in the creation of quantum entanglement. We recall from Chapter 2 in Section 2.3 that when one measures an observable that has a degenerate eigenvalue, then that measurement outcome has the result of projecting the state on the subspace spanned by the sum of the projectors, resulting in coherence left in the postmeasurement state. This same basic effect is also the case when we generalize to multiple systems, even when separated in space.

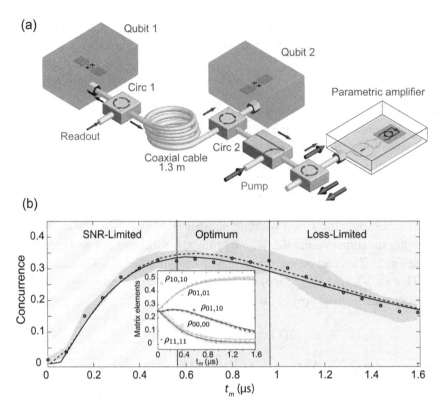

Figure 9.7 Experimental setup and data for measurement-induced entanglement creation (Roch et al., 2014). (a) Two qubits in separate cavities, connected by 1.3 m of coaxial cable, can be entangled with each other, despite there being no direct interaction. A microwave signal is sent sequentially in and out of both cavities, interacting with both qubits separately. The resulting signal is amplified with a phase-sensitive amplifier, realizing a partial-parity measurement. (b) Average concurrence, a measurement of entanglement, is plotted versus time in units of microseconds, when both qubits begin the separable $|x = 1\rangle$ eigenstates. The concurrence rises initially as the measurement begins to distinguish the odd parity signal from the noise (SNR-Limited), creating entanglement. After the characteristic measurement time of the detection process, the entanglement saturates (the Optimum). (c) Intrinsic decoherence processes take over, destroying the entanglement (Loss-Limited). Adapted with permission from the American Physical Society.

Let us consider the simplest case, where there are two qubits, A and B, each prepared in an equal coherent superposition of $|0\rangle$ and $|1\rangle$, which are initially separable. The initial state is then

$$|\Psi\rangle = \frac{1}{2}(|0\rangle_A + |1\rangle_A) \otimes (|0\rangle_B + |1\rangle_B) = \frac{1}{2}(|00\rangle + |01\rangle + |10\rangle + |11\rangle). \quad (9.61)$$

Here we used the shorthand $|\alpha\beta\rangle = |\alpha\rangle_A \otimes |\beta\rangle_B$. We now define a *parity measurement*, where the outcome of the measurement indicates the parity of the qubit $\alpha + \beta$ modulo 2. Thus, if the result of the measurement is odd, we know the state must span the subspace defined by the basis kets $|00\rangle$ and $|11\rangle$, while if the result of the measurement is odd, we know the state must span the subspace defined by the basis kets $|01\rangle$ and $|10\rangle$. A projective parity measurement is mathematically given with two projection operators $\hat{\Pi}_e$ and $\hat{\Pi}_o$,

$$\hat{\Pi}_e = |00\rangle\langle00| + |11\rangle\langle11|, \quad \hat{\Pi}_o = |01\rangle\langle01| + |10\rangle\langle10|. \tag{9.62}$$

The result of applying either projection operator on the initially separable state is given by

$$\hat{\Pi}_e|\Psi\rangle = \frac{1}{\sqrt{2}}\frac{|00\rangle + |11\rangle}{\sqrt{2}}, \quad \hat{\Pi}_o|\Psi\rangle = \frac{1}{\sqrt{2}}\frac{|01\rangle + |10\rangle}{\sqrt{2}}. \tag{9.63}$$

Both of the resulting states are fully entangled, and either outcome (even or odd) occurs with probability of $P_{e,o} = \langle\Psi|\hat{\Pi}_{e,o}|\Psi\rangle = 1/2$. Surprisingly, the entanglement takes place through the collective measurement of the two parties, and not necessarily through any direct interaction.

Direct Methods: Joint or Sequential Interaction

We can devise appropriate qubit observables to implement these collective measurements. The physical system must be constructed so as to engineer a two-qubit Hamiltonian of the form

$$\hat{H}_p = \Delta E \hat{\sigma}_z^A \hat{\sigma}_z^B. \tag{9.64}$$

The eigenvalues $\pm\Delta E$ of this Hamiltonian are doubly degenerate and correspond to the even subspace (00) or (11) and the odd subspace (01) or (10).

Often in the laboratory it is easier to engineer a "partial parity" measurement having three outcomes. The partial parity measurement is associated with a modified observable, $\hat{H}_{pp} = \Delta E'(\hat{\sigma}_z^A + \hat{\sigma}_z^B)$, the eigenvalues of which correspond to the projection operators $\hat{\Pi}_{00}, \hat{\Pi}_{11}, \hat{\Pi}_o$.

Circuit QED implementations of these two types of measurements are shown in panels (a) and (b) of Fig. 9.8. In panel (a), radiation in the microwave frequency field is sequentially interacted with qubit A and qubit B via a dispersive coupling, located in separate cavities. In this case, we can model the two dispersive interactions with the Hamiltonian,

$$\hat{H} = \chi_A \hat{\sigma}_z^A a^\dagger a + \chi_B \hat{\sigma}_z^B \hat{a}^\dagger \hat{a} + \hbar\omega_c \hat{a}^\dagger \hat{a} + \epsilon_A \hat{\sigma}_z^A + \epsilon_B \hat{\sigma}_z^B. \tag{9.65}$$

Here the electromagnetic mode is quantized with creation and annihilation operators $\hat{a}, \hat{a}^\dagger$. The two qubit interactions are described by the Jaynes–Cummings model in the dispersive approximation, Eq. (5.25), where we have introduced the more compact notation, $\chi = g^2/\Delta$, for each interaction. The combined interaction can be viewed as a shift of the oscillator frequency, $\hbar\omega_c \rightarrow \hbar\omega_c + \chi_A\hat{\sigma}_z^A + \chi_B\hat{\sigma}_z^B$, which is dependent on the states of the two qubits.

The quantum dynamics of this model can be solved in the Heisenberg equation of motion picture,

$$\dot{\hat{a}} = \frac{i}{\hbar}[\hat{H}, \hat{a}] = i\hat{\omega}\hat{a}, \tag{9.66}$$

where $\hat{\omega} = \omega_c + \chi_A\hat{\sigma}_z^A + \chi_B\hat{\sigma}_z^B$, which we can treat as four different c-numbers in general, when the two qubits are in their eigenstates. The injected states of electromagnetic radiation are typically in coherent states, so we can apply the equation of motion to a coherent state $|\alpha\rangle$ to obtain $\dot{\alpha} = i\tilde{\Omega}\alpha$, which has the solution $\alpha(t) = \exp(i\tilde{\Omega}t)\alpha(0)$. Light will exit the cavities after the decay $\kappa_{A,B}^{-1}$, which sets the interaction time in the preceding analysis. By the linearity of quantum mechanics, given an initial separable state of the bi-qubit/ cavity system, the postinteraction state is of the form

$$|\psi\rangle|\alpha\rangle \rightarrow \sum_j c_j|j\rangle|\alpha_j\rangle, \tag{9.67}$$

where qubit states $|j\rangle$ take on the states $|00\rangle, |01\rangle, |10\rangle, |11\rangle$, as $j = 1, 2, 3, 4$. Here, the coefficients c_j are determined from the initial two-qubit state, $\sum_j c_j|j\rangle$. The coherent states $|\alpha_j\rangle$ are of the form

$$|\alpha_j\rangle = |\alpha e^{i\phi_j}\rangle, \tag{9.68}$$

where the phases $\phi_j = \pm 2\chi_A/\kappa_A \pm 2\chi_B/\kappa_B$, and the pluses and minuses correspond to the eigenvalues of the measured collective operator. Notice that the dispersive coupling only changes the phase of the coherent state, and not the amplitude (expected photon energy).

After amplification of the coherent state (see the discussion in Chapter 8), the result of the readout implements an effective measurement of the two-qubit operator:

$$\hat{O} = (2\chi_A/\kappa_A)\hat{\sigma}_z^A + (2\chi_B/\kappa_B)\hat{\sigma}_z^B. \tag{9.69}$$

As a result, we can implement the partial parity measurement by setting $\chi_A/\kappa_A = \chi_B/\kappa_B$. Such a choice leads to the possibility of three acquired phase shifts, $-2\phi, 0, 2\phi$, where $\phi = 2\chi/\kappa$. This choice implements the "lollipop" diagram in Fig. 9.8(c), where the coherent states corresponding to the $|01\rangle$ and $|10\rangle$ states overlap. Phase-sensitive amplification then takes place along the x_2 axis to distinguish the three possibilities. Like in the single-qubit measurement case, it takes time to distinguish the various possibilities. The signal-to-noise ratio to distinguish

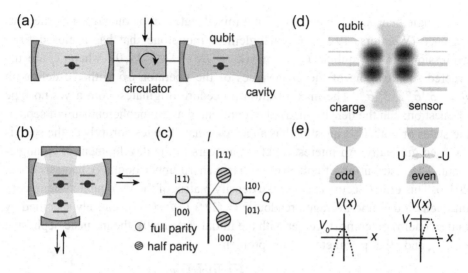

Figure 9.8 Illustrations of different types of physical entangling measurements. (a) Design of a superconducting circuit measurement of "half parity." Microwave frequency states interact with two superconducting circuits in separate cavities, sequentially. The net dispersive phase shift on the coherent state of the radiation is such that the joint two-qubit states $|01\rangle, |10\rangle$ are indistinguishable from each other, as shown on the phase space quadrature plots. (b) A full parity measurement is implemented with a dispersive phase shift much larger than the cavity linewidth. Ideal phase responses of a coherent tone reflected on each cavity are plotted versus frequency detuning for different qubit states. The parity probe tones are centered on the odd-parity resonances. (c) The phase space (IQ) plots show the ideal steady-state reflected tone for the shown qubit configuration. Dashed circles are centered on all possible steady-state responses. (d) Implementation of a parity measurement for two quantum dots measured by a common quantum point contact. (e) The different potential landscapes of the quantum point contact, common to the odd and even configurations of the electrons in the two quantum dots.

the three possibilities is related to the number of photons in the cavity, as well as the other parameters of the measurement, Eq. (5.11). This method was used to entangle superconducting qubits living in separate cavities in Roch et al. (2014), as shown in Fig. 9.8(a).

The creation of entanglement during the measurement process does not always happen as a continuous process, despite the measurement being continuous. The effect of "entanglement genesis" can be seen in individual runs of the experiment, where an initially mixed state becomes suddenly entangled at a certain point in time during the continuous measurement, along the way to a fully entangled odd-parity state (Williams and Jordan, 2008; Saira et al., 2014). This effect is related to the opposite, dissipative effect, where entanglement suddenly dies in the finite time (Yu and Eberly, 2009).

Entanglement for a general two-qubit mixed state can be quantified by the concurrence (Wootters, 1998), an entanglement monotone that has a closed form expression for two qubits: $C = \max(\lambda_1 - \lambda_2 - \lambda_3 - \lambda_4, 0)$, where λ_i are the ordered square roots of the eigenvalues of the operator $\hat{\rho}\tilde{\rho}$ with the definition $\tilde{\rho} = \hat{\sigma}_y^A \hat{\sigma}_y^B \hat{\rho}^* \hat{\sigma}_y^A \hat{\sigma}_y^B$. The maximization procedure originates from a yes/no type of question: can the density matrix be represented as an incoherent sum of separable states or not? This question has an answer that changes abruptly as the state is changed gradually. An interesting effect appears in the development of entanglement. The distribution of concurrence has a sharp upper bound (Chantasri et al., 2016). This effect occurs because the density matrix of the two-qubit system, conditioned on the time-averaged readout $\bar{r} = (1/t)\int_0^t dt' r(t')$, is entirely specified by that (random) outcome, together with the initial state $\rho_{ij}(0)$, the intrinsic dephasing rate γ_{ij}, and other parameters of the problem:

$$\rho_{ij}(t) = \rho_{ij}(0)\frac{\sqrt{p(\bar{r}|i)p(\bar{r}|j)}e^{-\gamma_{ij}t}}{\sum_j \rho_{kk}(0)p(\bar{r}|k)}. \tag{9.70}$$

This equation for the continuous state collapse follows immediately from the quantum Bayesian approach, or from implementing POVM theory with a measurement operator,

$$\hat{\Omega}(\bar{r}) = (|01\rangle\langle 01| + |10\rangle\langle 10|)\sqrt{p(\bar{r}|\text{odd})} + |00\rangle\langle 00|\sqrt{p(\bar{r}|00)} + |11\rangle\langle 11|\sqrt{p(\bar{r}|11)}, \tag{9.71}$$

where each of the distributions $p(\bar{r}|j)$ is Gaussian, whose mean and width are set by the interaction parameters and amplification physics, as described in Chapters 5 and 8. We assume no additional drive is present.

The concurrence of the measured state, considered as a function of the time-averaged readout $\bar{r}$, has a maximum when $\bar{r} = 0$, corresponding to the readout where the state is most likely in the odd-parity subspace. We consider an initial state discussed earlier where each qubit is in an $|+x\rangle$ eigenstate, and assuming a symmetric system, this upper bound on concurrence takes the form

$$C_{max} = \frac{e^{-\gamma t} - e^{-\Gamma_m t}}{1 + e^{-\Gamma_m t}}. \tag{9.72}$$

Here Γ_m is the measurement rate, given by the ratio of the square of the mean of the distribution to its width, the inverse of the characteristic measurement time Eq. (5.54). We notice a competition between the measurement rate and the intrinsic dephasing rate, γ. If $\gamma = 0$, the maximum concurrence value begins at 0 at the initial time, as is correct for initially separable states, and then climbs with time to asymptotically approach 1, the maximum value for perfect entanglement. However, given finite intrinsic dephasing, the maximum attainable value of entanglement drops exponentially with the dephasing rate. The average concurrence for the experiment is plotted in Fig. 9.7, demonstrating the process of creation of entanglement, followed by its gradual decay. Both the theoretical and experiment probability distribution have a sharp upper cutoff (Chantasri et al., 2016). Physically, this indicates that there is an upper limit on how fast entanglement can be created by the continuous measurement in this situation, even for rare events of the measurement process.

Full Parity Measurement

In order to make a full parity measurement, the response of the detector to the $|00\rangle$ and $|11\rangle$ state must also be the same, in addition to the odd parity subspace. In the circuit QED case, this can be accomplished by making the dispersive interaction strength χ of both qubits larger than the cavity linewidth κ.

To see this in detail, we recall the reflection scattering coefficient from a cavity, discussed in Eq. (7.25), is given by

$$r = -\frac{1 + i\theta/2}{1 - i\theta/2},$$ (9.73)

where $\theta = 4QA\hat{z}$ is the phase shift acquired by the reflection coefficient for small θ in the notation of Chapter 7. Let us adopt the preceding notation defining $\chi = A\Omega$ as the dispersive shift, and recall $\kappa = \Omega/Q$ is the cavity decay rate. The dispersive frequency shift times the Wigner delay gives us the relationship $\theta = 4(\chi/\kappa)\hat{z}$. We are now interested in the reflection phase for large values of θ. Direct calculation shows that the magnitude of r is 1, indicating that this one-sided cavity reflects all the light entering it. We can thus express $r = \exp(i\Phi)$, a pure phase. Taking the logarithm of the expression for r (9.73) gives

$$\Phi = \pi + 2\arctan(\theta/2) = \pi + 2\arctan\left(\frac{2\chi\hat{z}}{\kappa}\right).$$ (9.74)

Dropping the reflection phase of π, we see this function ranges from $-\pi$ to π as $2\chi/\kappa$ varies over its range. The phases $-\pi$ and π are identified, so the same phase response is obtained if there is a large positive or negative detuning of the cavity, larger than κ.

In the context of the full parity measurement, the operator $\hat{z}$ is replaced by the operator $\hat{O}$ defined in Eq. (9.69). Thus we see that if the two dispersive coupling and decay rates are set equal to each other, then the phase responses of the odd parity states are equal to each other, and the phase responses of the even parity states are equal to each other as well. This situation is sketched in Fig. 9.8(b). The resulting measurement operator is then

$$\hat{\Omega}(\bar{r}) = (|01\rangle\langle01| + |10\rangle\langle10|)\sqrt{p(\bar{r}|\text{odd})} + (|00\rangle\langle00| + |11\rangle\langle11|)\sqrt{p(\bar{r}|\text{even})},$$ (9.75)

where now there are only two Gaussian distributions, $p(\bar{r}|\text{even}), p(\bar{r}|\text{odd})$, corresponding to the two-qubit state being either in the odd or even parity subspace.

The quantum dot case is similar. If two DQDs are present on either side of the QPC, then the QPC current can resolve when the electrons in the quantum dots are both in the outer dots, or both in the inner dots, but cannot distinguish when one is in the inner dot and one is in the outer dot, realizing a partial parity measurement. A full parity measurement can be designed by tuning the detector to a point where the linear response vanishes, and the response is purely quadratic (Mao et al., 2004). Such a detector then gives the same response to the two situations when the two electrons are either both close or both far. Another possibility is shown in Figure 9.8, panel (d). By turning the dots parallel to the QPC two odd parity subspaces correspond to the same scattering potential of the scattering electrons in the QPC and thus give the same response (Trauzettel et al., 2006). The two even parity cases shift the scattering potential in one direction or another, relative to the symmetric point, creating a charge dipole. The heights of the two possible barriers are the same, so the electrical current will be the same in both cases, giving identical measurement outcomes. However, the difference of the dipole corresponding to a spatial shift of the position of the scattering barriers results in an acquired phase in the even parity subspace, that will be conditional on the electrical current realization. The acquired phase, although stochastic, can be inferred from the measurement readout, and in fact, can be reversed with feedback as been experimentally demonstrated in an analogous experiment in the Leonardo DiCarlo group (Riste et al., 2013). Related parity measurement designs using the phase periodicity effect have been proposed in Ruskov et al. (2006) and Haack et al. (2010) using Mach–Zender-type mesoscopic detectors.

Indirect Methods: Entanglement through Which-Path Interference

Another important method to generate entanglement is through which-path interference. The method is commonly used for long-distance Bell inequality tests when entanglement between two biphotons produced via spontaneous parametric down conversion result in a coincidence count after passing through a beam splitter. The result of the coincidence count is to transfer the entanglement between the two detected photons with the two undetected photons, such that now the two *undetected* photons are entangled, even though they never interacted in the past (Zukowski et al., 1993). This process is known as "entanglement swapping" and originates in the multiparticle interference effect, caused by indistinguishability of the different photon paths.

Here we describe how to create entanglement between two quantum emitters, such as atoms, nitrogen-vacancy centers in diamond, or nuclear or electron spins,

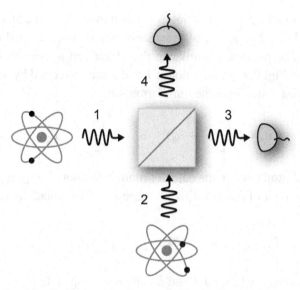

Figure 9.9 Two identical atoms, both prepared in the first excited state, will spontaneously decay, emitting a photon. Combining the (possible) emission of the atoms on a 50/50 beam splitter, the subsequent click of a detector indicates a photon was emitted. The fact that it is impossible to tell which atom emitted the photon results in an entangled state of the atoms, despite never having interacted. Replacing the click detectors with homodyne measurement with two local oscillators, out of phase by $\pi/2$, results in a continuous entanglement creation process.

that can spontaneously decay, emitting a photon. The effect is perhaps even more surprising, because there is no entanglement initially, and uses only what is usually seen as a source of decoherence and dissipation – spontaneous emission. If we consider two of these emitters of the same energy splitting, that can be spatially far apart, and combine these two possible photons on a beam splitter, then the detection of a single photon can create entanglement between these two distant emitters (Wiegner et al., 2010), as shown in Fig. 9.9. This method is remarkable in that the two entangled objects never had any interaction with each other, and the emitted photon comes from only one of the atoms – but it is impossible to tell which one. For free atoms, this is a challenging experiment, because light emits in all directions. However, high-efficiency detection of photons can be made in superconducting circuits, because the emitted radiation can be captured inside cavities and transferred to detectors via transmission lines (Campagne-Ibarcq et al., 2016; Naghiloo et al., 2016; Narla et al., 2016).

To see how this works in more detail, we consider the two-atom state in an initially separable state $|\psi\rangle = |ee\rangle$, where e indicates an excited atom and g indicates an atom in the ground state. One of the atoms emits a photon, decaying from e to g and creating a single photon at the frequency of the atomic transition, $\hbar\Omega = E_e - E_g$. Let γ be the rate of spontaneous emission, assumed to be identical for each atom. The physics is similar to that described in Section 6.4. Adopting the numbering in Fig. 9.9, photon modes 1 and 2 are converted by the linear beam splitter into modes 3 and 4 via the transformation,

$$\hat{a}_3^\dagger = \frac{1}{\sqrt{2}}(\hat{a}_1^\dagger + \hat{a}_2^\dagger), \quad \hat{a}_4^\dagger = \frac{1}{\sqrt{2}}(\hat{a}_1^\dagger - \hat{a}_2^\dagger). \tag{9.76}$$

By detecting a photon click in the detector monitoring port 3 or port 4, corresponding to a measurement of $\hat{a}_3^\dagger\hat{a}_3$ or $\hat{a}_4^\dagger\hat{a}_4$, the associated Lindblad operators are given by

$$\hat{L}_3 = \sqrt{\gamma}(\hat{\sigma}_-^A + \hat{\sigma}_-^B), \quad \hat{L}_4 = i\sqrt{\gamma}(\hat{\sigma}_-^A - \hat{\sigma}_-^B), \tag{9.77}$$

where the two atoms are labeled A and B. The postjump state is therefore given by Eq. (6.12), so the resulting conditional states are given by

$$\hat{L}_3|ee\rangle \propto |eg\rangle + |ge\rangle, \quad \hat{L}_4|ee\rangle \propto |ge\rangle - |eg\rangle. \tag{9.78}$$

Thus the two odd-parity Bell states are produced in the process. This click happens with a rate γ. A second click signals the emission of both photons, resulting in the ground state $|gg\rangle$. Note that $\hat{L}_3\hat{L}_4|ee\rangle = \hat{L}_4\hat{L}_3|ee\rangle = 0$, so if one detector fires, it must also fire a second time when the second photon is detected. This dynamical process of entanglement creation and death is shown in Fig. 9.10(a), for different realizations of the quantum jump trajectories, taking into account both the jump and no-jump dynamics.

Instead of photon detectors, we can replace the detectors with other beam splitters and combine the emission with two local oscillators that have different phase relations in general. This corresponds to homodyne detection of each port of the interferometer (Viviescas et al., 2010). There is a subtlety here, because in order to efficiently generate entanglement, the relative phase of the two local oscillators must be 90° out of phase in order to erase any quantum information about which atom the radiation originated from. The preceding two Lindblad operators are still applicable, but now we have two different homodyne outputs we label r_3, r_4. These outputs correspond to the two continuous outputs of the amplification of the fields with the two homodyne detection schemes. Repeating the derivation leading to Eq. (6.10) for the most general case of m readouts with Lindblad operators $\hat{L}_\nu$, Wiener increments dW_ν and detection efficiencies η_ν gives us the most general form of the stochastic master equation of a normalized density matrix in Itô form,

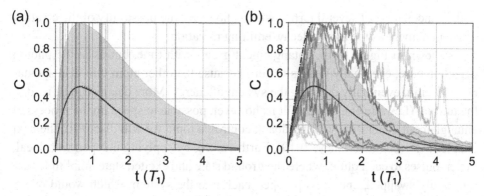

Figure 9.10 Remote entanglement production: concurrence is plotted versus time, in units of the inverse decay rate, $T_1 = 1/\gamma$. (a) Continuous monitoring with two click detectors. Different quantum jump trajectories are shown, indicating the perfect creation of entanglement with the first click, and its death in the second. (b) Continuous monitoring with two homodyne detectors. Different realizations of the diffusive quantum trajectories are shown. The dot-dashed line indicates the fastest entanglement can be created with this technique. The ensemble averages are shown in the solid curve in both plots, and coincide. All trajectories begin in the separable state $|ee\rangle$ (Lewalle et al., 2020).

$$d\hat{\rho} = -i[\hat{H}, \hat{\rho}] + \sum_{\nu=1}^{m} \left(\hat{L}_\nu \hat{\rho} \hat{L}_\nu^\dagger - \frac{1}{2}(\hat{L}_\nu^\dagger \hat{L}_\nu \hat{\rho} + \hat{\rho} \hat{L}_\nu^\dagger \hat{L}_\nu) \right. \tag{9.79}$$

$$\left. + \sqrt{\eta_\nu} \left[(\hat{L}_\nu \hat{\rho} + \hat{\rho} \hat{L}_\nu) - \text{Tr}[\hat{L}_\nu \hat{\rho} + \hat{\rho} \hat{L}_\nu^\dagger] \hat{\rho} \right] dW_\nu \right).$$

The corresponding shifted and rescaled readouts are given by

$$r_\nu = \sqrt{\eta_\nu} \text{Tr}[(\hat{L}_\nu + L_\nu^\dagger)\hat{\rho}] + \frac{dW_\nu}{dt}, \tag{9.80}$$

where the index $\nu = 1, \ldots, m$.

Applied to our case of $m = 2$, for the Lindblad operators given earlier, we have the stochastic master equation in the case of efficient measurement, given by

$$d\hat{\rho} = -i[\hat{H}, \hat{\rho}] + 2\gamma \hat{\sigma}_-^B \hat{\rho} \hat{\sigma}_+^A + 2\gamma \hat{\sigma}_-^A \hat{\rho} \hat{\sigma}_+^B - \gamma(\hat{\Pi}_e^A + \hat{\Pi}_e^B)\hat{\rho} - \gamma \hat{\rho}(\hat{\Pi}_e^A + \hat{\Pi}_e^B)$$
$$+ \sqrt{\gamma}[[(\hat{\sigma}_-^A + \hat{\sigma}_-^B)\hat{\rho} + \hat{\rho}(\hat{\sigma}_-^A + \hat{\sigma}_-^B)]dW_3 + i[(\hat{\sigma}_+^A - \hat{\sigma}_+^B)\hat{\rho} - \hat{\rho}(\hat{\sigma}_+^A - \hat{\sigma}_+^B)]dW_4]$$
$$- \sqrt{\gamma}\text{Tr}[(\hat{\sigma}_x^A + \hat{\sigma}_x^B)\hat{\rho}]\hat{\rho} dW_3 - \sqrt{\gamma}\text{Tr}[(\hat{\sigma}_y^A - \hat{\sigma}_y^B)\hat{\rho}]\hat{\rho} dW_4, \tag{9.81}$$

with the two readouts

$$r_3 = \text{Tr}[(\hat{\sigma}_x^A + \hat{\sigma}_x^B)\hat{\rho}] + dW_3/dt, \quad r_4 = \text{Tr}[(\hat{\sigma}_y^A - \hat{\sigma}_y^B)\hat{\rho}] + dW_4/dt. \tag{9.82}$$

Here we introduced the Pauli matrices for each atom, as well as the projectors on the excited states $\hat{\Pi}_e$ for both atoms. The symmetry of Lindblad operators results in the cancellation of several terms, which allows for the possibility of the creation of entanglement as the atoms decay, emitting radiation.

As shown in Fig. 9.10(b), entanglement grows with time, but as in the photon detection case, eventually decays to zero because the other atom will also spontaneously emit, eventually draining the system of energy. More details can be read in the tutorial (Lewalle et al., 2021). It is, however, possible to speed up the process of entanglement generation with feedback control, a topic we will discuss in the next chapter (Martin and Whaley, 2019). Furthermore, by applying independent, local, fast π-pulses to each qubit, where the ground state and excited state amplitudes are repeatedly swapped, energy is pumped back into the system, which would otherwise be lost by the dissipation. This provides a way to keep the two atoms entangled indefinitely, despite being far away (Lewalle et al., 2020).

Exercises

Exercise 9.4.1 Show that the quantum partial-collapse described by (4.14) applied to the state $\hat{\sigma}_x \hat{\rho}' \hat{\sigma}_x$ recovers the state $\hat{\sigma}_x \hat{\rho} \hat{\sigma}_x$. A final π-pulse restores the original state $\hat{\rho}$.

Exercise 9.4.2 Calculate the uncollapse probability P_{uc} using state $\hat{\sigma}_x \hat{\rho}' \hat{\sigma}_x$, to find the result (9.1).

Exercise 9.4.3 Show that the stochastic Hamiltonian is a constant of motion for the most likely paths.

Exercise 9.4.4 In the dispersive measurement example, solve for the most likely paths (x, y, z, r) exactly as a function of time in the special case where $\epsilon = \Delta = 0$. Take the initial and final conditions as x_i, y_i, z_i and the final conditions x_f, y_f, z_f. Can all final boundary conditions be met? Hint: Show that $\dot{r} = 0$ for the most likely readout and use that constant of motion, together with the stochastic Hamiltonian, to construct the solution.

Exercise 9.4.5 Show that the equations of motion for the most likely path for the dispersive measurement with no dephasing are given by (9.18, 9.21, 9.22).

Exercise 9.4.6 Suppose the most likely measurement result r is substituted into the equations of motion for θ, p_θ. Show the stochastic Hamiltonian (9.23) describes the resulting dynamics.

Exercise 9.4.7 Start with the original stochastic action involving the readout r. Integrate out the measurement result r using a Gaussian integral at each time step. Show the resulting stochastic Hamiltonian is the same as that in the previous exercise.

Exercise 9.4.8 Work out the Itô form of the stochastic master equation when one detector is measuring $\hat{\sigma}_z$ and the other is measuring $\sigma_\delta = (\cos \delta)\hat{\sigma}_z + (\sin \delta)\hat{\sigma}_x$. What happens in the limiting cases where $\delta = 0, \pi/4, \pi/2$?

Exercise 9.4.9 Suppose a two-qubit system is initialized in a separable state where each qubit is prepared in the state $(|0\rangle + |1\rangle)/\sqrt{2}$. Given the measurement operator (9.71) resulting in result $\bar{r}$, find the postmeasurement state.

Exercise 9.4.10 For the previous problem, calculate the concurrence of the postmeasurement state. Plot the concurrence versus r.

10

Feedback and Control

In the previous chapters of this book, you have learned how to describe and predict the results of measurements done on quantum systems. We have focused on the probabilities of the results and the subsequent changes of the quantum state in order to correctly predict the odds of subsequent measurements. In this chapter we move from observing to acting. You are now masters of quantum mechanics, so it is time to control the quantum state. Control can be obtained by the use of unitary or nonunitary processes, monitored or unmonitored processes, or any combination thereof. We will focus here on measurement-based feedback control. This type of control involves observing some event (or nonevent) and then making a choice to act on that information. A feedback loop is created, where we observe and consequently change the physics of a quantum system, then operate on that system based on what we learned; this process is then repeated. We will begin with a general description of the theory of feedback, and then give three examples of how this is done theoretically as well as in practice.

10.1 General Theory of Feedback

When implementing feedback control, the first thing that must be done is to decide on the goal or objective of the feedback control. This could be to stabilize the quantum state to the destabilizing nature of the measurement process; it could be to make a better or more accurate measurement; or it could be to implement a quantum information task. In all cases, we can define measures of how successful the control is: Is the time to the goal shortened? Is the fidelity of the quantum process or state improved? Did the error rate go down? and so on.

We here focus on Hamiltonian control of the quantum system – given some tunable parameters X_i in the system Hamiltonian $\hat{H}(X_i)$, we can make these parameters functions of the measurement readouts r_j, where $j = 1, \ldots, m$. In the feedback loop, the measurement results are then used to control the Hamiltonian of the system, so there is an interplay between the measurement results, the quantum state disturbance, and the unitary dynamics (Wiseman, 1994). This type of control is different from pure unitary control – for example, when a time-dependent Hamiltonian is applied to coherently control the state to meet a final target or maximize sensitivity, as is the case, for example, in shortcuts to adiabaticity (Guéry-Odelin et al., 2019), or time-dependent quantum metrology (Pang and Jordan, 2017).

Let us begin with discrete generalized measurements, as described in Chapter 3. Taking the measurement operators to be the set $\{\hat{\Omega}_j\}$, we obtain some measurement result, j, with probability $P_j = \langle\psi|\hat{\Omega}_j^\dagger\hat{\Omega}_j|\psi\rangle$, where $|\psi\rangle$ is the initial state. The postmeasurement state is given by

$$|\psi_j\rangle = \frac{\hat{\Omega}_j|\psi\rangle}{\sqrt{P_j}},\tag{10.1}$$

as we saw in Chapter 3. As the observer, we can now apply control with a time-dependent Hamiltonian $H(t)$ to implement a control unitary operation $\hat{U}_j$ on the quantum state. We now know the result j of the measurement, so this unitary can also be made to depend on j, as is indicated by the subscript. Thus the new quantum state is

$$|\psi_j'\rangle = \frac{\hat{U}_j\hat{\Omega}_j|\psi\rangle}{\sqrt{P_j}}.\tag{10.2}$$

For a sequence of measurement results, $j_1, j_2, \ldots j_n$, the unitary control will then take the form of a sequence of operations applied after each measurement, $\hat{U}_1, \hat{U}_2, \ldots, \hat{U}_n$, so the final state takes the form of a nested series of operators,

$$|\psi_f\rangle \propto \hat{U}_n\hat{\Omega}_{j_n}\ldots\hat{U}_2\hat{\Omega}_{j_2}\hat{U}_1\hat{\Omega}_{j_1}|\psi\rangle.\tag{10.3}$$

Let us give a simple example of this procedure, applied to stabilize a quantum state of a qubit at a particular pure state, in the presence of a $\hat{\sigma}_z$ generalized measurement (Jordan and Korotkov, 2006). Recall from our earlier analysis that if no unitary operation is applied during a continuous $\hat{\sigma}_z$ measurement for a time t with characteristic measurement time τ_m, the resulting quantum state is given by

$$\begin{pmatrix}\tilde{\alpha}\\\tilde{\beta}\end{pmatrix} \propto \begin{pmatrix}e^{S/2}\alpha\\e^{-S/2}\beta\end{pmatrix},\tag{10.4}$$

the results from Eq. (5.58), where we defined $S = (1/\tau_m)\int_0^t dt' r(t')$, which is a random variable. Suppose we wish to restore the state to its initial condition, after the measurement has stopped. Parameterizing the pure quantum state as $(\alpha, \beta) = (\cos[\theta/2], \sin[\theta/2])$, and setting the azimuthal angle to 0 for simplicity, we see that by taking the ratio of the coefficients, the new polar angle is related to the old polar angle by

$$\tan[\tilde{\theta}/2] = e^{-S} \tan[\theta/2]. \tag{10.5}$$

Thus, a control pulse can be applied to the qubit, causing a rotation by angle $\tilde{\theta} - \theta$ in order to restore the state. This control pulse will be different after every measurement in general, but will perfectly restore the initial state. Notice that if the measurement result S is 0, then the final and initial angles are the same, so no pulse need be applied. In the case of very large measurement result $S \to \pm\infty$, the new angle corresponds to either 0 or π, the collapsed state points.

Often, we wish to implement continuous feedback control. In this case, there is typically a delay between the measurement and the feedback control operation. In such a situation, the system has usually gone on to change further since the measurement, resulting in dated information. The feedback then effectively introduces a memory effect, depending on the state of the system at some previous time. While such a realistic situation can be handled, the mathematics gets quite complicated, so we will focus on the idealized situation where this delay time can be neglected.

10.2 Continuous Quantum State Analog Stabilization

We now start to describe experiments where the observer transitions from simply watching quantum effects evolve in time to controlling them in an active fashion. Let us consider the case of stabilizing a single qubit undergoing continuous Rabi flopping between the ground and excited states (Vijay et al., 2012). In an ensemble-average, the probability of finding the system in, say, the excited state simply oscillates in a sinusoidal fashion with time. The amplitude of this oscillation decays with time as the system dephases upon interacting with a noisy environment.

We can think of such a noisy qubit as an oscillator whose frequency jitters due to dephasing. In a classical circuit, if a clock has timing jitter, one can improve stability by phase locking it to a more precise reference such as a local rubidium frequency standard or a fancy atomic clock. Further, the qubit is essentially a quantum voltage-controlled oscillator: the frequency of Rabi oscillations Ω_R is proportional to the drive amplitude of the pulse applied. This motivates the following weak-measurement-based stabilization protocol (Korotkov, 2005). We can probe the state of the qubit, which is coupled to a resonant cavity of linewidth κ, via a homodyne measurement and feed the signal directly to an analog multiplier at room temperature. There, the measurement signal is multiplied with a classical reference oscillator set to operate at the nominal Rabi frequency of the qubit. The product then controls the amplitude of the Rabi drive being delivered to the qubit. The system is configured such that if the qubit lags the reference oscillator in phase, the Rabi drive is increased, resulting in a faster quantum oscillator. Conversely, the instantaneous Rabi rate is decreased if the phase leads the classical reference.

For such a scheme to work, there is an important hierarchy of several timescales that must be respected. First, the dephasing rate due to the environment Γ_ϕ must be significantly smaller than the dephasing induced by the measurement process Γ_m. Here we are assuming that the dominant decoherence process is dephasing. This ensures that information from the quantum system is not transferred to inaccessible degrees of freedom. Next, we must not measure so strongly that we collapse the wavefunction and suppress the Rabi oscillations themselves. Thus, $\Gamma_m < \Omega_R$. Relatedly, if we want to treat the system as a qubit coupled to a Markovian bath, thereby suppressing any memory effects, we choose the linewidth $\kappa > \Omega_R$ so that the photons escaping the cavity represent the real-time state of the qubit. Similarly, the bandwidth of the superconducting near-quantum-noise-limited Josephson Parametric Amplifier (JPA) used to probe the cavity response must be larger than κ.

Figure 10.1 details the key results of this feedback experiment. In panel (a), we see the ensemble-averaged Rabi oscillations, which decay on the timescale of about a microsecond. If we take the Fourier transform of these oscillations, we observe a peak at the Rabi frequency, as shown in panel (b). The linewidth there is set by the dephasing rate. In this experiment, the homodyne measurement is arranged such that a single quadrature contains the information about the qubit state. Probing the other (squeezed) quadrature shows no signal. The strength of the measurement is varied by changing the average photon number $\bar{n}$ in the cavity. In panel (c), we see the expected linear dependence of the dephasing rate on $\bar{n}$, with the y-intercept yielding the decoherence rate associated with the environment. As desired, we can readily enter a regime where the measurement is the dominant source of dephasing, and the measurement rate is slower than the 3 MHz Rabi frequency used in the experiment. Note that this regime is accessed by having less than one microwave photon, on average, in the cavity! Thus, a very low-noise amplifier is needed to efficiently record such weak signals.

When the feedback loop is activated, the linewidth of the Rabi oscillations significantly narrows, commensurate with an oscillation with a high degree of phase coherence. This is readily seen in panel (b). Correspondingly, in the time domain, we see in panel (d) that Rabi oscillations in this closed-loop feedback cycle persist indefinitely! There are many basic features of quantum mechanics that are corroborated by this result. We have the idea that acquiring information from a quantum system necessarily perturbs it, and this results in the decay of an ensemble property. Moreover, the purity of the quantum state is not a priori degraded, and one can correct for an unwanted evolution based on information extracted from a weak measurement that does not fully project the system onto an eigenstate.

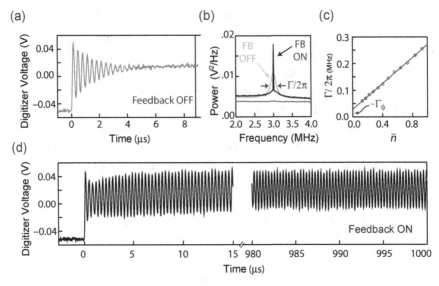

Figure 10.1 (a) Ensemble-averaged Rabi oscillations for a transmon qubit. The decay results from environmental decoherence. (b) Averaged Fourier transforms of individual measurement traces from (a), with and without the feedback protocol applied. The linewidth is related to the dephasing rate Γ. Measurement of the noninformational quadrature (flat line) is frequency-independent. (c) Γ is plotted as a function of the measurement strength as parameterized by the average number of photons $\bar{n}$ in the readout resonator. (d) Feedback-stabilized Rabi oscillations, which persist well beyond the decay time observed in (a) (Vijay et al., 2012). Adapted with permission from Springer Nature.

A quantitative analysis of this experiment can be made by taking the simplest case of no-environmental dephasing, so we parameterize the state with the polar angle θ, so the equation of motion is given by (9.50), where we take $\tau_x \to \infty$, so only a z measurement is occurring. The desired evolution is simply $z = \cos(\Delta_d t), y = \sin(\Delta_d t)$. We define the error signal $\delta\theta = \theta(t) - \Delta_d t$, which will be the feedback signal to minimize. The preceding experiment can then be modeled with a Rabi drive that is state-dependent, given by

$$\Delta_f(\theta) = \Delta_d(1 - F\delta\theta), \tag{10.6}$$

where F is the feedback factor that controls how strong the feedback is (Ruskov and Korotkov, 2002). This is a simplified version of the preceding experiment, where an estimate of the phase is made from filtered readout signals directly.

In the diffusive Rabi limit, $\Delta_d \gg 1/\tau_m$, so $\delta\theta$ may be treated as a slow variable, where the Rabi period $T = 2\pi/\Delta_d$ is the fast timescale. We consider the time-averaged effect of the noise on this slow variable by defining a new noise term, $\tilde{\xi}$, given by

$$\tilde{\xi} = -\frac{1}{T} \int_t^{t+T} dt' \sin[\delta\theta(t') + \Delta_d t'] \frac{\xi(t')}{\sqrt{\tau_m}}. \tag{10.7}$$

This new noise term has zero mean and variance given by $\langle \tilde{\xi}^2 \rangle = 1/(2T\tau_m)$. The resulting equation of motion is given by

$$\delta\dot{\theta} = -F\Delta_d\delta\theta + \tilde{\xi}. \tag{10.8}$$

The average dynamics can be found and takes the simple form

$$\langle \delta\theta(t) \rangle = e^{-F\Delta_d t} \langle \delta\theta(0) \rangle, \tag{10.9}$$

showing an exponential approach to the desired solution. The auto-correlation function $K_z(\tau) = \langle z(t + \tau)z(t) \rangle$ can be calculated with either the Fokker–Planck equation (Ruskov and Korotkov, 2002), or with the stochastic path integral approach (Chantasri and Jordan, 2015) to find the result

$$K_z(\tau) = \frac{\cos(\Delta_d\tau)}{2} \exp\left(\frac{e^{-F\Delta_d\tau} - 1}{4\tau_m F\Delta_d} \right). \tag{10.10}$$

This solution shows much tighter correlations with the feedback on than with it off, controlled by the feedback parameter F.

As a final thought about this experiment, we note that this is an example of direct feedback where no complex classical data processing of the measurement was needed to condition the future evolution of the qubit. We will now discuss experiments where fast digital electronics are used to process measurement data and generate conditioned control sequences in real time.

10.3 Canonical Phase Measurement

Consider now an experiment where the problem at hand involves measuring the phase of an electromagnetic wave with quantum-limited accuracy. If we consider a bosonic mode with n quanta, we can readily express the generalized position and momentum quadrature operators in terms of the creation and annihilation operators $\hat{a}^\dagger$ and $\hat{a}$, respectively: $\hat{X} = (\hat{a}^\dagger + \hat{a})$ and $\hat{P} = (\hat{a}^\dagger - \hat{a})/i$. We can measure these field quadratures with a homodyne or heterodyne measurement, as described in Section 8.1, using linear optical components such as beam splitters and mixers. We can also readily count quanta using a photomultiplier tube in the optical domain or superconducting circuits at microwave frequencies, gaining direct access to $\hat{n} = \hat{a}^\dagger\hat{a}$. However, we do not have a measurement device that directly measures the phase that, for all practical purposes, can be thought of as the quantity canonically conjugate to the number of quanta n. While defining a true quantum phase operator is tricky, especially given subtleties related to 2π periodicity, we can associate an operator that returns a phase measurement ϕ and leaves the system in a uniform superposition of definite number states,

$$|\phi\rangle\langle\phi| = \sum_{n=0}^{\infty} e^{in\phi}|n\rangle\langle n|. \qquad (10.11)$$

This is compatible with the picture of Kraus operators and POVMs developed earlier if we associate the Kraus operator $\hat{\Omega}_\phi = |\phi\rangle\langle\phi|$ with this type of canonical phase measurement.

We can graphically visualize the different measurements discussed above by plotting the usual quantum optics phasor diagram, where the length of the vector represents the number of quanta, and the projections on the axes represent the values of the generalized position and momentum. The phase has the natural meaning of the angle of the vector with respect to a reference axis. The conventional method to determine this angle is a heterodyne measurement where one measures the projection along an axis that is rapidly rotating. In the absence of any knowledge of the orientation of the vector, heterodyne measurement is a robust procedure, and all possible measurement axes are sampled. However, to reach near-unity efficiency determination of θ, one must measure in a direction orthogonal to the vector, namely in the circumferential direction. Measurements that yield information about the radial coordinate probe the number of quanta and reduce the efficiency of a phase measurement. Furthermore, in the limit of a very weak signal, such as a single quantum of excitation, fluctuations that cause the phase to jitter are significant, thus resulting in a time-varying phase.

A scheme that can achieve quantum-limited measurement sensitivity in the limit of a single-photon-level signal is the "adaptivedyne measurement protocol" (Wiseman, 1995). One starts with a weak continuous measurement of an electromagnetic signal. After each increment of partial state information is obtained, the observer changes the measurement axis in real time to be orthogonal to the best guess of where the vector points (see Fig. 10.2). To test out this protocol, Martin et al. (2020) prepare a superconducting qubit in a superposition state whose a priori unknown phase is imprinted on the photon emitted from the cavity coupled to the qubit. The photon is detected by a JPA whose pump phase is adjusted in real time to perform an adaptivedyne measurement.

Let us now describe the more subtle aspects of the experiment in greater detail. The emitter is a transmon qubit dispersively coupled to a microwave frequency cavity. In this experiment, the $\sim$ GHz qubit is Rabi flopped at $\Omega_R = 20$ MHz. In this rotating frame, the bare ground and excited states of the qubit $|g\rangle$ and $|e\rangle$, respectively, become the dressed states $|\pm\rangle = \frac{1}{\sqrt{2}}|e\rangle \pm i|g\rangle$. The transition rate between

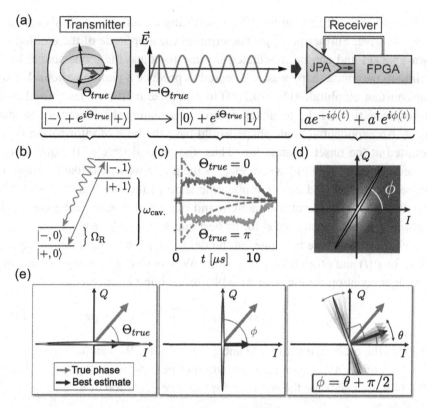

Figure 10.2 (a) Schematic diagram of an atom in a cavity with phase Θ_{true} encoded into its dipole moment. The atom continuously decays and emits a photon into a one-dimensional waveguide with phase encoded into the quadrature components of the electric field (e) as shown. The JPA receives the photon and measures an amplitude quadrature of the field, which is continuously updated by an FPGA as the photon arrives. A sideband cooling scheme (b) converts a qubit excitation to a cavity excitation, which is then emitted as a single photon at the cavity frequency with the engineered mode shape shown in (c). The dashed gray line shows the mode shape if a constant cooling rate was used instead. The output of the JPA is shown in (d), amplified along ϕ and squeezed along the other axes. The feedback process is shown in (e) where the controller initially picks an arbitrary measurement axis phase and then rapidly rotates the measurement axis, attempting to maintain the phase measurement condition $\phi = \theta + \pi/2$ (Martin et al., 2020). Adapted with permission from Springer Nature.

these states can be controlled by applying a microwave frequency drive detuned by the Rabi frequency from the cavity resonance ω_{cav}. In this case, a microwave tone applied at $\omega_{cav} + \Omega_R$ promotes the qubit from the $|+\rangle$ state with no photons in the cavity to the $|-\rangle$ state with one photon in the cavity. This latter state then decays by emitting a photon to the $|-\rangle$ state with an empty cavity; this state does not couple to the sideband modulation. This process can be used to modulate the stimulated emission rate of the dressed qubit $\gamma(t)$ by adjusting the amplitude of this additional microwave drive. Furthermore, $\gamma(t)$ determines the amplitude of the emitted photon wavepacket and can be modulated in time to adjust the mode shape. In this experiment, the mode shape was engineered to produce a photon wavepacket with a near-constant amplitude (Fig. 10.2(c)) to minimize inefficiencies caused by the finite response time of the feedback controller actuating the adaptivedyne measurement. An exponential mode shape would have the bulk of information rapidly transmitted at the onset of emission. Thus, the general state of the qubit can be described as $|-\rangle + e^{i\Theta_{true}}|+\rangle$ where Θ_{true} is the phase we would like to determine. This phase is transferred onto the photonic state of the cavity as $|0\rangle + e^{i\Theta_{true}}|1\rangle$ where $|0\rangle$ and $|1\rangle$ represent the vacuum and one photon state of the cavity. The detector now returns a value of ϕ for its best guess of the phase Θ_{true}.

It is worth noting here how one can model the system using a master equation to solve for $\gamma(t)$ and other related quantities. We can start by writing the stochastic Schrödinger equation for the pure state $|\tilde{\Psi}\rangle$ of the qubit emitter as

$$\frac{d}{dt}|\tilde{\Psi}\rangle = [-\frac{1}{2}\gamma(t)\hat{\sigma}_-^\dagger\hat{\sigma}_- + \sqrt{\gamma(t)}e^{-i\phi(t)}\hat{\sigma}_-r(t)]|\tilde{\Psi}\rangle. \qquad (10.12)$$

Here $r(t)$ is the measurement record and $\hat{\sigma}_- = |-\rangle\langle+|$. We can further express the state of the emitter as a general superposition of $|\pm\rangle$: $|\tilde{\Psi}\rangle = c_-|-\rangle + c_+|+\rangle$. Upon substitution, we can write the equation for the time evolution of c_+ as

$$\frac{d}{dt}c_+(t) = -\frac{1}{2}\gamma(t)c_+(t). \qquad (10.13)$$

We can formally solve this to yield

$$c_+(t) = c_+(0)e^{-\frac{1}{2}\int_0^t \gamma(s)ds}, \qquad (10.14)$$

which makes sense for a system that decays from $|+\rangle$ to $|-\rangle$. Now, to connect this decay rate with the mode shape of the photon, we can define the latter as $u(t)$ and equate it to the instantaneous emitted intensity: $u(t) = \gamma(t)|c_+(t)|^2$. If we constrain $u(t)$ to be a constant (i.e. flat photon mode), then $\gamma(t) = 1/(\tau - t)$ where τ is the duration of the photon wavepacket. Note here that experimentally the strength of the sideband drive has to be increased as time progresses to maintain a constant output of information.

The other half of the story is how to optimally implement a feedback protocol to estimate the phase of the qubit dipole source and by extension the emitted photon. We can compute the equation of motion of c_-. Following the procedure given earlier for c_+, we obtain

$$c_-(t) = c_-(0) + c_+(0)\int_0^t e^{-i\phi(s)}\sqrt{u(s)}r(s)ds. \qquad (10.15)$$

Again starting with $c_0 = 0, c_+ = 1$, the phase θ of the emitter is simply the relative phase between c_+ and c_-^*. This can be expressed as the argument of a complex number R,

$$\theta = \arg(R), \ R = \int_0^t e^{-i\phi(s)} \sqrt{u(s)} r(s) ds. \tag{10.16}$$

Thus, in operation, the measurement apparatus determines R based on the measurement record r and sets the phase of the detector to be $\phi = \theta + \pi/2$; see Fig. 10.2(e). Furthermore, it is possible to show that this combination results in a canonical phase measurement with no information about the number of quanta in the state.

The job of detection in this experiment is carried out by a JPA whose local oscillator pump phase is continuously changed to maintain the optimal phase measurement condition defined earlier. For this task, a fast, field programmable gate array (FPGA) is used to modulate the pump tone applied to the JPA; the total feedback delay time is 374 ns. The JPA is operated with only 6 dB of power gain so as to achieve wide-band operation – the signal bandwidth in this regime is 45 MHz. The JPA is followed by a traveling wave amplifier, resulting in a total measurement efficiency of approximately 0.4.

The efficacy of this adaptivedyne protocol can be evaluated in a few different ways. We can see in Fig. 10.3(b) that the difference between the desired quadrature phase and the best guess of the measurement setup is smaller for a adaptivedyne protocol and does not rapidly saturate as it does for a heterodyne routine. From a signal processing point of view, adaptivedyne produces a 15 percent lower standard error in comparison to a heterodyne measurement carried out in the same setup, and this is in good agreement with theoretical predictions. More subtly, one can look at the dynamics of the emitter in this setup to see if one is truly implementing a phase measurement where photon number information is minimally probed. We can see this in two ways. One, if we examine the quantum trajectories (see Fig. 10.3(a,c) for adaptivedyne and heterodyne measurement), we see that for the former, backaction along the polar access is suppressed and the trajectories form a ring shape. For heterodyne, this is not the case and the trajectories do span the polar cap. Furthermore, we can look at the uncertainty $d\theta$ caused by the measurement backaction. The backaction should be greatest in the θ variable for a phase measurement. We can see (Fig. 10.3(d)) that as the adaptivedyne measurement proceeds, $d\theta$ also grows in time as we approach a canonical phase measurement, and is larger than in the heterodyne case, indicating that more phase information is being extracted. We therefore see an improvement in the resolution of a phase measurement and can use the quantum entanglement of the emitter and the receiver to verify that these improvements are commensurate with a protocol that more closely approaches a canonical phase measurement.

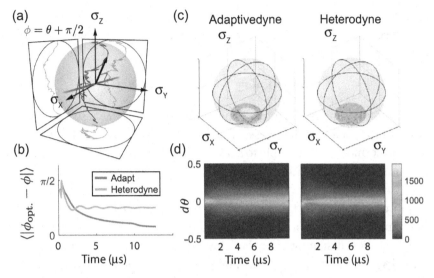

Figure 10.3 (a) A single adaptivedyne quantum trajectory with the qubit initialized in the $|+\rangle$ state. In (b) the tracking results for both heterodyne and adaptivedyne detection with $\phi_{opt} = \theta(t) + \pi/2$ in the optimal measurement axis are plotted. Adaptivedyne detection considerably outperforms heterodyne detection and comes close to the ideal phase by $t = 13\,\mu s$. The distribution of trajectories at $t = 10\,\mu s$ is shown in (c). Suppression of the photon number backaction causes the adaptivedyne trajectories to cluster in a ringlike shape at late time points. (d) Statistics of the phase backaction $d\theta$ for adaptivedyne and heterodyne detection. On average, the phase backaction is considerably larger for adaptivedyne detection, indicating that more phase information is acquired (Martin et al., 2020). Adapted with permission from Springer Nature.

10.4 Continuous Error Correction

In the previous section, we described a feedback experiment to improve the measurement resolution of a single observable, namely the phase of an electromagnetic signal. Here, we highlight the use of similar hardware to implement a version of continuous quantum error correction (Livingston et al., 2022). The measurement protocol involves determining and comparing the parity of two pairs of qubits. As such, in this case, we couple cavities to qubits in such a way that the dispersive shifts induced by the qubits do not reveal state information about the qubits, but rather inform us about a joint property. Quantum error correction is an important task in quantum information processing. The key idea is to measure when an error occurs without destroying the quantum information. This is accomplished by storing the logical qubit of information in a collective state of several qubits and doing collective measurements on the combined system (Shor, 1995; Lidar and Brun, 2013).

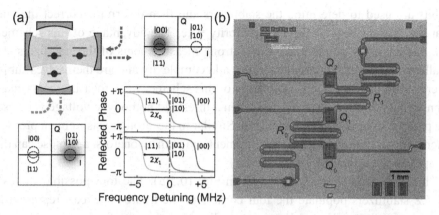

Figure 10.4 (a) A continuous-measurement implementation of the three-qubit, bit-flip error correction code consisting of three qubits in two cavities. Each cavity implements a full parity measurement. The ideal phase responses of a coherent tone reflected off each cavity for different possible qubit states are shown in the lower right graph. The parity probe tones are centered on the odd-parity resonances. The phase space (IQ) plots show the ideal steady-state reflected tone for the shown qubit configuration. The dashed circles are centered on all possible steady-state responses. (b) An electron micrograph of the superconducting chip with three transmon qubits (Q_0, Q_1, Q_2) and two joint read-out resonators (R_0, R_1) (Livingston et al., 2022). Adapted with permission from Springer Nature.

The experimental setup involves coupling three superconducting qubits (Q_0, Q_1, Q_2) to two resonators (R_0, R_1). The physical layout of the chip is shown in Fig. 10.4(b) where R_0 is capacitively connected to both Q_0 and Q_1, and R_1 to Q_1 and Q_2. Each pair of qubits is described by the four states $|00\rangle, |01\rangle, |10\rangle$, and $|11\rangle$. The dispersive shifts induced by either of the odd parity combinations, $|01\rangle, |10\rangle$, are engineered to be identical, as shown in Fig. 10.4(a). The even combinations, $|00\rangle$ and $|11\rangle$, do induce potentially distinguishable frequency shifts χ_0, χ_1 on the order of 4 MHz. The linewidth κ of each resonator is intentionally made to be smaller than χ and is less than 1 MHz. Thus, biasing either resonator at a center frequency of the resonance when the odd states are occupied results in a change of resonator population if the system switches to an even parity state, but the specific combination $|00\rangle$ or $|11\rangle$ cannot be pinpointed by spectroscopy. This setup therefore implements a full parity measurement on two qubits.

We can assign the qubit state to correspond to the Pauli $\hat{\sigma}_z$ operator, and the experimental sequence returns the stabilizers $\hat{\sigma}_{z,0}\hat{\sigma}_{z,1}$ and $\hat{\sigma}_{z,1}\hat{\sigma}_{z,2}$. To implement continuous error correction, we can consider a generalization of the three-qubit bit-flip code. Here, logical zero and one are encoded as $|0_L\rangle = |000\rangle$ and $|1_L\rangle = |111\rangle$, respectively. In a traditional, digital implementation of this code, additional ancillae qubits are used to determine the stabilizers and then determine/correct the qubit that has undergone a bit flip by a majority vote. The advantage of this scheme is that one can in principle make rapid strong measurements of the stabilizers. The challenge is that the additional ancillae add complexity and are themselves subject to errors! In the continuous adaptation of this scheme, weak measurements enable monitoring of both R_0 and R_1, and when a threshold homodyne voltage is crossed, the parity is assumed to have changed. Similar to the strong measurement protocol, after comparing the results of measurements on both resonators, a π pulse can then be applied to correct the errant qubit.

This process is shown graphically in Fig. 10.5(a). The four possible values of the ZZ stabilizers populate the four corners of the figure. The axes represent the homodyne voltages V_0 and V_1 corresponding to the result of a weak measurement of resonators R_0 or R_1, respectively. If the system is initialized in the state $|000\rangle$, then $\langle \hat{\sigma}_{z,0}\hat{\sigma}_{z,1}\rangle$ and $\langle \hat{\sigma}_{z,1}\hat{\sigma}_{z,2}\rangle$ are both $+1$. Flipping one bit turns one or the other of these measurements outcomes to -1; flipping both bits turns them both to -1. Plotted are individual quantum trajectories that originate at positive values for both V_0 and V_1, corresponding to the $|000\rangle$ state, and then evolve in time away from this configuration as errors occur. The dashed squares in each corner represent threshold values that quantify when a voltage excursion is large enough to qualify as a change in parity. As a given trajectory crosses one of the error thresholds, one of three possible π pulses are applied to flip the appropriate errant bit.

The efficacy of this error correction protocol can be evaluated in a few different ways. A very direct one is to flip a bit to see if the controller corrects it properly. This procedure is shown to work about 90 percent of the time, with the remnant error due to the finite T_1 of the qubits. Perhaps a more dramatic benchmark is to see an enhancement of the decay time of the logical states. This error correction protocol is designed to correct bit flips of the individual bare qubits, but will also protect the logical states from decay. When a logical qubit decays, we lose phase information. For example, when the state $1/\sqrt{2}(|0_L\rangle + |1_L\rangle)$ undergoes a decay, it is restored to a mixed state with the same probability distribution in the computational basis. In this case, we would get a density matrix $1/2(|0_L\rangle\langle 0_L| + |1_L\rangle\langle 1_L|)$. The decay constant to reach this steady state can be defined as the T_1 of the logical state. In Fig. 10.5(b), this logical T_1 time is plotted with and without feedback applied. A conventional $T_1 \sim 22\ \mu s$ decay of one of the bare qubits is shown for comparison. With the continuous error correction protocol in action, the logical qubit T_1 is improved by two to three times the bare qubit lifetime!

An important detail in this experiment is how to optimally process the raw homodyne voltage date acquired from each resonator. The continuous parity measurements (the rescaled homodyne voltages) take the form of $r_0 = \langle \hat{\sigma}_{z,0}\hat{\sigma}_{z,1}\rangle + \sqrt{\tau_0}\xi_0$ and $r_1 = \langle \hat{\sigma}_{z,1}\hat{\sigma}_{z,2}\rangle + \sqrt{\tau_1}\xi_1$, where $\tau_{0,1}$ are the characteristic measurement

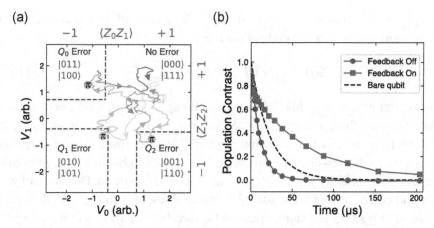

Figure 10.5 (a) Sample experimental voltage traces of the feedback controller correcting induced bit flips with the system starting in $|000\rangle$. If no errors occur, both voltages (V_0 and V_1) remain positive. When an error occurs, one or both of the voltages flip and a predetermined threshold is exceeded, triggering the controller to send a corrective π pulse to bring the system back to the original state. (b) Population decay of the excited logical state, with and without feedback. With feedback on, the lifetime of the logical basis state is longer than that of an individual bare qubit (Livingston et al., 2022). Adapted with permission from Springer Nature.

times for the resonators R_0, R_1. Ideally, the measurement times are made as short as possible in order to reveal any change of the parities, signaling the onset of an error. Single errors are relatively easy to correct, because any kind of time-average of the signal will reveal a bit flip from a parity of $+1$ to a parity of -1. However, the logical errors come from bit flips on different qubits, separated by a short time. For example, a bit flip on qubit 1, followed by a bit flip on qubit 2 results in a flip of parities: $\langle\hat{\sigma}_{z,1}\hat{\sigma}_{z,2}\rangle$ changes first to -1, and then back to $+1$, while the parity $\langle\hat{\sigma}_{z,2}\hat{\sigma}_{z,3}\rangle$ makes first no change, and then changes to -1. Without knowledge of the first parity flip, this last situation appears to be a bit flip on qubit 3, so an erroneous correction pulse would be sent to flip qubit 3. Thus, fast time resolution is needed to identify and correct errors with little time in between them. The most general approach to estimation with Markovian physics is to assign each of the eight possible encodings (generated by applying every possible bit flip to the first encoding) a probability, and to apply a Bayesian filter to track the probability assignments of each possibility. At any time, the encoding with the most probability is returned as our best estimate of the state of the system. The system can then be actively corrected, or passively tracked, with correction only applied to the end. The Bayesian filter gives the optimal error identification (Mabuchi, 2009; Mohseninia et al., 2020). However, application of the filter involves computation on the fly. A simpler method that is slightly worse in identifying errors but improves processing time is to simply apply a moving exponential filter, where the past data is averaged, such that the most recent data is weighted the most:

$$S(t) = (1/\tau) \int_{-\infty}^{t} dt' r(t - t') e^{-(t-t')/\tau}. \tag{10.17}$$

In the digital processing, this can be efficiently applied with a time-series of data r_j with the smoothing relation, $S_j = S_{j-1} + \alpha(r_j - S_{j-1})$, where $\alpha \approx \Delta t/\tau$, and Δt is the time step, is the smoothing factor. This procedure then requires only two time steps of memory. The timescale τ is chosen to be long enough to allow parity distinguishability while still allowing fast detection times. Decisions of when to act and correct errors are made by applying thresholds to the smoothed parity signals, so only smoothed signals passing the thresholds activate a correction pulse. The thresholds cut down on false negative events due to rare noise fluctuations (Livingston et al., 2022).

11

Epilogue: What Does It All Mean?

This chapter is not a scientific chapter. It is a philosophical chapter, where we tell you not only what we think is behind all of this physics, but also what other people think, as well. The content touches on metaphysics – principles and ideas above physics. That is, this topic cannot be directly tested. Nevertheless, we can marshal evidence to support our view, as well as tell you why other scientists think otherwise. We return to our mantra in the first chapter – the best interpretation is a fruitful interpretation. A good view of quantum mechanics (or any other scientific area of investigation) should not only explain known phenomena, it should predict unknown phenomena. This is, in our opinion, where so many "interpretations" of quantum mechanics fail. To the extent that different interpretations can lead to new discoveries, they are good, even if they may not be the ultimate truth. We will also give an outlook on the field as a whole – where it is going, what the most interesting new directions are – and even consider the possibility of post-quantum science.

11.1 What Quantum Mechanics Is and Is Not

Science is fundamentally a descriptive activity. If anyone tells you "science" says this or that, they are a liar. Natural philosophy is a better and historically accurate way of expressing this activity. Science can tell you about patterns in nature and clever ways of organizing those patterns, often involving mathematics. In many cases, the patterns enable us to predict the future behavior of certain phenomena, and even allow us to develop technology that utilize those patterns of behavior for different ends. If you are looking for the meaning of life or the ultimate truths, science will disappoint if you go deep enough. It can help you discover amazing phenomena and uncover patterns in those phenomena if enough repetitions are done, but that is all. In quantum mechanics, we are doubly lucky. Initial conditions made identical to the best of our ability usually give rise to different final outcomes, no matter how carefully they are arranged or how simple the system is. Neverthe-less, patterns in the aggregate emerge. Quantum mechanics, as a theory, explains these patterns. We should not forget that the elements of the theory are products of people like you and me. That is, they are the product of minds, and demand minds to be able to comprehend and understand them. This is likely the most con-troversial aspect of quantum mechanics – it needs an observer to even formulate the theory. The reason for this is that the quantum state is not a state external to us – it is a state internal to us. One cannot consistently speak of an objective quantum state. Perhaps one can speak of objective reality existing in the quantum realm, but that is not quantum theory. To speak of quantum states divorced from an observer is nonsense. To do otherwise is to give reality to our calculational devices. Just as probability distributions only exist in the mind of statisticians, quantum states only exist in the minds of quantum mechanics. This is not to say that everything is subjective – we have gone to great lengths in this book to detail the very precise predictions of quantum measurement theory. There can be facts of nature that all observers agree on. There are structural aspects of the theory that are observer inde-pendent – for example, the energy spectrum of a system, the dimension of a Hilbert space, and the fact that wavefunctions for bosons (fermions) must be symmet-ric (anti-symmetric). Given repeated experiments on identically prepared systems, eventually all observers will agree on the correct quantum state assignment, except ones with highly biased priors, who can be identified beforehand (Srednicki, 2005). Our point of view does not rule out new physics. In the future, it is quite possible that a new and improved physical theory will emerge that is better able to explain the facts around us than quantum mechanics.

11.2 Charting the Quantum Technological Frontier

Inasmuch as quantum mechanics propels us to think of the philosophy of the world around us, it is at the same time rigorous and allows us to describe physical phe-nomena with unparalleled acumen. The theory is arguably one the most successful triumphs of science, and no deviations from it have been seen . . . yet. It is in this spirit that the pursuit of quantum information technologies is, at its most inner core, an exploration and a test of the most fundamental tenets of quantum theory. The tools that we have now developed promise entry to a realm where quantum entan-glement is prepared on demand in a many-body system and tapped for a practical purpose. This is different than unearthing coherent effects in systems that all have roughly the complexity of a single atom or photon – the domain for which the quantum solution was produced – but simply takes on different microscopic and macroscopic physical forms. It is in this sense that the quest to build a quantum computer, for example, is the ultimate test of quantum mechanics for systems with many degrees of freedom. Relatedly, quantum computing protocols give us powerful, efficient, new probes to explore the complex world around us. Ultimately, we are endeavoring to synthesize matter that has a degree of complexity that is not readily accessible in the physical world around us. This pursuit is the seamless juxtaposition of philosophy, physics, and practicality.

As such, we should not view the potential applications of quantum technology solely from the lens of what we have thus far proposed. Only with the benefit of hindsight can we connect the dots that led us from the first vacuum tube to the first transistor to the trillions of modern transistors on a computer chip. We are still in search of the quantum coherent integrated circuit. Its discovery will no doubt be accompanied by mini-revolutions in materials science, manufacturing, and information processing. We should take solace in the fact that accessing a new physics resource, in this case quantum entanglement, is bound to produce new tools. There is vast intellectual capital present in the quantum tool chest. Steady progress and an open mind will let us open it fully, and not just peek inside through a corner.

11.3 What We Have Learned and the Road Ahead

In this book, you have learned about many of the advances made in the theory and practice of quantum measurement, well beyond what Niels Bohr knew about. The cardinal properties of textbook quantum measurement have been broken and remade: You have learned that measurements can take time, that it is possible to reverse them, and that they are not necessary described as a mathematical projection operator. You have learned precise statements concerning using quantum amplification to make classical signals, and how to realize them with cutting-edge technology. You have learned about quantum trajectory theory and how to predict the physics of quantum state diffusion and jumps using advanced mathematical tools. We have surveyed a variety of new phenomena that have been discovered and demonstrated experimentally, including how to use feedback to control a quantum system.

To the extent that we have scientifically supported extensions of quantum mechanics (field theory, for example), these extensions rely and extend the aspects of the theory we are presently concerned with. The view of the authors is that the founders of quantum theory – Bohr, principally – had it basically right, and that the Copenhagen interpretation is able to scientifically explain the phenomena we see in the laboratory. As a scientific theory, quantum mechanics provides a set of rules and equations to apply in a variety of circumstances to give probabilistic predictions about what events occur. It self-consistently captures the patterns in nature. That is all we can expect of a scientific theory, so properly speaking, quantum mechanics needs nothing more. For the philosophically minded, however, more is desired. Science cannot give it to us, but it can provide some clues, based on what we have learned in this book.

We have learned a number of scientific phenomena associated with "wavefunction collapse" or "state reduction." It can be nonlocal – that is, distance and time are no obstacle in the state collapse. It can be controlled by the experimenter to happen slow or fast. It can be jumpy or diffusive depending on the kind of measurement that is done, even on the same physical system. It seems to always be correlated with the degree of information the observer can extract from the system – even in principle. That last fact is brought home dramatically by quantum measurement reversal – the wavefunction can be uncollapsed if the information acquired about the system is contradicted by a subsequent measurement.

All these facts lead us to the conclusion that what is collapsing is in fact not anything external to the observer, but rather what is internal to the observer: the wavefunction is in the eye of the beholder. One way of expressing this, due to Joe Eberly, is that the quantum state of the particle does not belong to the particle, it belongs to you. If we adopt this perspective, it explains many of the preceding puzzling facts – the quantum state is in fact a reflection of your state of knowledge, and new information acquired from the world simply allows you to adjust this state of knowledge to correctly predict the statistics of future events (Caves et al., 2002; Leifer and Spekkens, 2013). Further, this reasoning can be reversed and an observer can retrodict past facts from current ones! It is important, however, to note that scientists and philosophers violently disagree about this topic. There are many rival interpretations about quantum physics – the Copenhagen interpretation, many-worlds, Bohmian mechanics, consistent histories, the transactional interpretation, QBism, the relational interpretation, the two-state-vector interpretation, and many more. Each has its strengths and weaknesses. To do justice to these interpretations, another book-length discussion is required. It is our view these are likely all wrong at the deepest level, but are useful to know about as alternative points of view. We have already given our preference for the "epistemic" point of view – that is, the quantum state is a reflection of our state of knowledge, which is the case in the Copenhagen interpretation, or the QBist approach, rather than the "ontic" point of view – that it (the wavefunction or quantum state) exists as a physical object in space and time, as is the case in the many-worlds, or the Bohmian interpretation (pilot wave), for example. Indeed, the likely best answer to what the meaning of the wavefunction is, is that it does not exist – only measurement outcomes do. This can be brought home forcefully by noting there are formulations of quantum mechanics that have neither states nor observables (Dressel and Jordan, 2013), and when states are rederived, generalized states can appear that are predictive, retrodictive, or the more exotic time bidirectional or interdictive states, depending on what quantities are being inferred. This fact reinforces our view that the notion of state is useful as a mathematical device to make inferences about events in the world but should not be taken very seriously as part of the world's ontology. Let us return to our mantra in the first chapter – the best interpretation is a fruitful interpretation. The topics of weak measurements and weak values were initially discovered by Yakir Aharonov and colleagues using the two-state vector formalism. It is the epistemic interpretation that has helped us to discover and think productively about the quantum effects described in this book. The more we know about quantum mechanics – that is, the more discoveries we make about how the quantum world works, which is happening every day – the more likely we can arrive at the right metaphysical understanding of the world.

This brings us to a discussion of the most exciting new frontiers in quantum physics. We have already mentioned the field of quantum complexity as a blooming area of quantum physics, where the need to create, preserve, control, and measure large-scale entanglement for the purpose of quantum computation opens the opportunity to test the extreme limits of quantum theory. The field of quantum thermodynamics has become an exciting area that explores topics ranging from rethinking thermodynamic concepts in single quantum systems, to designing new kinds of quantum-based engines and refrigerators. The interplay between energy exchange and the information acquired by quantum measurement is a fascinating area investigated by Eugene Wigner that continues to the present day (Wigner, 1952). These concepts build on seminal works in classical thermodynamics and statistical mechanics, such as Maxwell's demon (Bennett, 1987), Szilard engines (Szilard, 1929), and Landauer's principle (Landauer, 1961). An often overlooked point is that if the observable being measured does not commute with the system Hamiltonian, then there is in general an exchange of energy between the system and the measuring apparatus. That is, measurement is not simply "looking" in quantum mechanics – it takes resources from the measurement apparatus in order to carry out the desired measurement. The same is true with other conserved quantities such as momentum and angular momentum. If an apparatus of fixed energy repeatedly measures in the preceding situation, the measurement apparatus will eventually no longer return faithful results about the system being measured because it will not be able to do so and conserve energy. This feature has led to conceptually new kinds of engines powered by quantum measurements; see, for example, Elouard et al. (2017); Elouard and Jordan (2018); Jordan et al. (2020); Bresque et al. (2021). In those works, the measurement result of an operator not commuting with the system Hamiltonian can be used to deduce the postmeasurement state and how energy has been stochastically transferred from the measurement apparatus to the system. Conditional operations can then be carried out in order to rectify this process and create a cyclic engine running on the "fuel" of the measurement process. This phenomenon of stochastic energy exchange without thermal reservoirs is core to the emerging field of *quantum energetics.*

A huge research area that is rapidly advancing is the field of quantum sensing and metrology, which we have only slightly touched in this book. The underlying principle in this field is to use the unique features of quantum mechanics, such as coherence and entanglement, to gain some advantage in measuring parameters, such as phase, frequency, rotation speed, coupling constants of the Hamiltonian, time, and any number of other important quantities. This area is likely to be of more immediate application to societal needs than quantum computing, simply because the resource requirements are more modest. More and more, quantum is becoming an adjective that can be applied to a wide variety of topics. It will be fascinating to see how the field develops and what applications catch on.

11.4 Beyond Quantum

Finally, we can't resist speculating on a theory beyond quantum mechanics. There are many challenging features of quantum mechanics that still resist a satisfying explanation. The dichotomies of the Schrödinger equation versus measurement, the quantum/classical boundary, unitary versus nonunitary dynamics remain vexing issues. It is our prediction that we need to go beyond quantum mechanics to finally iron out these difficulties and make a seamless theory. However, a word of warning is needed. Many people assume that if the next theory can come along, our previous notions of classical realism will be restored, and indeed, much of the motivation of quantum interpretations is to recover this feature. However, it is our prediction that if anything, such a future theory will be even more bizarre than quantum mechanics and will challenge other cherished beliefs, such as the nature of time and locality. It is important that in our pursuit for scientific truth that we do not take a step backwards but push forward to follow where our experiments take us. Metaphysics is a hidden motivation in science, and should not be forgotten or ignored. It can drive innovation, creating challenging alternative perspectives and different points of view about our current and future theories. The great advance in quantum physics is to let go of perfect knowledge and embrace ignorance. In fact, if we do embrace selective ignorance – as we have seen in this book – such a choice is a resource that can power a new quantum information revolution.

Appendix A
Review of Classical Probability Theory

Since quantum mechanics is a generalized theory of probability, it is important to have a firm grasp of classical probability theory. However, in our experience, students and the population in general have a hard time understanding basic concepts in probability. While the topic is really outside the scope of this book, we deem it so important that we give this appendix to review some of the key results of this discipline.

Classical Probability Theory

Classical probability theory is a systematized way to reason about ignorance. This field of mathematics no doubt began with gambling. Only by making informed decisions about ignorance could the gambler win on average and increase his winnings.

Let us begin with a sample space of events. What are the possibilities? For a flipped coin, there are two options, while for a conventional die there are six. For a continuous variable, like position, an infinite sample space exists (ignoring detector resolution), which requires some mathematical care. We can then assign probabilities to that sample space, $\{p_i\}$, where $i = 1, \ldots, N$. Operationally, these can be assigned as the frequencies of results in the limit of a large number of trials. For example, for a fair coin under repeated flips, we assign probabilities of $1/2$ to both heads and tails, although for any finite number of trials, the frequencies will usually deviate from that to some degree that probability theory can also quantify.

We can give a formal mathematical foundation to probability theory via the Kolmogorov axioms (Kolmogorov and Bharucha-Reid, 2018), first published in 1933. These are as follows:

- Probability is defined as a nonnegative real number $p_i \geq 0$, for i in the sample space.

- The probability of some event occurring in the sample space is 1.
- Consider a set of disjoint, or mutually exclusive events A and B; their probability satisfies $P(A \cup B) = P(A) + P(B)$, where $\cup$ indicates the union of the events.

The events A and B can be sets of elementary events. We can read $P(A \cup B)$ as the "probability of either A or B occurring." The events are disjoint if they have no overlap on a Venn diagram.

From these axioms, we can derive other well-known results, such as $\sum_i p_i = 1$. This normalization of total probability can be shown from axioms 2 and 3, applied to all events.

Let us introduce the notation $P(A \cap B)$, where $\cap$ is the intersection symbol, as the "probability of A and B occurring." The intersection would be the region in common to A and B on a Venn diagram. Calling the complement of A with the notation A^c, we can then prove these results.

- $P(A^c) = 1 - P(A)$
- The probability of the empty set is 0.
- For any event A, $P(A) \leq 1$.
- $P(A \cup B) = P(A) + P(B) - P(A \cap B)$
- If $A \subset B$, then $P(A) < P(B)$.

Some very important concepts in probability theory are those of joint and conditional probability. The *conditional probability* $P(A|B)$ is defined as the probability of event A, given event B has occurred. The fundamental rule of probability is

$$P(A|B)P(B) = P(A \cap B). \tag{A.1}$$

We can make sense of this formula as follows. Suppose event B occurs with probability $P(B)$. Given that B occurred, $P(A|B)$ is then the probability of getting A. So, the probability of getting *both* A and B is the product of the two. It is important to stress that the joint probability, $P(A \cap B)$, is symmetric – it is both the probability of getting A and B, as well as the probability of getting B and A. We can also just express this *joint* probability of the events as $P(A, B) = P(A \cap B)$.

The *marginal probability* is just the probability of one event regardless of the other. So, if we only want to know the probability of A occurring, and we don't care about B, this can be calculated as

$$P(A) = \sum_j P(A, B = B_j), \tag{A.2}$$

where the sum is over all values of B.

The notion of *statistical independence* is when one variable is uncorrelated with another. The expression "What does that have to do with the price of tea in China?" is a folksy encapsulation of this idea. Formally, we say events A and

B are statistically independent if $P(A, B) = P(A)P(B)$. In this case, the marginal probability of A is just $P(A)$.

We can now use the fundamental rule of probability together with the concepts of marginal and conditional probabilities to state the *law of total probability*,

$$P(A) = \sum_j P(A, B = B_j) = \sum_j P(A|B = B_j)P(B_j). \tag{A.3}$$

Our last rule of probability is to relate the joint probability of events A and B to the reversed conditioned state,

$$P(A, B) = P(A|B)P(B) = P(B|A)P(A). \tag{A.4}$$

From the last equality in this chain of reasoning, we can express one conditioning in terms of the reverse conditioning:

$$P(A|B) = \frac{P(B|A)P(A)}{P(B)}. \tag{A.5}$$

This important relation is called "Bayes' rule" or "Bayes' theorem," named after the Reverend Thomas Bayes and published posthumously in 1763 (Bayes, 1763). It is very useful in calculating probabilities of events, in particular the "inverse probability," as he called it. Sometimes students are confused by the difference between A given B, versus B given A. Another folksy way of seeing the difference is if we ask, "What is the probability of being pregnant, given that you're a women?" versus "What is the probability of being a woman, given that you're pregnant?"

Bayes' rule is very helpful for what is called "Bayesian inference" – a method of starting from complete ignorance and converging to the optimal probability assignment of events. From a Bayesian view of probability theory, we view $P(A)$ as a degree of belief that A is true. Say, for example, A is that it will rain tomorrow. This belief can even be formed about nonrepeatable events, such as the possibility of you dying tomorrow. This is called the *prior*, or *a priori* belief. The weather forecaster gives some number $P(A) = 1/10$ as his belief in the proposition. However, new information or data can come in to the weather forecaster, such as new satellite data, or dark clouds forming overhead. Call that new data B. The probability $P(B|A)$ is the probability of the data occurring, assuming that rain will happen tomorrow. The probability $P(A|B)$ is called the *posterior*, or *a posteriori* belief in A, given that B has occurred. The ratio $P(B|A)/P(B)$ is then interpreted as the quantified support for the belief A, given the evidence or data B. So, for example, if the weather forecaster notices that every time dark clouds form, rain falls tomorrow 90 percent of the time, but the unconditional formation of dark clouds is $P(B) = 1/5$, then the a posteriori belief that it will rain tomorrow is

$$P(\text{rain tomorrow} | \text{dark clouds}) = P(\text{prior})\frac{P(\text{dark clouds} | \text{rain tomorrow})}{P(\text{dark clouds})}$$

$$= 0.1\frac{0.9}{0.2} = 0.45. \tag{A.6}$$

Consequently, the belief of the weather forecaster that it will rain tomorrow has increased from a 10 percent chance of rain to a 45 percent chance of rain. This comes from the new information "dark clouds have appeared."

Appendix B

Mixed Quantum States

B.0.1 The Density Matrix

This appendix concerns the subjects of mixed quantum states. While many readers of this text will already be familiar with them, some are new to this subject. The basic object of study is to replace the quantum state $|\psi\rangle$ with a density operator, or matrix, $\hat{\rho}$. Some physicists think that only pure quantum states exist, and only our lack of knowledge leads us to discuss mixed states. Whether that is true or not, the density matrix is a useful and powerful concept. Let us introduce it with a thought experiment. Consider two parties, Alice and Bob. Alice can prepare single photons in either the state $|\psi_1\rangle$, or in some other state $|\psi_2\rangle$. Now Alice flips a fair coin and sends a photon in state $|\psi_1\rangle$ to Bob if the coin is heads, and sends a photon in state $|\psi_2\rangle$ to Bob if the coin is tails.

The question at hand is what state should Bob assign to the photon if he has no knowledge of the results of Alice's coin-flipping game, but knows the basic setup of the game? The mixed quantum state is applicable to this situation. For a "pure state" $|\psi\rangle$ the corresponding density operator is simply its outer product, $|\psi\rangle\langle\psi|$. The generalization of the projection probability (2.11) is given by

$$P_j = \text{Tr}[\hat{\rho}\hat{\Pi}_j], \tag{B.1}$$

where Tr is the trace symbol. It is defined by summing the diagonal matrix elements in any basis. If we choose to trace in the basis where the projection operator is diagonalized, then we straightforwardly recover our earlier result, so $P_j = \text{Tr}[|\psi\rangle\langle\psi||j\rangle\langle j|] = |\langle j|\psi\rangle|^2$.

If we now have a statistical mixture of states $\hat{\rho}_1 = |\psi_1\rangle\langle\psi_1|$ and $\hat{\rho}_2 = |\psi_2\rangle\langle\psi_2|$, with equal probability of occurring, then the correct state is given by

$$\hat{\rho} = \frac{1}{2}\hat{\rho}_1 + \frac{1}{2}\hat{\rho}_2. \tag{B.2}$$

The probability of the result j is still given by (B.1),

$$P_j = \frac{1}{2}|\langle j|\psi_1\rangle|^2 + \frac{1}{2}|\langle j|\psi_2\rangle|^2. \tag{B.3}$$

This is just the law of total probability for classical statistics (A.3). The result j could come from two possibilities: on one hand, the photon could be in state $|\psi_1\rangle$, which produces result j with probability $|\langle j|\psi_1\rangle|^2$. On the other hand, the photon could be in state $|\psi_2\rangle$, which produces result j with probability $|\langle j|\psi_2\rangle|^2$. Consequently, we weight these probabilities together with the respective probabilities of their being produced – 1/2 each in this case. Indeed, this reasoning from the results then justifies the density operator definition (B.2).

More generally, suppose the photon is prepared in a set of possible states $\{|\psi_1\rangle, \ldots, |\psi_N\rangle\}$, with probabilities $\{p_1, \ldots, p_N\}$ respectively, such that $0 \le p_j \le 1$, and $\sum_j p_j = 1$, as usual. Then Bob's best state is to define a density operator

$$\hat{\rho} = \sum_j p_j |\psi_j\rangle\langle\psi_j|. \tag{B.4}$$

This corresponds to a weighted average of density operators $\hat{\rho}_j = |\psi_j\rangle\langle\psi_j|$. Given a Hermitian observable $\hat{O}$, the expectation value of $\hat{O}$ is given by

$$\langle\hat{O}\rangle = \mathrm{Tr}[\hat{\rho}\hat{O}]. \tag{B.5}$$

Notice this does not correspond to the objective truth in this case – Alice can always beat Bob in a game of chance about the state of the photon because she knows which of the two states is prepared. Nevertheless, because Bob is making statistical predictions, this is the best he can do with the information at hand.

Notice the density matrix is Hermitian. Each term in the series is of the form $|\psi_i\rangle\langle\psi_i|$, which is a positive operator; that is, the quadratic form $\langle\chi|\psi_i\rangle\langle\psi_i|\chi\rangle = |\langle\chi|\psi_i\rangle|^2 \ge 0$ for every state $|\chi\rangle$. Therefore $\hat{\rho}$ is also a positive operator, so we can diagonalize $\hat{\rho}$ with real eigenvalues $w_i \ge 0$, and orthonormal eigenkets $\{|i\rangle\}$.

To proceed further, notice that

$$\mathrm{Tr}[\hat{\rho}] = \mathrm{Tr}\left[\sum_i p_i |\psi_i\rangle\langle\psi_i|\right] = \sum_i p_i \mathrm{Tr}\left[|\psi_i\rangle\langle\psi_i|\right]. \tag{B.6}$$

For each term in the preceding sum, we have $\mathrm{Tr}[|\psi_i\rangle\langle\psi_i|] = \sum_j \langle j|\psi_i\rangle\langle\psi_i|j\rangle$, where $\{|j\rangle\}$ is an arbitrary orthonormal basis. We may switch the ordering of the two terms and extract a complete set of states, leaving 1. Equivalently, each term in the preceding expression may be interpreted as the probability of finding result j given state $|\psi_i\rangle$, and therefore summing over all results gives 1. Since the probability distribution $\{p_i\}$ is normalized, we therefore have

$$\text{Tr}[\hat{\rho}] = 1. \tag{B.7}$$

This is the statement of state normalization for mixed states. By tracing in the eigenbasis of $\hat{\rho}$, it immediately follows that

$$\sum_i w_i = 1. \tag{B.8}$$

We have already established that w_i are real and nonnegative, so they must now satisfy $0 \leq w_i \leq 1$ for all i. Let us reconsider the expectation value of an arbitrary operator $\hat{O}$ (B.5) where we choose to trace in the eigenbasis of $\hat{\rho}$,

$$\langle \hat{O} \rangle = \sum_i w_i \langle i|\hat{O}|i\rangle. \tag{B.9}$$

Here, we interpret the equation as a weighted average of the diagonal matrix elements of $\hat{O}$, in the eigenbasis that diagonalizes $\hat{\rho}$, weighted by the corresponding eigenvalues of $\hat{\rho}$.

We will now discuss the differences between the quantum state described as a Dirac ket and the density operator (or matrix). Notice that, for any element of the weighted sum (B.4) in the definition of the density matrix, we have the property

$$\hat{\rho}_j^2 = (|\psi_j\rangle\langle\psi_j|)(|\psi_j\rangle\langle\psi_j|) = |\psi_j\rangle\langle\psi_j| = \hat{\rho}_j, \tag{B.10}$$

which followed from the normalization of the state, or equivalently from the fact that $\hat{\rho}_j$ is a projection operator on the state $|\psi_j\rangle$. However, what about a general density operator? Let us consider its diagonal representation,

$$\hat{\rho}^2 = \left(\sum_i w_i|i\rangle\langle i|\right)\left(\sum_j w_j|j\rangle\langle j|\right) = \sum_{ij} w_i w_j|i\rangle\langle i|j\rangle\langle j| = \sum_i w_i^2|i\rangle\langle i|. \tag{B.11}$$

The last equality follows from the orthonormality of the basis states. Thus, we see $\hat{\rho}^2 = \hat{\rho}$ only if the condition $w_i^2 = w_i$ is satisfied for all i. This last condition is true only if w_i is equal to 0 or 1. Further, since the sum of all w_i must be 1, only states that have one of the $w_i = 1$ and the rest are all 0 satisfy this criteria. Consequently, only if $\hat{\rho}$ is a projection operator, $|\phi\rangle\langle\phi|$ for some $|\phi\rangle$, is this condition satisfied. Generally, therefore, $\hat{\rho}^2 \neq \hat{\rho}$, which we define as a *mixed state*; otherwise, it is defined as a *pure state*.

A density operator (or matrix) is also an accurate description for a fully quantum system if we are ignorant of or cannot access a part of the quantum system. Consider a system and environment, where we can control and measure the system, but not the environment. Let us define orthonormal bases for the system and environment as $\{|j\rangle_S\}$ and $\{|k\rangle_E\}$ respectively. Then a general pure state of the combined system and environment may be represented as

$$|\Psi\rangle = \sum_{j,k} c_{jk} |j\rangle_S \otimes |k\rangle_E, \tag{B.12}$$

where c_{jk} are complex coefficients as usual. Let an arbitrary operator $\hat{A}$ be restricted to act only on the system states, but not the environment states. That is,

$$\hat{A}(|a\rangle_S \otimes |b\rangle_E) = (\hat{A}|a\rangle_S) \otimes |b\rangle_E. \tag{B.13}$$

Consider the expectation value of the operator $\hat{A}$ in the state $|\Psi\rangle$, given by

$$\langle\Psi|\hat{A}|\Psi\rangle = \sum_{j,k} \sum_{j',k'} c_{jk} c_{j'k'}^* \langle j'| \otimes \langle k'|\hat{A}|j\rangle \otimes |k\rangle, \tag{B.14}$$

$$= \sum_{j,j'} \left(\sum_k c_{jk} c_{j'k}^* \right) \langle j'|\hat{A}|j\rangle. \tag{B.15}$$

Here, we have passed the environment states past the system operator $\hat{A}$ and invoked their orthonormality to eliminate one of the k sums. Let us make a shorthand for the coefficients appearing inside the j, j' sum to be $\rho_{j,j'} = \sum_k c_{jk} c_{j'k}^*$. $\rho_{j,j'}$ are defined to be the density matrix elements of the operator $\hat{\rho}$. That is, $\rho_{jj'} = \langle j|\hat{\rho}|j'\rangle$. With this definition, the expectation value of A takes on the simple form

$$\langle\Psi|A|\Psi\rangle = \sum_{j,j'} \langle j|\hat{\rho}|j'\rangle\langle j'|A|j\rangle = \mathrm{Tr}[\hat{\rho}A], \tag{B.16}$$

where we have used the completeness of the basis and the definition of the trace.

A complementary point of view is to consider the density operator of the entire system and trace out the environmental degrees of freedom,

$$\mathrm{Tr}_E|\Psi\rangle\langle\Psi| = \mathrm{Tr}_E\left[\sum_{j,k} \sum_{j',k'} c_{jk} c_{j'k'}^* |j\rangle_S \otimes |k\rangle_{ES}\langle j'|_E\langle k'| \right], \tag{B.17}$$

$$= \sum_{j,j'} \left(\sum_k c_{j'k}^* c_{jk} \right) |j\rangle_{SS}\langle j'|, \tag{B.18}$$

where we have traced in the environmental basis $\{|k\rangle_E\}$. We notice that the coefficients in the preceding expansion are nothing else but the density matrix elements, $\rho_{jj'}$. Consequently, we find

$$\mathrm{Tr}_E|\Psi\rangle\langle\Psi| = \sum_{j,j'} \rho_{j,j'} |j\rangle_{SS}\langle j'| = \hat{\rho}. \tag{B.19}$$

The density operator for the system is therefore nothing more than the density operator of the composite system with the environmental degrees of freedom traced out.

The elements of the density matrix have a simple interpretation. Recalling the probability of a projective measurement (B.1) for an outcome j, we can trace in the basis of a complete set of projectors to find

$$P_j = \text{Tr}[\hat{\rho}\hat{\Pi}_j] = \langle j|\hat{\rho}|j\rangle = \rho_{j,j}. \tag{B.20}$$

Thus the probability of outcome j is simply the diagonal density matrix element in the basis of the measurement operator. These are sometimes called the *populations* of the system. The off-diagonal density matrix elements $\rho_{i,j}$ for $i \neq j$ are sometimes called the *coherences*. These results can readily be extended to continuous variable systems using the methods described in Section 2.4.

Dynamics

Let us now consider the problem of time-development of the density matrix. For simple closed systems with a time-independent Hamiltonian, we have the state evolution of the eigenstates of $\hat{\rho}$ to be

$$|i(t)\rangle = e^{-i\hat{H}t/\hbar}|i(0)\rangle, \tag{B.21}$$

where we consider the system Hamiltonian only. Consequently, the time-development of $\hat{\rho}$ is given by

$$\hat{\rho}(t) = \sum_i w_i e^{-i\hat{H}t/\hbar}|i\rangle\langle i|e^{i\hat{H}t/\hbar} = e^{-i\hat{H}t/\hbar}\hat{\rho}e^{i\hat{H}t/\hbar}. \tag{B.22}$$

We can derive a differential equation for $\hat{\rho}$ by taking a time derivative to find

$$\frac{d\hat{\rho}}{dt} = -i[\hat{H}, \hat{\rho}]. \tag{B.23}$$

This is the analog of the Schrödinger equation for density matrices.

Mixedness and Entropy

It is useful to consider the degree of "mixedness" of the state $\hat{\rho}$. We previously discussed how for the general mixed state that $\hat{\rho}^2$ does not equal $\hat{\rho}$. We therefore define the *purity* of a quantum state as $\mathcal{P} = \text{Tr}[\hat{\rho}^2]$. The purity can be expressed in terms of the eigenvalues of $\hat{\rho}$ as

$$\mathcal{P} = \sum_i w_i^2, \tag{B.24}$$

by simply summing the diagonal values of $\hat{\rho}^2$. The condition $\sum_i w_i = 1$, together with $0 \leq w_i \leq 1$, implies that $0 \leq \mathcal{P} \leq 1$. For one extreme limit of a pure state, the purity is $\mathcal{P} = 1$. On the other extreme, the system is maximally mixed. For an N-dimensional system, that corresponds to $w_i = 1/N$. Therefore, we have the maximally mixed state to have a purity of $\mathcal{P} = \sum_i N^{-2} = 1/N$. Another useful measure of purity is the *von Neumann entropy*. It is defined as the Shannon entropy of the density matrix eigenvalues, $\mathcal{S} = -\sum_j w_j \ln w_j$. The Shannon entropy is a measurement of the amount of information concerning a random variable. This quantity can be made basis independent through the definition,

$$\mathcal{S} = -\text{Tr}[\hat{\rho} \ln \hat{\rho}]. \tag{B.25}$$

For a pure state, when one of the $w_i = 1$ and the rest are zero, $w_i \ln w_i \rightarrow 0$ both for $w_i = 0, 1$. Consequently, $\mathcal{S} = 0$ for a pure state – it has no entropy. On the opposite extreme, for a maximally mixed state, $w_i = 1/N$, so $\mathcal{S} = -\sum_{i=1}^{N}(1/N)\ln(1/N) = \ln N$. From exercise B.0.2, the entropy cannot increase in time from system dynamics alone, without measurement or interactions from outside the system.

As a final result for this introduction to mixed states, we can inquire about the postmeasurement state following a projective measurement on the system. We can still apply the measurement operator $\hat{\Pi}_j$ but on both sides of the density operator, so the state conditioned on result j is given as

$$\hat{\rho} \rightarrow \frac{\hat{\Pi}_j \hat{\rho} \hat{\Pi}_j}{\text{Tr}[\hat{\rho}\hat{\Pi}_j]} = \frac{|j\rangle\langle j|\langle j|\hat{\rho}|j\rangle}{\langle j|\hat{\rho}|j\rangle} = |j\rangle\langle j|. \tag{B.26}$$

That is, the density matrix of the system is simply the projection operator on state $|j\rangle$, which is equivalent to state collapse onto the state $|j\rangle$. This has a simple interpretation as asking the system (whose state you are uncertain about) what state it is in by measuring some observable. The answer comes back j, so the system must be in the pure state $|j\rangle$.

Exercises

Exercise B.0.1 Show that the expectation value of the observable $\mathcal{O}$ is simply the weighted average of $\langle \hat{O}_i \rangle$, where the latter is the expected value of $\hat{O}$ in the state $|\psi_i\rangle$.

Exercise B.0.2 Show that any function of $\hat{\rho}(t)$ has a trace that is time-independent.

References

Aharonov, Y., and Bohm, D. 1959. Significance of electromagnetic potentials in the quantum theory. *Physical Review*, **115**(3), 485–491.

Aharonov, Y., and Botero, A. 2005. Quantum averages of weak values. *Physical Review A*, **72**(5), 052111.

Aharonov, Y., Albert, D. Z., and Vaidman, L. 1988. How the result of a measurement of a component of the spin of a spin-1/2 particle can turn out to be 100. *Physical Review Letters*, **60**(14), 1351–1354.

Aristotle. *Physics*. Book VI, chapter 5, 239b5-32. In J. Barnes (ed.), *The Complete Works of Aristotle*. The Revised Oxford Translation, Vol. 1, 1991. Princeton University Press.

Arthurs, E., and Kelly, J. L., Jr. 1965. BSTJ briefs: On the simultaneous measurement of a pair of conjugate observables. *The Bell System Technical Journal*, **44**(4), 725–729.

Aspect, A., Grangier, P., and Roger, G. 1981. Experimental tests of realistic local theories via Bell's theorem. *Physical Review Letters*, **47**(7), 460–463.

Aspect, A., Dalibard, J., and Roger, G. 1982. Experimental test of Bell's inequalities using time-varying analyzers. *Physical Review Letters*, **49**(25), 1804.

Aumentado, J. 2020. Superconducting parametric amplifiers: The state of the art in Josephson parametric amplifiers. *IEEE Microwave Magazine*, **21**, 45–59.

Autler, S. H., and Townes, C. H. 1955. Stark effect in rapidly varying fields. *Physical Review*, **100**(2), 703–722.

Averin, D. V. 2001. Continuous weak measurement of the macroscopic quantum coherent oscillations of magnetic flux. *Physica C: Superconductivity*, **352**(1–4), 120–124.

Averin, D. V. 2003. Linear quantum measurements. Pages 229–239 of Nazarov, Y. V. (ed.), *Quantum Noise in Mesoscopic Physics*. Springer.

Averin, D. V., and Sukhorukov, E. V. 2005. Counting statistics and detector properties of quantum point contacts. *Physical Review Letters*, **95**(12), 126803.

Barchielli, A., and Gregoratti, M. 2009. *Quantum Trajectories and Measurements in Continuous Time*. Springer.

Bayes, T. 1763. LII. An essay towards solving a problem in the doctrine of chances. By the late Rev. Mr. Bayes, FRS communicated by Mr. Price, in a letter to John Canton, AMFR S. *Philosophical Transactions of the Royal Society of London*, 370–418.

Belinfante, F. J. 2016. *Measurements and Time Reversal in Objective Quantum Theory*. International Series in Natural Philosophy, **75**. Elsevier.

Bell, J. 1990. Against "measurement." *Physics World*, **3**(8), 33–41.

Bell, J. S. 1964. On the Einstein–Podolsky–Rosen paradox. *Physics Physique Fizika*, **1**(3), 195–200.

Bennett, C. H., and Brassard, G. "Quantum cryptography: Public key distribution and coin tossing." In *Proceedings of IEEE International Conference on Computers, Systems and Signal Processing*, volume 175, page 8. New York, 1984.

Bennett, C. H. 1987. Demons, engines and the second law. *Scientific American*, **257**(5), 108–117.

Bennett, C. H., and Brassard, G. 2014. Quantum cryptography: Public key distribution and coin tossing. *Theoretical Computer Science*, **560**(1), 7–11.

Bergeal, N., Vijay, R., Manucharyan, V. E. et al. 2010a. Analog information processing at the quantum limit with a Josephson ring modulator. *Nature Physics*, **6**(4), 296–302.

Bergeal, N., Schackert, F., Metcalfe, M. et al. 2010b. Phase-preserving amplification near the quantum limit with a Josephson ring modulator. *Nature*, **465**(7294), 64–68.

Bergquist, J. C., Hulet, R. G., Itano, W. M., and Wineland, D. J. 1986. Observation of quantum jumps in a single atom. *Physical Review Letters*, **57**, 1699–1702.

Bhattacharyya, A. 1943. On a measure of divergence between two statistical populations defined by their probability distributions. *Bulletin of the Calcutta Mathematical Society*, **35**, 99–109.

Blais, A., Huang, R.-S., Wallraff, A., Girvin, S. M., and Schoelkopf, R. J. 2004. Cavity quantum electrodynamics for superconducting electrical circuits: An architecture for quantum computation. *Physical Review A*, **69**, 062320.

Blais, A., Grimsmo, A. L., Girvin, S. M., and Wallraff, A. 2021. Circuit quantum electrodynamics. *Reviews of Modern Physics*, **93**(2), 025005.

Blanter, Y. M., and Büttiker, M. 2000. Shot noise in mesoscopic conductors. *Physics Reports*, **336**(1–2), 1–166.

Blatt, R., Ertmer, W., Zoller, P., and Hall, J. L. 1986. Atomic-beam cooling: A simulation approach. *Physical Review A*, **34**(4), 3022.

Bohm, D. 1952a. A suggested interpretation of the quantum theory in terms of "hidden" variables. I. *Physical Review*, **85**(2), 166–179.

Bohm, D. 1952b. A suggested interpretation of the quantum theory in terms of "hidden" variables. II. *Physical Review*, **85**, 180–193.

Bohm, D. 2012. *Quantum Theory*. Dover Publications.

Bohm, D., and Aharonov, Y. 1957. Discussion of experimental proof for the paradox of Einstein, Rosen, and Podolsky. *Physical Review*, **108**, 1070–1076.

Bohr, N. 2016. On the constitution of atoms and molecules. Pages 13–33 of O. Darrigol, B. Duplantier, J.-M. Raimond, and V. Rivasseau, eds., *Niels Bohr, 1913–2013*. Springer.

Bohr, N. 2010. *Atomic Physics and Human Knowledge*. Dover Publications.

Born, M. 1926. Quantum mechanics of collision processes. *Zeitschrift für Physik*, **38**, 803.

Boutin, S., Toyli, D. M., Venkatramani, A. V. et al. 2017. Effect of higher-order nonlinearities on amplification and squeezing in Josephson parametric amplifiers. *Physical Review Applied*, **8**(5), 054030.

Braginsky, V. B., and Khalili, F. Y. 1995. *Quantum Measurement*. Cambridge University Press.

Bresque, L., Camati, P. A., Rogers, S. et al. 2021. Two-qubit engine fueled by entanglement and local measurements. *Physical Review Letters*, **126**(12), 120605.

Breuer, H.-P., and Petruccione, F. 2002. *The Theory of Open Quantum Systems*. Oxford University Press.

Brun, T. A. 2002. A simple model of quantum trajectories. *American Journal of Physics*, **70**(7), 719–737.

Brune, M., Haroche, S., Lefevre, V., Raimond, J. M., and Zagury, N. 1990. Quantum non-demolition measurement of small photon numbers by Rydberg-atom phase-sensitive detection. *Physical Review Letters*, **65**, 976–979.

Brune, M., Hagley, E., Dreyer, J. et al. 1996. Observing the progressive decoherence of the "meter" in a quantum measurement. *Physical Review Letters*, **77**, 4887–4890.

Büttiker, M. 1986. Four-terminal phase-coherent conductance. *Physical Review Letters*, **57**, 1761–1764.

Büttiker, M. 1990. Quantized transmission of a saddle-point constriction. *Physical Review B*, **41**(11), 7906–7909.

Büttiker, M., Prêtre, A., and Thomas, H. 1993. Dynamic conductance and the scattering matrix of small conductors. *Physical Review Letters*, **70**, 4114–4117.

Campagne-Ibarcq, P., Six, P., Bretheau, L. et al. 2016. Observing quantum state diffusion by heterodyne detection of fluorescence. *Physical Review X*, **6**, 011002.

Carmichael, H. J. 1993. *An Open Systems Approach to Quantum Optics*. Springer.

Caves, C. M. 1982. Quantum limits on noise in linear amplifiers. *Physical Review D*, **26**(8), 1817–1839.

Caves, C. M., and Milburn, G. J. 1987. Quantum-mechanical model for continuous position measurements. *Physical Review A*, **36**(12), 5543–5555.

Caves, C. M., Fuchs, C. A., and Schack, R. 2002. Quantum probabilities as Bayesian probabilities. *Physical Review A*, **65**(2), 022305.

Chantasri, A., and Jordan, A. N. 2015. Stochastic path-integral formalism for continuous quantum measurement. *Physical Review A*, **92**(3), 032125.

Chantasri, A., Dressel, J., and Jordan, A. N. 2013. Action principle for continuous quantum measurement. *Physical Review A*, **88**(4), 042110.

Chantasri, A., Kimchi-Schwartz, M. E., Roch, N., Siddiqi, I., and Jordan, A. N. 2016. Quantum trajectories and their statistics for remotely entangled quantum bits. *Physical Review X*, **6**(4), 041052.

Chantasri, A., Atalaya, J., Hacohen-Gourgy, S. et al. 2018. Simultaneous continuous measurement of noncommuting observables: Quantum state correlations. *Physical Review A*, **97**(1), 012118.

Clarke, J., and Wilhelm, F. 2008. Superconducting quantum bits. *Nature*, **453**(07), 1031–1042.

Clauser, J. F. 1976. Experimental investigation of a polarization correlation anomaly. *Physical Review Letters*, **36**(21), 1223–1226.

Clauser, J. F., Horne, M. A., Shimony, A., and Holt, R. A. 1969. Proposed experiment to test local hidden-variable theories. *Physical Review Letters*, **23**(15), 880–884.

Clerk, A. A., Devoret, M. H., Girvin, S. M., Marquardt, F., and Schoelkopf, R. J. 2010. Introduction to quantum noise, measurement, and amplification. *Reviews of Modern Physics*, **82**, 1155–1208.

Clerk, A. A., Girvin, S. M., and Stone, A. D. 2003. Quantum-limited measurement and information in mesoscopic detectors. *Physical Review B*, **67**(16), 165324.

Cohen-Tannoudji, C., Grynberg, G., and Dupont-Roc, J. 1992. *Atom–Photon Interactions: Basic Processes and Applications*. New York: Wiley.

Collett, M. J., and Gardiner, C. W. 1984. Squeezing of intracavity and traveling-wave light fields produced in parametric amplification. *Physical Review A*, **30**(3), 1386–1391.

Cook, R. J., and Kimble, H. J. 1985. Possibility of direct observation of quantum jumps. *Physical Review Letters*, **54**, 1023–1026.

Davies, E. B. 1976. *Quantum Theory of Open Systems*. Academic Press.

De Broglie, L. 1927. La mécanique ondulatoire et la structure atomique de la matière et du rayonnement. *Journal De Physique Et Le Radium*.

Deutsch, D. 1985. Quantum theory, the Church–Turing principle and the universal quantum computer. *Proceedings of the Royal Society of London. A. Mathematical and Physical Sciences*, **400**(1818), 97–117.

Deutsch, D., and Jozsa, R. 1992. Rapid solution of problems by quantum computation. *Proceedings of the Royal Society of London. Series A: Mathematical and Physical Sciences*, **439**(1907), 553–558.

Devoret, M. H., Martinis, J. M., and Clarke, J. 1985. Measurements of macroscopic quantum tunneling out of the zero-voltage state of a current-biased Josephson junction. *Physical Review Letters*, **55**, 1908–1911.

Dixon, P. B., Starling, D. J., Jordan, A. N., and Howell, J. C. 2009. Ultrasensitive beam deflection measurement via interferometric weak value amplification. *Physical Review Letters*, **102**(17), 173601.

Doyle, A. C. 1894. *The Memoirs of Sherlock Holmes*. G. Newnes Ltd., UK.

Dressel, J., and Jordan, A. N. 2012a. Contextual-value approach to the generalized measurement of observables. *Physical Review A*, **85**(2), 022123.

Dressel, J., and Jordan, A. N. 2012b. Significance of the imaginary part of the weak value. *Physical Review A*, **85**(1), 012107.

Dressel, J., and Jordan, A. N. 2013. Quantum instruments as a foundation for both states and observables. *Physical Review A*, **88**(2), 022107.

Dressel, J., Agarwal, S., and Jordan, A. N. 2010. Contextual values of observables in quantum measurements. *Physical Review Letters*, **104**(24), 240401.

Dressel, J., Lyons, K., Jordan, A. N., Graham, T. M., and Kwiat, P. G. 2013. Strengthening weak-value amplification with recycled photons. *Physical Review A*, **88**(2), 023821.

Dressel, J., Malik, M., Miatto, F. M., Jordan, A. N., and Boyd, R. W. 2014. Colloquium: Understanding quantum weak values: Basics and applications. *Reviews of Modern Physics*, **86**(1), 307–316.

Eichler, C., and Wallraff, A. 2014. Controlling the dynamic range of a Josephson parametric amplifier. *EPJ Quantum Technology*, **1**(1), 1–19.

Einstein, A., Podolsky, B., and Rosen, N. 1935. Can quantum-mechanical description of physical reality be considered complete? *Physical Review*, **47**, 777–780.

Elouard, C., and Jordan, A. N. 2018. Efficient quantum measurement engines. *Physical Review Letters*, **120**(26), 260601.

Elouard, C., Herrera-Martí, D. A., Clusel, M., and Auffèves, A. 2017. The role of quantum measurement in stochastic thermodynamics. *npj Quantum Information*, **3**(1), 1–10.

Esposito, M., Ranadive, A., Planat, L., and Roch, N. 2021. Perspective on traveling wave microwave parametric amplifiers. *Applied Physics Letters*, **119**(12), 120501.

Esteve, D., Devoret, M. H., and Martinis, J. M. 1986. Effect of an arbitrary dissipative circuit on the quantum energy levels and tunneling of a Josephson junction. *Physical Review B*, **34**, 158–163.

Feizpour, A., Xing, X., and Steinberg, A. M. 2011. Amplifying single-photon nonlinearity using weak measurements. *Physical Review Letters*, **107**(13), 133603.

Feynman, R. P. 1982. Simulating physics with computers. *International Journal of Theoretical Physics*, **21**(6/7), 467–488.

Feynman, R. P. 1948. Space-time approach to non-relativistic quantum mechanics. *Reviews of Modern Physics*, **20**(2), 367–387.

Feynman, R. P., Leighton, R. B., and Sands, M. 2011. *The Feynman Lectures on Physics, Vol. III: The New Millennium Edition: Quantum Mechanics*. The Feynman Lectures on Physics. Basic Books.

Frattini, N. E., Vool, U., Shankar, S. et al. 2017. 3-wave mixing Josephson dipole element. *Applied Physics Letters*, **110**(22), 222603.

Freedman, S. J., and Clauser, J. F. 1972. Experimental test of local hidden-variable theories. *Physical Review Letters*, **28**(14), 938–941.

Friedrich, B., and Herschbach, D. 1998. Space quantization: Otto Stern's lucky star. *Daedalus*, **127**(1), 165–191.

Fry, E. S., and Thompson, R. C. 1976. Experimental test of local hidden-variable theories. *Physical Review Letters*, **37**(8), 465–468.

Gardiner, C. W. 1986. Inhibition of atomic phase decays by squeezed light: A direct effect of squeezing. *Physical Review Letters*, **56**, 1917–1920.

Gardiner, C. W., and Zoller, P. 2004. *Quantum Noise: A Handbook of Markovian and Non-Markovian Quantum Stochastic Methods with Applications to Quantum Optics*. Springer.

Gardiner, C. W. 1985. *Handbook of Stochastic Methods*. Vol. 3. Springer.

Gerlach, W., and Stern, O. 1922. Der experimentelle nachweis der richtungsquantelung im magnetfeld. *Zeitschrift für Physik*, **9**(1), 349–352.

Giustina, M., Versteegh, M. A. M., Wengerowsky, S. et al. 2015. Significant-loophole-free test of Bell's theorem with entangled photons. *Physical Review Letters*, **115**(25), 250401.

Gleick, J. 1986. Physicists finally get to see quantum jump with own eyes. *New York Times*, Oct. 1.

Goan, H.-S., Milburn, G. J., Wiseman, H. M., and Sun, H. B. 2001. Continuous quantum measurement of two coupled quantum dots using a point contact: A quantum trajectory approach. *Physical Review B*, **63**(12), 125326.

Goldstein, H. 2011. *Classical Mechanics*. Pearson Education India.

Gorini, V., Kossakowski, A., and Sudarshan, E. C. G. 1976. Completely positive dynamical semigroups of N-level systems. *Journal of Mathematical Physics*, **17**(5), 821–825.

Gottfried, K. 1989. Does quantum mechanics describe the collapse of the wavefunction? *Presented at 62 Years of Uncertainty, Erice, 5–14 August*.

Granger, G, Taubert, D., Young, C. E., et al. 2012. Quantum interference and phonon-mediated back-action in lateral quantum-dot circuits. *Nature Physics*, **8**(7), 522–527.

Grover, L. K. 1996. A fast quantum mechanical algorithm for database search. Pages 212–219 of *Proceedings of the Twenty-Eighth Annual ACM Symposium on Theory of Computing*. https://dl.acm.org/doi/10.1145/237814.237866.

Guéry-Odelin, D., Ruschhaupt, A., Kiely, A. et al. 2019. Shortcuts to adiabaticity: Concepts, methods, and applications. *Reviews of Modern Physics*, **91**(4), 045001.

Gurvitz, S. A. 1997. Measurements with a noninvasive detector and dephasing mechanism. *Physical Review B*, **56**(23), 15215.

Haack, G., Förster, H., and Büttiker, M. 2010. Parity detection and entanglement with a Mach–Zehnder interferometer. *Physical Review B*, **82**(15), 155303.

Hacohen-Gourgy, S., Martin, L. S., Flurin, E. et al. 2016. Quantum dynamics of simultaneously measured non-commuting observables. *Nature*, **538**(7626), 491–494.

Hallaji, M., Feizpour, A., Dmochowski, G., Sinclair, J., and Steinberg, A. M. 2017. Weak-value amplification of the nonlinear effect of a single photon. *Nature Physics*, **13**(6), 540–544.

Haroche, S., and Raimond, J.-M. 2006. *Exploring the Quantum: Atoms, Cavities, and Photons*. Oxford University Press.

Hatridge, M., Vijay, R., Slichter, D. H., Clarke, J., and Siddiqi, I. 2011. Dispersive magnetometry with a quantum limited SQUID parametric amplifier. *Physical Review B*, **83**(13), 134501.

Heisenberg, W. 1949. *The Physical Principles of the Quantum Theory*. Courier Corporation.

Heisenberg, W. 1985. Über den anschaulichen Inhalt der quantentheoretischen Kinematik und Mechanik. Pages 478–504 of Blum, W., Rechenberg, H., and Dürr, H. P. (eds.), *Original Scientific Papers Wissenschaftliche Originalarbeiten*. Springer.

Hensen, B., Bernien, H., Dréau, A. E. et al. 2015. Loophole-free Bell inequality violation using electron spins separated by 1.3 kilometres. *Nature*, **526**(7575), 682–686.

Holt, R. A. 1973. Atomic Cascade Experiments. *PhDT*.

Hong, C. K., Ou, Z. Y., and Mandel, L. 1987. Measurement of subpicosecond time intervals between two photons by interference. *Physical Review Letters*, **59**, 2044–2046.

Hosten, O., and Kwiat, P. 2008. Observation of the spin Hall effect of light via weak measurements. *Science*, **319**(5864), 787–790.

Jacobs, K. 2014. *Quantum Measurement Theory and Its Applications*. Cambridge University Press.

Jacobs, K., and Steck, D. A. 2006. A straightforward introduction to continuous quantum measurement. *Contemporary Physics*, **47**(5), 279–303.

Jaynes, E. T., and Cummings, F. W. 1963. Comparison of quantum and semiclassical radiation theories with application to the beam maser. *Proceedings of the IEEE*, **51**(1), 89–109.

Jordan, A. N. 2013. Watching the wavefunction collapse. *Nature*, **502**(7470), 177–178.

Jordan, A. N., and Büttiker, M. 2005. Continuous quantum measurement with independent detector cross correlations. *Physical Review Letters*, **95**(22), 220401.

Jordan, A. N., and Korotkov, A. N. 2006. Qubit feedback and control with kicked quantum nondemolition measurements: A quantum Bayesian analysis. *Physical Review B*, **74**(8), 085307.

Jordan, A. N., and Korotkov, A. N. 2010. Uncollapsing the wavefunction by undoing quantum measurements. *Contemporary Physics*, **51**(2), 125–147.

Jordan, A. N., Martínez-Rincón, J., and Howell, J. C. 2014. Technical advantages for weak-value amplification: When less is more. *Physical Review X*, **4**(1), 011031.

Jordan, A. N., Chantasri, A., Rouchon, P., and Huard, B. 2016. Anatomy of fluorescence: Quantum trajectory statistics from continuously measuring spontaneous emission. *Quantum Studies: Mathematics and Foundations*, **3**, 237–263.

Jordan, A. N., Elouard, C., and Auffèves, A. 2020. Quantum measurement engines and their relevance for quantum interpretations. *Quantum Studies: Mathematics and Foundations*, **7**(2), 203–215.

Josephson, B. 1974. The discovery of tunnelling supercurrents. *Reviews of Modern Physics*, **46**, 251–254.

Jozsa, R. 2007. Complex weak values in quantum measurement. *Physical Review A*, **76**(4), 044103.

Karmakar, T., Lewalle, P., and Jordan, A. N. 2022. Stochastic path-integral analysis of the continuously monitored quantum harmonic oscillator. *PRX Quantum*, **3**(1), 010327.

Katz, N., Ansmann, M., Bialczak, R. C. et al. 2006. Coherent state evolution in a superconducting qubit from partial-collapse measurement. *Science*, **312**(5779), 1498–1500.

Katz, N., Neeley, M., Ansmann, M. et al. 2008. Reversal of the weak measurement of a quantum state in a superconducting phase qubit. *Physical Review Letters*, **101**(20), 200401.

Kimble, H. J., and Mandel, L. 1976. Theory of resonance fluorescence. *Physical Review A*, **13**(6), 2123–2144.

Knill, E., Laflamme, R., and Milburn, G. J. 2001. A scheme for efficient quantum computation with linear optics. *Nature*, **409**(6816), 46–52.

Koch, J., Yu, T. M., Gambetta, J. et al. 2007. Charge-insensitive qubit design derived from the Cooper pair box. *Physical Review A*, **76**(4), 042319

Kok, P., Lee, H., and Dowling, J. P. 2002. Creation of large-photon-number path entanglement conditioned on photodetection. *Physical Review A*, **65**, 052104.

Kolmogorov, A. N., and Bharucha-Reid, A. T. 2018. *Foundations of the Theory of Probability: Second English Edition*. Dover Publications.

Korotkov, A. N. 1999. Continuous quantum measurement of a double dot. *Physical Review B*, **60**(8), 5737–5742.

Korotkov, A. N. 2001. Selective quantum evolution of a qubit state due to continuous measurement. *Physical Review B*, **63**(11), 115403.

Korotkov, A. N. 2005. Simple quantum feedback of a solid-state qubit. *Physical Review B*, **71**, 201305.

Korotkov, A. N., and Averin, D. V. 2001. Continuous weak measurement of quantum coherent oscillations. *Physical Review B*, **64**, 165310.

Korotkov, A. N., and Jordan, A. N. 2006. Undoing a weak quantum measurement of a solid-state qubit. *Physical Review Letters*, **97**(16), 166805.

Krafczyk, C., Jordan, A. N., Goggin, M. E., and Kwiat, P. G. 2021. Enhanced weak-value amplification via photon recycling. *Physical Review Letters*, **126**(22), 220801.

Kraus, K. 1981. Measuring processes in quantum mechanics I: Continuous observation and the watchdog effect. *Foundations of Physics*, **11**(7–8), 547–576.

Kraus, K. 1985. Measuring processes in quantum mechanics. II: The classical behavior of measuring instruments. *Foundations of Physics*, **15**(6), 717–730.

Kraus, K., Böhm, A., Dollard, J. D., and Wootters, W. H. 1983. States, effects, and operations: Fundamental notions of quantum theory. Lectures in mathematical physics at the University of Texas at Austin. *Lecture Notes in Physics*, **190**.

Lamb, W. E., and Retherford, R. C. 1947. Fine structure of the hydrogen atom by a microwave method. *Physical Review*, **72**(3), 241–243.

Landauer, R. 1961. Irreversibility and heat generation in the computing process. *IBM Journal of Research and Development*, **5**(3), 183–191.

Leggett, A. J., and Garg, A. 1985. Quantum mechanics versus macroscopic realism: Is the flux there when nobody looks? *Physical Review Letters*, **54**, 857–860.

Leifer, M. S., and Spekkens, R. W. 2013. Towards a formulation of quantum theory as a causally neutral theory of Bayesian inference. *Physical Review A*, **88**(5), 052130.

Lewalle, P., Chantasri, A., and Jordan, A. N. 2017. Prediction and characterization of multiple extremal paths in continuously monitored qubits. *Physical Review A*, **95**(4), 042126.

Lewalle, P., Steinmetz, J., and Jordan, A. N. 2018. Chaos in continuously monitored quantum systems: An optimal-path approach. *Physical Review A*, **98**, 012141.

Lewalle, P., Elouard, C., and Jordan, A. N. 2020. Entanglement-preserving limit cycles from sequential quantum measurements and feedback. *Physical Review A*, **102**(6), 062219.

Lewalle, P., Elouard, C., Manikandan, S. K. et al. 2021. Entanglement of a pair of quantum emitters via continuous fluorescence measurements: A tutorial. *Advances in Optics and Photonics*, **13**(3), 517–583.

Lidar, D. A., and Brun, T. A. 2013. *Quantum Error Correction*. Cambridge University Press.

Lindblad, G. 1976. On the generators of quantum dynamical semigroups. *Communications in Mathematical Physics*, **48**(2), 119–130.

Livingston, W. P., Blok, M. S., Flurin, E. et al. 2022. Experimental demonstration of continuous quantum error correction. *Nature Communications*, **13**(1), 1–7.

Lupu-Gladstein, N., Yilmaz, Y. B., Arvidsson-Shukur, D. R. M. et al. 2022. Negative quasiprobabilities enhance phase estimation in quantum-optics experiment. *Physical Review Letters*, **128**(22), 220504.

Lyons, K., Dressel, J., Jordan, A. N., Howell, J. C., and Kwiat, P. G. 2015. Power-recycled weak-value-based metrology. *Physical Review Letters*, **114**(17), 170801.

Mabuchi, H. 2009. Continuous quantum error correction as classical hybrid control. *New Journal of Physics*, **11**(10), 105044.

Macklin, C., O'Brien, K., Hover, D. et al. 2015. A near–quantum-limited Josephson traveling-wave parametric amplifier. *Science*, **350**(6258), 307–310.

Madelung, E.. 1926. Eine anschauliche Deutung der Gleichung von Schrödinger. *NW*, **14**(45), 1004. https://link.springer.com/article/10.1007/BF01504657.

Madelung, E. 1927. Quantentheorie in hydrodynamischer Form. *Zeitschrift für Physik*, **40**(3–4), 322–326.

Magaña-Loaiza, O. S., Mirhosseini, M., Rodenburg, B., and Boyd, R. W. 2014. Amplification of angular rotations using weak measurements. *Physical Review Letters*, **112**(20), 200401.

Mahan, G. D. 2013. *Many-Particle Physics*. Springer Science & Business Media.

Mandel, L., and Wolf, E. 1995. *Optical Coherence and Quantum Optics*. Cambridge University Press.

Manzano, D. 2020. A short introduction to the Lindblad master equation. *AIP Advances*, **10**(2), 025106.

Mao, W., Averin, D. V., Ruskov, R., and Korotkov, A. N. 2004. Mesoscopic quadratic quantum measurements. *Physical Review Letters*, **93**, 056803.

Martin, L. S., and Whaley, K. B. 2019. Single-shot deterministic entanglement between non-interacting systems with linear optics. *arXiv preprint arXiv:1912.00067*.

Martin, L. S., Livingston, W. P., Hacohen-Gourgy, S., Wiseman, H. M., and Siddiqi, I. 2020. Implementation of a canonical phase measurement with quantum feedback. *Nature Physics*, **16**(10), 1046–1049.

Martinis, J. M., Devoret, M. H., and Clarke, J. 1985. Energy-level quantization in the zero-voltage state of a current-biased Josephson junction. *Physical Review Letters*, **55**, 1543–1546.

Martinis, J. M., Nam, S., Aumentado, J., and Urbina, C. 2002. Rabi oscillations in a large Josephson-junction qubit. *Physical Review Letters*, **89**, 117901.

Mensky, M. B. 1993. *Continuous Quantum Measurements and Path Integrals*. Institute of Physics.

Mensky, M. B. 1994. Continuous quantum measurements: Restricted path integrals and master equations. *Physics Letters A*, **196**(3–4), 159–167.

Minev, Z. K., Mundhada, S. O., Shankar, S. et al. 2019. To catch and reverse a quantum jump mid-flight. *Nature*, **570**(7760), 200–204.

Mohseninia, R., Yang, J., Siddiqi, I., Jordan, A. N., and Dressel, J. 2020. Always-on quantum error tracking with continuous parity measurements. *Quantum*, **4**, 358.

Mollow, B. R. 1975. Pure-state analysis of resonant light scattering: Radiative damping, saturation, and multiphoton effects. *Physical Review A*, **12**(5), 1919.

Murch, K. W., Weber, S. J., Beck, K. M., Ginossar, E., and Siddiqi, I. 2013a. Reduction of the radiative decay of atomic coherence in squeezed vacuum. *Nature*, **499**(7456), 62–65.

Murch, K. W., Weber, S. J., Macklin, C., and Siddiqi, I. 2013b. Observing single quantum trajectories of a superconducting quantum bit. *Nature*, **502**(7470), 211–214.

Mygind, J., Pedersen, N. F., Soerensen, O. H., Dueholm, B., and Levinsen, M. T. 1979. Low-noise parametric amplification at 35 GHz in a single Josephson tunnel junction. *Applied Physics Letters*, **35**(1), 91–93.

Naghiloo, M., Foroozani, N., Tan, D., Jadbabaie, A., and Murch, K. W. 2016. Mapping quantum state dynamics in spontaneous emission. *Nature Communications*, **7**(1), 1–7.

Naghiloo, M., Tan, D., Harrington, P. M. et al. 2017. Quantum caustics in resonance-fluorescence trajectories. *Physical Review A*, **96**, 053807.

Nagourney, W., Sandberg, J., and Dehmelt, H. 1986. Shelved optical electron amplifier: Observation of quantum jumps. *Physical Review Letters*, **56**, 2797–2799.

Narla, A., Shankar, S., Hatridge, M., et al. 2016. Robust concurrent remote entanglement between two superconducting qubits. *Physical Review X*, **6**, 031036.

O'Brien, K., Macklin, C., Siddiqi, I., and Zhang, X. 2014. Resonant phase matching of Josephson junction traveling wave parametric amplifiers. *Physical Review Letters*, **113**(15), 157001.

Odom, B., Hanneke, D., d'Urso, B., and Gabrielse, G. 2006. New measurement of the electron magnetic moment using a one-electron quantum cyclotron. *Physical Review Letters*, **97**(3), 030801.

Ott, E. 2002. *Chaos in Dynamical Systems*. Cambridge University Press.

Pang, S., and Jordan, A. N. 2017. Optimal adaptive control for quantum metrology with time-dependent Hamiltonians. *Nature Communications*, **8**(1), 1–9.

Pang, S., Alonso, J. R. G., Brun, T. A., and Jordan, A. N. 2016. Protecting weak measurements against systematic errors. *Physical Review A*, **94**(1), 012329.

Pedersen, M. H., Van Langen, S. A., and Büttiker, M. 1998. Charge fluctuations in quantum point contacts and chaotic cavities in the presence of transport. *Physical Review B*, **57**(3), 1838–1846.

Percival, I. 1998. *Quantum State Diffusion*. Cambridge University Press.

Pilgram, S., and Büttiker, M. 2002. Efficiency of mesoscopic detectors. *Physical Review Letters*, **89**(20), 200401.

Plenio, M. B., and Knight, P. L. 1998. The quantum-jump approach to dissipative dynamics in quantum optics. *Reviews of Modern Physics*, **70**(1), 101.

Pryde, G. J., O'Brien, J. L., White, A. G., Ralph, T. C., and Wiseman, H. M. 2005. Measurement of quantum weak values of photon polarization. *Physical Review Letters*, **94**, 220405.

Ranadive, A., Esposito, M., Planat, L. et al. 2022. Kerr reversal in Josephson meta-material and traveling wave parametric amplification. *Nature Communications*, **13**(1), 1737.

Raussendorf, R., Browne, D. E., and Briegel, H. J. 2003. Measurement-based quantum computation on cluster states. *Physical Review A*, **68**, 022312.

Riste, D., Dukalski, M., Watson, C. A. et al. 2013. Deterministic entanglement of superconducting qubits by parity measurement and feedback. *Nature*, **502**(7471), 350–354.

Ritchie, N. W. M., Story, J. G., and Hulet, R. G. 1991a. Realization of a measurement of a "weak value." *Physical Review Letters*, **66**, 1107–1110.

Ritchie, N. W. M., Story, J. G., and Hulet, R. G. 1991b. Realization of a measurement of a "weak value." *Physical Review Letters*, **66**(9), 1107.

Roch, N., Schwartz, M. E., Motzoi, F. et al. 2014. Observation of measurement-induced entanglement and quantum trajectories of remote superconducting qubits. *Physical Review Letters*, **112**(17), 170501.

Rosenfeld, W., Burchardt, D., Garthoff, R. et al., 2017. Event-ready Bell test using entangled atoms simultaneously closing detection and locality loopholes. *Physical Review Letters*, **119**(1), 010402.

Ruskov, R., and Korotkov, A. N. 2002. Quantum feedback control of a solid-state qubit. *Physical Review B*, **66**(4), 041401.

Ruskov, R., Korotkov, A. N., and Mizel, A. 2006. Quantum Zeno stabilization in weak continuous measurement of two qubits. *Physical Review B*, **73**, 085317.

Ruskov, R., Korotkov, A. N., and Mølmer, K. 2010. Qubit state monitoring by measurement of three complementary observables. *Physical Review Letters*, **105**(10), 100506.

Saira, O.-P., Groen, J. P., Cramer, J. et al. 2014. Entanglement genesis by ancilla-based parity measurement in 2D circuit QED. *Physical Review Letters*, **112**(7), 070502.

Salart, D., Baas, A., Branciard, C., Gisin, N., and Zbinden, H. 2008. Testing the speed of "spooky action at a distance." *Nature*, **454**(7206), 861–864.

Sauter, Th., Neuhauser, W., Blatt, R., and Toschek, P. E. 1986. Observation of quantum jumps. *Physical Review Letters*, **57**, 1696–1698.

Schilpp, P. A. 1949. *The Library of Living Philosophers, Volume 7. Albert Einstein: Philosopher-Scientist*. Open Court.

Schrödinger, E. 1935. Die gegenwärtige Situation in der Quantenmechanik. *Naturwissenschaften*, **23**(48), 807–812.

Shalm, L. K., Meyer-Scott, E., Christensen, B. G. et al. 2015. Strong loophole-free test of local realism. *Physical Review Letters*, **115**(25), 250402.

Shor, P. W. 1994. Algorithms for quantum computation: Discrete logarithms and factoring. Pages 124–134 of *Proceedings 35th Annual Symposium on Foundations of Computer Science*. IEEE.

Shor, P. W. 1995. Scheme for reducing decoherence in quantum computer memory. *Physical Review A*, **52**(4), R2493.

Sinclair, J., Hallaji, M., Steinberg, A. M., Tollaksen, J., and Jordan, A. N. 2017. Weak-value amplification and optimal parameter estimation in the presence of correlated noise. *Physical Review A*, **96**(5), 052128.

Slichter, D. H., Müller, C., Vijay, R. et al. 2016. Quantum Zeno effect in the strong measurement regime of circuit quantum electrodynamics. *New Journal of Physics*, **18**(5), 053031.

Smith, F. T. 1960. Lifetime matrix in collision theory. *Physical Review*, **118**(1), 349.

Song, M., Steinmetz, J., Zhang, Y. et al. 2021. Enhanced on-chip phase measurement by inverse weak value amplification. *Nature Communications*, **12**(1), 1–7.

Srednicki, M. 2005. Subjective and objective probabilities in quantum mechanics. *Physical Review A*, **71**(5), 052107.

Starling, D. J., Dixon, P. B., Jordan, A. N., and Howell, J. C. 2009. Optimizing the signal-to-noise ratio of a beam-deflection measurement with interferometric weak values. *Physical Review A*, **80**(4), 041803.

Starling, D. J., Dixon, P. B., Jordan, A. N., and Howell, J. C. 2010. Precision frequency measurements with interferometric weak values. *Physical Review A*, **82**(6), 063822.

Steinberg, Aephraim M. 2010. A light touch. Nature, **463**(7283), 890-891.

Steinberg, A. M. 1995. How much time does a tunneling particle spend in the barrier region? *Physical Review Letters*, **74**(13), 2405.

Sudarshan, E. C. G., Mathews, P. M., and Rau, J.. 1961. Stochastic dynamics of quantum-mechanical systems. *Physical Review*, **121**(3), 920–924.

Szilard, L. 1929. Über die Entropieverminderung in einem thermodynamischen System bei Eingriffen intelligenter Wesen. *Zeitschrift für Physik*, **53**(11), 840–856.

Trauzettel, B., Jordan, A. N., Beenakker, C. W. J., and Büttiker, M. 2006. Parity meter for charge qubits: An efficient quantum entangler. *Physical Review B*, **73**(23), 235331.

Trimmer, J. D. 1980. The present situation in quantum mechanics: A translation of Schrödinger's "cat paradox" paper. *Proceedings of the American Philosophical Society*, **124**(5), 323–338.

Urbina, C., Esteve, D., Martinis, J. M. et al. 1991. Measurement of the latency time of Macroscopic Quantum Tunneling. *Physica B: Condensed Matter*, **169**(1), 26–31.

Van Erven, T., and Harremos, P. 2014. Rényi divergence and Kullback–Leibler divergence. *IEEE Transactions on Information Theory*, **60**(7), 3797–3820.

Van Kampen, N. G. 1981. Itô versus Stratonovich. *Journal of Statistical Physics*, **24**(1), 175–187.

Vijay, R., Macklin, C., Slichter, D. H. et al. 2012. Stabilizing Rabi oscillations in a superconducting qubit using quantum feedback. *Nature*, **490**(7418), 77–80.

Vijay, R., Slichter, D. H., and Siddiqi, I. 2011. Observation of quantum jumps in a superconducting artificial atom. *Physical Review Letters*, **106**, 110502.

Vion, Denis. 2004. Course 14: Josephson quantum bits based on a Cooper pair box. Pages 521–559 of Estève, D., Raimond, J.-M., and Dalibard, J. (eds.), *Quantum Entanglement and Information Processing*. Les Houches, vol. 79. Elsevier.

Viviescas, C., Guevara, I., Carvalho, A. R. R., Busse, M., and Buchleitner, A. 2010. Entanglement dynamics in open two-qubit systems via diffusive quantum trajectories. *Physical Review Letters*, **105**(21), 210502.

Viza, G. I., Martínez-Rincón, J., Alves, G. B., Jordan, A. N., and Howell, J. C. 2015. Experimentally quantifying the advantages of weak-value-based metrology. *Physical Review A*, **92**(3), 032127.

von Neumann, J. 1955. *Mathematical Foundations of Quantum Mechanics: New Edition*. Princeton University Press.

Wahlsten, S., Rudner, S., and Claeson, T. 1978. Arrays of Josephson tunnel junctions as parametric amplifiers. *Journal of Applied Physics*, **49**(7), 4248–4263.

Wallraff, A., Schuster, D. I., Blais, A., et al. 2004. Strong coupling of a single photon to a superconducting qubit using circuit quantum electrodynamics. *Nature*, **431**(7005), 162–167.

Walls, D. F. 1983. Squeezed states of light. *Nature*, **306**(5939), 141–146.

Weber, S. J., Chantasri, A., Dressel, J. et al. 2014. Mapping the optimal route between two quantum states. *Nature*, **511**(7511), 570–573.

Wheeler, J. A., and Zurek, W. H. 2014. *Quantum Theory and Measurement*. Princeton Series in Physics, vol. 53. Princeton University Press.

Wiegner, R., Thiel, C., Von Zanthier, J., and Agarwal, G. S. 2010. Creating path entanglement and violating Bell inequalities by independent photon sources. *Physics Letters A*, **374**(34), 3405–3409.

Wigner, E. P. 1952. Die Messung quantenmechanischer Operatoren. *Zeitschrift für Physik A Hadrons and Nuclei*, **133**(1), 101–108.

Wigner, E. P. 1955. Lower limit for the energy derivative of the scattering phase shift. *Physical Review*, **98**(1), 145.

Williams, N. S., and Jordan, A. N. 2008. Entanglement genesis under continuous parity measurement. *Physical Review A*, **78**(6), 062322.

Wineland, D. J., Bollinger, J. J., Itano, W. M., Moore, F. L., and Heinzen, D. J. 1992. Spin squeezing and reduced quantum noise in spectroscopy. *Physical Review A*, **46**, R6797–R6800.

Wiseman, H. M. 1994. Quantum theory of continuous feedback. *Physical Review A*, **49**(3), 2133–2150.

Wiseman, H. M. 1995. Adaptive phase measurements of optical modes: Going beyond the marginal Q distribution. *Phys. Rev. Lett.*, **75**, 4587–4590.

Wiseman, H. M., and Milburn, G. J. 2010. *Quantum Measurement and Control.* Cambridge University Press.

Wong, E., and Zakai, M. 1965a. On the convergence of ordinary integrals to stochastic integrals. *The Annals of Mathematical Statistics*, **36**(5), 1560–1564.

Wong, E., and Zakai, M. 1965b. On the relation between ordinary and stochastic differential equations. *International Journal of Engineering Science*, **3**(2), 213–229.

Wootters, W. K. 1998. Entanglement of formation of an arbitrary state of two qubits. *Physical Review Letters*, **80**(10), 2245.

Yu, T., and Eberly, J. H. 2009. Sudden death of entanglement. *Science*, **323**(5914), 598–601.

Yurke, B., Corruccini, L. R., Kaminsky, P. G. et al. 1989. Observation of parametric amplification and deamplification in a Josephson parametric amplifier. *Physical Review A*, **39**(5), 2519–2533.

Zhao, S., and Withington, S. 2021. Quantum analysis of second-order effects in superconducting travelling-wave parametric amplifiers. *Journal of Physics D: Applied Physics*, **54**(36), 365303.

Zhou, X., Schmitt, V., Bertet, P. et al. 2014. High-gain weakly nonlinear flux-modulated Josephson parametric amplifier using a SQUID array. *Physical Review B*, **89**, 214517.

Zoller, P., Marte, M., and Walls, D. F. 1987. Quantum jumps in atomic systems. *Physical Review A*, **35**(1), 198.

Zukowski, M., Zeilinger, A., Horne, M. A., and Ekert, A. K. 1993. "Event-ready-detectors" Bell experiment via entanglement swapping. *Physical Review Letters*, **71**(26), 4287–4290.

Zurek, W. H. 1981. Pointer basis of quantum apparatus: Into what mixture does the wave packet collapse? *Physical Review D*, **24**(6), 1516–1525.

Zurek, W. H. 1982. Environment-induced superselection rules. *Physical Review D*, **26**(8), 1862–1880.

Index

Printed in the United States
by Baker & Taylor Publisher Services

Printed in the United States
by Baker & Taylor Publisher Services